MAMMALIAN PALEOECOLOGY

MAMMALIAN PALEOECOLOGY

Using the Past to Study the Present

FELISA A. SMITH

JOHNS HOPKINS UNIVERSITY PRESS   BALTIMORE

Johns Hopkins University Press
2715 North Charles Street
Baltimore, Maryland 21218-4363
www.press.jhu.edu

Library of Congress Cataloging-in-Publication Data

Names: Smith, Felisa A., author.
Title: Mammalian paleoecology : using the past to study the present /
    Felisa A. Smith.
Description: Baltimore : Johns Hopkins University Press, 2021. |
    Includes bibliographical references and index.
Identifiers: LCCN 2020045433 | ISBN 9781421441405 (hardcover) |
    ISBN 9781421441412 (ebook)
Subjects: LCSH: Mammals, Fossil. | Mammal remains (Archaeology) |
    Paleoecology.
Classification: LCC QE881 .S64 2021 | DDC 569—dc23
LC record available at https://lccn.loc.gov/2020045433

A catalog record for this book is available from the British Library.

*Special discounts are available for bulk purchases of this book. For more
information, please contact Special Sales at specialsales@jh.edu.*

Johns Hopkins University Press uses environmentally friendly book
materials, including recycled text paper that is composed of at least
30 percent post-consumer waste, whenever possible.

To the fierce, intelligent, and funny women in my life:
my aunt, Rosemary Pellegrini Quarantotti, and
my mother, Maria del Carmen Calvo Herrero,
both of whom I lost over the past year and greatly miss;
and especially,
to my ever-so-talented daughters, who inspire me daily:

Emma A. Elliott Smith, M.S., Ph.D.
Rosemary E. Elliott Smith, M.S.

Contents

Acknowledgments

This project was supported in part by two sabbatical semesters from the University of New Mexico, which were greatly appreciated. Research support was provided by the Integrating Macroecological Pattern and Process across Scales (IMPPS) NSF Research Coordination Network (DEB-0541625) and the National Science Foundation, Division of Environmental Biology grant (DEB-1555525), Beyond Causation: Characterizing the Local, Regional, and Global Impacts of the Late Pleistocene Megafaunal Extinction. As a roving Sabbatarian, I wrote in a number of places, including the Rocky Mountain Biological Laboratory in Gothic, Colorado, which provided a quiet and scenically lovely place to kick-start this project; the University of Wyoming and National Park Service Research Station in Grand Teton National Park, where the views and wildlife were amazing; and both the University of Chicago and University of California. I would be remiss if I forgot the many coffee shops where I lingered for countless hours, and on one occasion knocked a cup of coffee into my laptop. Thank you, all.

I am extremely grateful to my lab group for understanding as I focused on finishing this book instead of other things; I will have more time this year, honestly. In particular, the past year was extremely difficult as I lost both my mother and aunt, both of whom I was very close to. I would especially like to thank Vince Burke at Johns Hopkins University Press, my original editor, for his endless patience as I struggled to prioritize work on this volume. His encouragement was instrumental; my new editor, Tiffany Gasbarrini, and her assistant, Esther Rodriguez, have been equally supportive. Thank you, all!

Finally, I thank my wonderful family (Emma, Rosy, and Scott) for their consistent encouragement, love, and support, and of course, all of my furry friends who have sat upon, chewed on, or merely ignored this volume as it was being written.

MAMMALIAN PALEOECOLOGY

1 Introduction

This is a book about ecology in the past and what it can tell us about the present and future. By past, I mean the literal definition as in *"gone by in time and no longer existing."* That is, this is a book about the interactions of organisms and their environment in the time before European settlement of the Americas. In fact, most of this book is concerned with the time before humans came to occupy the Americas, or in some cases, before our hominin lineage even existed on Earth at all.

More specifically, this is a book primarily about *mammalian* ecology of the past. Restricting most—although not all—of the discussion in this text to mammals is a strategic move, motivated by their abundant and well-resolved fossil record, the ability to readily identify species and lineages, and the ability to characterize life histories and ecologies. And, I confess, because of an intrinsic bias: mammals are cool. No other group contains the vast array of sizes and life forms they do. Mammals range from the miniscule pygmy shrew (*Suncus etruscus*), which at 1.8 g doesn't even weigh as much as a single American dime (2.268 g, according to the US Mint), to the largest animal of all time, the blue whale (*Balaenoptera musculus*), which at almost 180 tons weighs just slightly less than 80,000,000 dimes (Smith and Lyons 2013; Fig. 1.1). And, I might add, the blue whale is also twice the size of any dinosaur that ever roamed the Earth. Not only do mammals vary in size

by more than eight orders of magnitude, but they also make their home in oceans and streams, in the soil, in the hottest deserts, in the wettest tropics, and in the coldest arctic, and they have successfully colonized the air. In all, mammals display a bewildering and impressive array of diverse life history and ecological characteristics. What's not to like?

The Past as Prologue

The past has much to teach us; while our current deluge of pressing environmental problems—ocean acidification, ecosystem restructuring, and in particular, climate change—may be unprecedented in human history, all have ancient analogs. After all, the Earth is dynamic and has been beset by many perturbations over the past 4.6 Ga. A case in point is anthropogenic climate change. Here, the late Quaternary provides abundant examples for examining the influence of changing abiotic conditions on the distribution, ecology, and evolution of organisms (Graham et al. 1996; Clark et al. 2001; Davis and Shaw 2001; Gavin et al. 2007; Williams and Jackson 2007; Birks and Birks 2008; Louys 2012 and references therein). Evidence from numerous paleoclimate proxies (e.g., pollen, cross-dated tree ring chronologies, ice cores, and other indicators, discussed in chapter 11) suggests abrupt climate shifts have occurred with regularity in the past (Allen and Anderson 1993; Dansgaard et al. 1993; Bond and Lotti

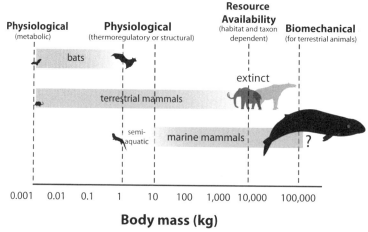

Fig. 1.1. Factors influencing the minimum and maximum size of mammals. Note that the minimum size of ~1.8 g is represented by both volant and nonvolant mammals; aquatic mammals have a much larger minimum body size that appears to be set by the thermoregulatory demands of living in an aquatic environment (see text). The largest terrestrial mammal, *Indricotherium*, reached masses reportedly in excess of 12-15 tons. Interestingly, this is about an order of magnitude smaller than the largest terrestrial dinosaurs and could reflect a difference between endothermic and exothermic animals if resources limit size in terrestrial environments. The upper limit to ocean life has not yet been determined, although at ~180 tons the blue whale is the largest mammal we know of. Redrawn from Smith and Lyons 2013. Animal silhouettes from PhyloPic (http://phylopic.org) under Creative Commons license.

1995; Alley 2000; Birks and Ammann 2000). Some, such as the terminus of the Younger Dryas cold episode around 11.5 ka, were sizable, with temperature warming of as much as 5-10°C reportedly occurring within a decade or two (Alley et al. 1993; Alley 2000). This rate of change was actually higher than that expected under all current scenarios of anthropogenic climate change (IPCC 2007) and must have posed significant challenges for organisms. Yet, virtually all species extant today were also present during this time and somehow successfully coped with the consequences of this abrupt warming through a combination of range shifts and adaptation. Indeed, we have recently shown that climate change over the past 65 Ma did not lead to increased rates of mammal extinction, probably because geographic ranges were larger and animals were free to move around (Smith et al. 2018, 2019). Thus, while the underlying cause of human-mediated climate change is different, the magnitude itself is not novel in Earth history.

Similarly, in conservation ecology, there is a renewed emphasis on determining appropriate baselines and in disentangling anthropogenic effects on ecosystems from natural ones (Donlan et al. 2005; Lyman 2006). Even determining whether a mammal is an exotic or native inhabitant, or differentiating an invasive species from a natural extirpation/recolonization event depends on a temporal perspective (Lyman 2006). Not surprisingly, over the past decade there has been renewed interest in bringing a historical perspective into contemporary ecological and conservation studies (Lyman 2006; Botkin et al. 2007, Gavin et al. 2007, Williams and Jackson 2007). After all, as Shakespeare wrote in his play the Tempest, *"What's past is prologue."* This highly relevant quote is engraved on the New Mexico State Archive building in my hometown of Santa Fe, as well as in the National Archive buildings in Washington, DC (Fig. 1.2). Thus, characterizing how mammals coped with earlier environmental challenges provides critical insights as we struggle to manage our changing Earth.

Space and Time

Replication is key in modern biology. In the lab, the same experimental design may be run multiple times. In the field, scientists tend to have multiple

Fig. 1.2. The New Mexico State Archive building in Santa Fe, New Mexico. Photo by author.

experimental plots. By examining ecological dynamics across gradients, scientists are able to confirm important findings and explore their limits. Most modern biologists strive to replicate experiments across spatial environmental gradients; exploring processes across geography allows better understanding about ecological function (Rull 2014). The implicit assumption is that by doing so they are incorporating the possible range of natural variation in important parameters and thus more accurately portraying their relationship (Fig. 1.3).

While using a broader spatial scale clearly increases the natural range of variation in both abiotic and biotic conditions, space is not synonymous with time (Fig. 1.3; Smith and Boyer 2012). A disparity may arise because of non-analog climatic conditions found in the past, leading to assemblages of mammal or vegetative communities not found together today

(e.g., Huntley 1990; Overpeck et al. 1992; Graham et al. 1996; Williams et al. 2001; Williams and Jackson 2007), or because the type or magnitude of change in the past dwarfs that represented along a modern spatial gradient (Jackson 2007), much like the terminus of the Younger Dryas cold episode. Moreover, it can take many generations for some processes to be manifest, which vastly exceeds the time scale of typical ecological studies (Smith and Boyer 2012; Rull 2014). For example, the rearrangement of ecological communities following the Last Glacial Maximum, or the terminal Pleistocene megafauna extinction, may have taken millennia (MacDonald et al. 2008).

A good example is the relationship of maximum body mass of mammals with area (Fig. 1.4). It turns out if the size of the largest-bodied mammal on an island is plotted versus the size of the island it occupies, you get a nice linear relationship in log

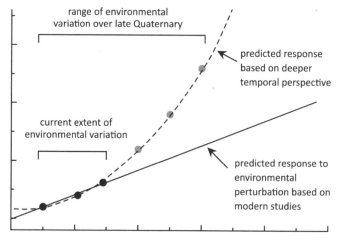

Fig. 1.3. A paleo perspective may allow the examination of a greater range of biotic responses to environmental perturbations. A focus of much of contemporary biology and conservation science is in predicting how organisms will respond to the ongoing changes in environmental conditions. Contemporary ecological studies, even when replicated across space, are typically restricted to a fairly narrow range of environmental conditions. When the temporal scale is increased, a different relationship between the perturbation and the biotic response may be revealed. Redrawn from Jablonski 2003.

space (e.g., a power relationship in linear space, see chapter 5). Indeed, about half the variation in maximum body size can be explained simply by the island area; this probably reflects limitations set by resources (Burness et al. 2001; Smith et al. 2010a). This is true even when habitat islands, continents, or ocean basins are included (Smith et al. 2010a; Smith and Boyer 2012). Thus, there appears to be a hard upper constraint on body size that is set by the size of the habitat it occupies. This could be pretty useful in management planning if, for example, re-introductions were under consideration.

However, these data are incomplete: many species of large-bodied mammals went extinct in the late Quaternary due to the exploits of humans (Smith et al. 2018). If this analysis is redone using the historic record of the largest-bodied mammal found in each area, the relationship still holds, but it is much stronger (about 70% of the variance is explained) and, importantly, both the intercept and the slope differ (Fig. 1.4). Thus, while an analysis based solely on modern data suggests the largest-bodied mammal an island of 1,000 km² can support is about 430 g, the more complete historical dataset predicts an animal more than five times larger—around 2.3 kg (Smith and Boyer 2012). If such incomplete data are used in analyses for conservation planning and management, poor choices might be made. Thus, ecological history matters.

Structure of This Volume

My goal in writing this book is to coalesce information from a wide variety of sources into a single—and hopefully lucid—volume providing an overview of how we obtain information from long-dead mammals and what this information can tell us of relevance to ongoing pressing environmental issues. My hope is that the book will prove useful to those interested in mammals and/or paleoecology and, especially, those interested in conservation paleoecology, which I believe is increasingly relevant in our anthropogenically altered world. While I have targeted this book largely toward graduate students and advanced undergraduates, I hope it might also be of interest to other academics or professionals seeking a better understanding of mammalian paleoecology. Throughout, I assume that the reader is familiar with fundamental ecological concepts and processes; those without a biological background might find it useful to refer to a standard ecology text from time to time.

The past few decades have seen the compilation of a number of large synthetic datasets in both ecology and paleontology (the Paleobiology Database, Neotoma, Neogene Mammals of the Old World, Metadatabase of Mammals, Paleoclim, etc.). These, along with great advances in computation power and in analytical and statistical techniques and approaches, have greatly facilitated progress in both disciplines and, importantly, allowed a change in emphasis from

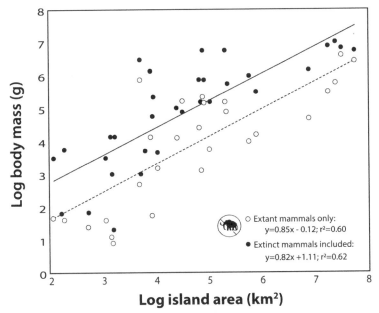

Fig. 1.4. The relationship between the maximum size of a mammal and the area of its habitat. For each island, continent, or ocean basin, the largest mammal was identified. Note that when the largest extant mammals are compiled with the largest extinct mammals (i.e., those found prior to human activities), there is only a slight difference in the slope but the intercept differs greatly. This reflects the intense size-selective extinctions caused by humans (Smith et al. 2018). Redrawn from Smith and Boyer 2012. Mammoth silhouette from PhyloPic (http://phylopic.org) under Creative Commons license.

descriptive to quantitative studies. Thus, we can now address a number of broad-scale questions, including the patterns of body size of all life over all time (Smith et al. 2016a) and the drivers of mammal evolution over the Cenozoic (e.g., Smith et al. 2010a). Much of this volume employs information from these and other broad-scale synthetic databases.

The book begins with some of the fundamental principles, or the "nuts and bolts," of paleoecology— that is, a bit about the fossil record, dating, taphonomy, and other issues of relevance when interpreting the geologic record. I imagine those with a robust physical science background may find some of these first chapters fairly basic. The second major section of the book is devoted to how we characterize the ecology of long-dead mammals. Thus, chapter 5 discusses the importance of mammal body size, how we can estimate size from teeth and post-cranial materials, and what size and shape reveals about the life history and ecology of long-dead organisms (Fig. 1.5). In chapter 6, I continue along this theme, discussing the structure, function, and utility of different types of mammal teeth. Teeth are the most durable mammal fossil and can reveal much about mammals, including their body size and general ecology. Chapters 7 and 8 highlight other important methods and proxies used

Fig. 1.5. The author in her natural habitat. Measuring fossil elements of late Pleistocene mammals at the Texas Memorial Museum in Austin, Texas. Photo courtesy of S. Kathleen Lyons.

in modern paleoecology, including stable isotope, ancient DNA, owl pellet, *Sporormiella*, and paleomidden analyses. Finally, the last section of the book is devoted to several case studies: areas where a historical or paleontological perspective greatly enhances our ability to untangle complex environmental issues. Here I focus on biodiversity and climate change—perhaps the most pressing environmental problems we face today.

Throughout I have attempted to strike a balance between providing sufficient references for the reader to follow up on important points and unnecessarily clogging the prose. I tried to provide some historical perspective where possible. I have undoubtably left out many important contributions that should be included; if so, I hope those authors will forgive me. Additionally, I provide a Further Reading section at the end of each chapter. These works are either seminal or particularly informative and should provide a platform for exploring the ideas presented. Finally, I note that this book has a decidedly North American focus. This reflects my own biases and knowledge base; many of the examples are drawn from my own research, much of it in the western United States. I have tried to provide examples from other continents where relevant.

Part One

General Principles of Paleoecology

Old Bones, Footprints, and Trace Evidence of Life

She sells seashells on the sea-shore. The shells she sells are seashells, I'm sure. For if she sells seashells on the sea-shore, then I'm sure she sells seashore shells.
—Terry Sullivan, British songwriter, 1908 (Winick 2017)

Making Sense of Odd Objects

You probably learned this tongue twister as a child. Like many popular sayings, there is a cultural myth behind the rhyme. The "she" in this well-worn phrase likely refers to Mary Anning, a working-class woman who lived in the little seaside town of Lyme Regis in Dorset, England, during the first half of the nineteenth century (Emling 2009; but see Winick 2017). Mary was a professional fossil collector and dealer who sold her "curios" in a shop attached to her home. In addition to seashells, Mary discovered the first ichthyosaur, plesiosaur, and pterosaur skeletons, as well as a number of important fossil fish. She collected primarily during the cold winter months because the heavy rains at this time caused landslides, which often exposed new finds. The coastal cliffs she worked alongside were quite unstable and dangerous; one sudden rock fall led to the death of her beloved dog (Fig. 2.1a). While not publicly acknowledged at the time because of her gender and low social class, Mary's discoveries and careful preparations of her specimens contributed profoundly to the transformation of paleontology from a cultural curiosity to an actual science (Torrens 1995; Emling 2009). Although she regularly corresponded with and sold fossils to many of the leading figures in geology and paleontology, she was by no means part of the scientific establishment (Torrens 1995; Emling 2009). Indeed, as

unconceivable as it seems now, most scientific descriptions of her specimens were published without acknowledgment of her role; nor did she receive any attribution in the many museums and private collections that housed her spectacular finds (Fig. 2.1b). Sadly, it wasn't until some 163 years after her death that she was publicly acknowledged by the Royal Society as one of the "ten British women who have most influenced the history of science" (https://royalsociety.org/topics-policy/diversity-in-science/influential-british-women-science).

But, while Mary's discoveries and her hard and sad life may now make her one of the most famous fossil hunters of her time (Torrens 1995), she was certainly by no means the first. What the earliest humans made of the strange artifacts they occasionally encountered we can only guess. There is some evidence that fossils (e.g., marine shells, corals, squid, and amber) were used for ornamentation as long as 30,000-40,000 years ago (White 1993), but whether early man recognized these for what they were is open to debate. Certainly by the time of the ancient Greeks, some scholars recognized that fossils might represent some sort of living thing. The Greek philosopher, poet, and critic Xenophanes described *"flat impressions of all sorts of marine life"* in Malta, and impressions of fish and seals in quarries in Syracuse. These discoveries led him to conclude that these areas had once been underwater (McKirahan 1994,

Fig. 2.1a. One of the first fossil hunters, Mary Anning. Her beloved dog Tray (who was later crushed in a landslide that narrowly missed Mary) is shown sleeping at her feet; in the background is the Golden Cap outcrop situated between Bridport and Charmouth in Dorset, England. The name comes from the distinctive golden greensand rock found at the cliff top. Portrait by B.J.M. Donne in 1847 or 1850 on display at Natural History Museum in London, courtesy of Wikimedia Commons.

p. 65). Other Greeks and Romans, including Herodotus, Theophrastus, Aristotle, Suetonius, and the emperor Augustus, were fascinated also by these strange and often fragmentary remains, although they differed in their interpretations of what they were. The Roman Pliny the Elder wrote, *"In the vicinity of Munda in Spain, the place where the Dictator Cæsar defeated Pompeius, there are stones found, which, when broken asunder, bear the impression of palm leaves"* (Pliny [77] 1855, chap. 29).

The attempts of early philosophers and naturalists to reconstruct animals from fossil fragments led to the design of many fanciful creatures. The legend of the griffin, for example, may have derived from fossils of *Protoceratops*, an early horned dinosaur found in Mongolia and China (Mayor 2000). *Protoceratops* was a quadrupedal dinosaur with a highly distinctive neck frill at the back of its skull and a massive frontal beak; these are morphological features similar to that attributed to the griffin (Fig. 2.2). Moreover, in mythology griffins are said to guard gold deposits; interestingly, *Protoceratops* fossils are often found in the Gobi Desert, an area of sedimentary deposits rich with gold (Mayor 2000).

Similarly, the fossil of a *Basilosaurus* whale may have inspired the story of the monster of Joppa, a voracious sea monster sent by Poseidon in revenge for a boast (Mayor 2000). The legend goes on to say that Andromeda was chained to a rock as a sacrificial offering to appease the monster and was ultimately

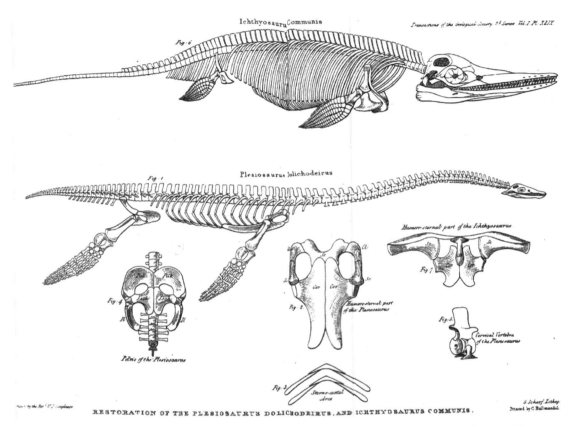

Fig. 2.1b. Illustrations of several of Mary Anning's most spectacular finds, including an ichthyosaur and a plesiosaur. These drawings appeared in William D. Conybeare's original description of the species. He was an English paleontologist and clergyman who made a particular study of marine reptile fossils. In keeping with the traditions of the day, no mention of Mary Anning appeared in these publications. Conybeare 1824.

(and fortuitously!) rescued by Perseus. Indeed, it has been suggested that much of classical Greek and Roman mythology actually stems from such attempts to interpret the fossil remains of long-dead animals (Mayor 2000). If true, how fascinating to consider that long-dead animals might have so profoundly influenced human cultural evolution.

The Nature of Fossils

The etymology of the term "fossil" is straightforward; it comes from the past participle of the Latin verb *fodere*, which means "to dig," in reference to the fact that fossils are mostly found buried in sedimentary deposits. The word was coined by Georgius Agricola, the founder of geology, in his *De Natura Fossilium* (*On the Nature of Fossils*), published in 1546. But a fossil, as Agricola saw it, was not necessarily alive; indeed there was considerable controversy

about whether they represented anything that had been once living. Agricola included minerals and gems in his book as types of fossils. Nonetheless, rather than classify on the basis of mystical powers or some other subjective basis as others had done, he used their simple physical properties, which could be readily quantified. "*Thus minerals have differences which we observe by color, taste, odor, place of origin, natural strength and weakness, shape, form, and size*" (Agricola 1546). In this, his work was a conceptual step forward and helped lay the foundation for modern geology.

The nature of fossils continued to be debated for centuries with little tangible progress. The German Jesuit scholar Athanasius Kircher, argued they were "sports of nature," that is, odd geometric shapes that had somehow been impressed onto rocks. Confusion partly arose because fossils were hard to

Fig. 2.2. The griffin was a mythical creature that was said to have the body of a lion and the head and wings of an eagle. Legend had it that these powerful and majestic creatures preyed on gold prospectors in distant Asia. Mayor (2000) argues that the source of this myth lies in the fossils that Scythian gold-miners found as they journeyed and prospected in the Gobi Desert. This region contains some of the richest dinosaur fields; they are in close proximity to mountains containing rich gold deposits. The most commonly encountered specimen is *Protoceratops*, thought to be the prototype of the legend. Ancient Greeks may have transmuted this myth. Drawing by Pearson Scott Foresman, courtesy of Wikimedia Commons.

Although some, such as the great French naturalist Comte de Buffon, had argued for this process much earlier, extinction would not be seriously considered until the 1800s when another French scholar named Cuvier collected compelling evidence to document it. Despite the prevailing confusion, some scientists did arrive at the correct conclusion. The original Renaissance man, Leonardo da Vinci, believed that fossils were the remains of ancient life; this led him to the heretical view that the Earth was older than stated in the Bible (Baucon 2010). Unfortunately, being too busy with his many other pursuits (art, engineering, anatomy, cartography, architecture, inventor, literature, etc.), he never managed to publish his careful studies of Italian fossils. Had he done so, the debate over their nature may have ended much earlier (Baucon 2010).

A major figure of this time was Niels Stensen, or Nicolai Stenonis (in keeping with the academic customs of the time, he Latinized his Danish name; he is often referred to as Nicolas Steno). He was a Danish Catholic bishop and scientist who specialized in anatomy. In October 1666, when Stenonis was working as an anatomist at the Santa Maria Nuova hospital in Florence, two Italian fishermen caught a huge female shark off the coast of Livorno. The Grand Duke of Tuscany, to whom the shark was given, had the head sent to Stenonis. As he dissected it, he noticed that the teeth were shaped very much like *glossopetrae* or "tongue stones," odd triangular pieces of rock that had long been known and that he had studied at university in Copenhagen (Kermit 2002). The ontogeny of these tongue stones was still very much in dispute. Earlier writers, such as Pliny the Elder, had argued that they fell from the sky; others (including Theophrastus and later Athanasius Kircher) believed that they "grew" in rocks. But, based on careful comparisons with the shark caught near Livorno, Stenonis argued tongue stones were actually the teeth of ancient giant sharks that had turned to stone when the "corpuscles" (we now call them molecules) had gradually been replaced by minerals.

This was somewhat radical. And Stenonis went further. His geological rambles over the countryside of Italy had suggested to him that rocks might be dynamic, forming layers over time that might then encapsulate the remains of ancient life. Such layers

identify—they were generally found in bits and pieces and sometimes did not seem to belong to anything remotely recognizable. And, indeed, sometimes they *were* simply an oddly shaped mineral or rock. Because there was no consensus on what a fossil was, many writers assumed that they were anything "dug up." This issue undoubtedly hampered analysis and classification. Moreover, scholars of this age tended to be generalists. While they might be conversant in a wide variety of scientific disciplines, only some obtained the expertise necessary to evaluate fossil finds. Certainly, without detailed knowledge of comparative anatomy, *anyone* would be hard pressed to figure out what some of these strange specimens were. And, the concept of extinction was largely unthinkable.

could provide a record of life over time, with the oldest rocks/fossils found on the bottom and the younger ones on top. This simple but elegant idea came to be called Steno's Law of Superposition; it is one of the foundational principles of stratigraphy (the study of rock layers). While not always true (rock strata can be disrupted by some geological process, or deposited on uneven surfaces), it continues to provide an important framework to interpret the relative age of rock formations.

As discoveries continued to mount, it eventually became clear that fossils were not only the remains of living things, but that some of these living things were, well, no longer living. This idea was troubling because it directly contradicted the religious beliefs of the day. Among those advocating the concept of extinction was Georges Cuvier, a French scientist with a particular expertise in anatomy. He was one of the scientists who corresponded with and purchased fossils from Mary Anning (after originally being quite skeptical of the validity of her specimens), and he went on to become the foundational figure in the development of vertebrate paleontology. Cuvier studied the fossil remains of what were clearly elephants found near Paris and realized that they were different from that of living species. The popular answer at the time to this puzzle was that these species had just not yet been discovered. The famous Swedish scientist Carl Linnaeus (who developed the system of biological nomenclature we use today), argued these missing animals were "hiding" in unexplored territories of the Earth (Watson 2006). And the American philosopher (and later president) Thomas Jefferson believed that mammoths and giant lions could be found in the interior of the vast North American continent. Indeed, this belief was one reason behind Jefferson's commissioning of the Lewis and Clark expedition of the early 1800s. Cuvier found the idea of giant mammoths or dinosaurs roaming unexplored regions of the Earth implausible. A much simpler explanation was that they were extinct.

Gradually the weight of fossil evidence—the exciting discovery of "terrible lizards" (dinosaurs), flying and large aquatic reptiles, and the like—changed scientific understanding both of the nature of fossils and the concept of extinction (Owen 1841;

Watson 2006). Interestingly, Mary Anning's discoveries were to prove pivotal in this debate. Her skeletons of inexplicably strange creatures such as ichthyosaurs and plesiosaurs were hard to explain; they were simply unlike any living animal on Earth. Paleontology had become a legitimate discipline. Fossil hunting eventually became a popular pastime. The fossils that working class people like Mary Anning found were eagerly bought by wealthy collectors to add to their private collections (Emling 2009). Paleontology even gave rise to a scientific soap opera: the intense and nasty rivalry between Edward Drinker Cope of the Academy of Natural Sciences in Philadelphia and Othniel Charles Marsh of the Peabody Museum of Natural History at Yale (Box 2.1). This feud, which involved all sorts of underhanded behavior like theft, bribery, and even the deliberate destruction of irreplaceable materials, came to be called the Bone Wars (Colbert 1984; Jaffe 2000). It was not paleontology's finest hour.

Thomas Jefferson and the Great Claw

When most people think of Thomas Jefferson, they naturally think of the Declaration of Independence or perhaps recall that he was the third president of the nascent United States. Few realize that he was also deeply interested in science, especially paleontology. His home at Monticello was a virtual natural history museum, with all kinds of bones and other strange objects hanging on the walls.

Jefferson privately financed several scientific expeditions and was also responsible for commissioning the transcontinental Lewis and Clark expedition that began in 1804. While the primary objective was economic and included exploring the Louisiana Purchase and establishing trade with the Native Americans, Jefferson was also keenly interested in observations of natural history. Because Jefferson's interest was well known, his friends often sent him interesting curiosities. The most famous of these were some bones sent by Colonel John Stuart from West Virginia. Discovered by workmen mining sodium nitrate, the collection consisted of parts of an upper and lower arm with extremely large claws at the end of each finger. Jefferson was fascinated. By careful comparisons with the anatomical accounts of other animals, he ultimately decided it was a huge

Box 2.1
The Bone Wars

Despite the Hollywood image, scientists are usually pretty normal. Most don't walk around in white lab coats with thick Coke-bottle glasses and a bad fashion sense. Thus, it is no surprise that scientists are just as prone to jealousies, rivalries, and ambition as the average person. Indeed, the backdrops behind some of the greatest scientific discoveries involve Shakespearian-like dramas of betrayal and dishonesty. Perhaps the most infamous of these was the intense competition between the paleontologists Edward Drinker Cope from the Academy of Natural Sciences in Philadelphia and Othniel Charles Marsh from the Peabody Museum of Natural History at Yale. This feud, which eventually became the stuff of legends, was referred to as the Bone Wars.

It started out fairly innocuously. The two men met in Berlin in 1863 and apparently liked each other at first. They spent several days together and even corresponded afterward for a while. At the time, Cope was traveling through Europe, perhaps partially to avoid military service in the ongoing American Civil War. He had spent the previous five years working part time at the Academy of Natural Sciences in Philadelphia, publishing a number of papers despite his lack of a formal university degree. Throughout his life, Cope had difficulty landing a permanent academic job. Instead, he often used his own funds to support his research endeavors. In contrast, Marsh was a graduate student studying paleontology and anatomy in Berlin when they met. Upon returning to the United States, Marsh took up an appointment as a professor of vertebrate paleontology at Yale, a position facilitated by his wealthy uncle, who donated money to build a museum. Cope and Marsh clearly had at least a brief friendship; shortly after returning to his scientific pursuits in the United States, Cope named an amphibian fossil (*Ptyonius marshii*) after Marsh, and Marsh quickly reciprocated by naming a fossil mosasaur in honor of Cope (*Mosasaurus copeanus*).

But the friendship was short-lived. It began to sour when Cope took Marsh to a Cretaceous fossil locality in New Jersey where a skeleton of *Hadrosaurus* had been found; Marsh quietly made a better deal with the owner of the site so that any new fossils would be sent to him at Yale. Cope was livid when he found out about Marsh's duplicity. But the death knell was Marsh's discovery of a mistake in Cope's reconstruction of a fossil plesiosaur (a large aquatic animal with four flippers and a long neck and tail). Cope had put together the extinct marine reptile with a short neck and a long tail. The men argued vehemently about the reconstruction and eventually called in the country's leading anatomist, Joseph Leidy, to arbitrate.

Later, when recounting this episode, Marsh wrote, *"The skeleton itself was arranged in the Museum of the Philadelphia Academy of Sciences, according to this restoration, and when Professor Cope showed it to me and explained its peculiarities, I noticed that the articulations of the vertebrae were reversed and suggested to him gently that he had the whole thing wrong end foremost. His indignation was great, and he asserted in strong language that he had studied the animal for many months and ought to at least know one end from the other"* (January 19, 1890, *New York Herald*). Leidy agreed that Cope had placed the skull on the wrong end of the animal. Cope was so humiliated by this mistake that he tried to buy back all copies of the journal where he had published the original description. He was only partially successful. Thus, the feud began.

Although paleontology arrived late to North America, it was now taking off with a vengeance. In 1867, Congress authorized a number of federally funded geological expeditions in the American West, including the King, Hayden, Powell, and Wheeler surveys. The completion of the Transcontinental Railroad a few years later not only allowed easier access to the West, but also exposed a number of new fossil localities. Both men went that direction, at first as members of the government surveys. Competition between Cope and Marsh was fierce. Spies were a constant concern; both Cope and Marsh took to bribing workers from opposing expeditions to report on discoveries and localities. They were so paranoid about being "scooped" that they sometimes destroyed or damaged fossils or even filled in excavations to prevent their opponent from recovering them. Reportedly, there were even occasions where opposing expeditions threw stones at each other. Fossils were stolen; shipments deliberately misdirected. Marsh's treatment of his assistants (who were forbidden to publish under their own names) was apparently quite harsh; many of his assistants later became his most outspoken critics.

And it got worse. Marsh used his growing political influence to become the chief paleontologist of the newly formed US Geological Survey, which not only gave him extensive access to government funding but also allowed him to prevent Cope from doing so. Marsh was enormously successful in his discoveries; he named and described over one hundred dinosaurs, including our most beloved genera: *Apatosaurus* (i.e., brontosaurus), *Stegosaurus*, *Allosaurus*, *Diplodocus*, and *Triceratops*. For a time, Cope relied on his inheritance to fund expeditions that were able to search for fossils year-round, but ultimately his money dried up. Desperate, he invested in silver mines in New Mexico, but these failed and Cope became impoverished. Unable to outcompete Marsh in the field, he focused on trying to ruin Marsh's professional credibility with his most potent weapon, his pen.

Cope's scientific output was legendary. During his thirty-seven-year career, he published over thirteen hundred articles describing and naming more than six hundred vertebrate species, including hundreds of fishes and dozens of dinosaurs—a truly astonishing achievement. Today Cope is probably best known for the rule that bears his name: the idea that lineages increase in size over geologic time. Cope's productivity was aided by his purchase in 1878 of the rights to the scientific journal the *American Naturalist*, which he edited until his death. Cope's animosity toward Marsh was not subtle. Indeed, he published many papers with titles such as "Remarks on discoveries recently made by Prof. O.C. Marsh," "Critical review of Marsh's contributions . . . ," "On some of Prof. Marsh's criticisms," "Reply to Marsh's criticism in *Nature*," "Criticism of Marsh's paper . . . ," and so forth.

But both men had made many mistakes. The rush to describe new discoveries led to many hastily written and inaccurate descriptions. When first describing one of the largest dinosaurs ever, *Apatosaurus*, Marsh mistakenly put the head of a *Camarasaurus* on it. When he later found a complete *Apatosaurus* skeleton, he described it as a new genus of dinosaur: *Brontosaurus*. The mistake was spotted within a few years, but the name lingered on for almost another hundred years. A fellow scientist of the time, Henry Osborn, characterized the situation as "nomenclatural chaos" and later "*joked that a 'trinomial system' emerged for naming prehistoric animals: an original Leidy name and the two Cope and Marsh names*" (Wallace 2000, p. 83). Consequently, there were many errors by both Marsh and Cope to highlight, and both men were merciless in their critiques.

The feud came to a boil in January of 1890 when a series of articles entitled "Scientists wage bitter warfare" appeared in the *New York Herald*. These articles and rebuttals by Cope and Marsh made harsh accusations of fraud, plagiarism, and scientific malfeasance. The debate that had formerly been confined to the scientific community was now public, and the ugliness of it tarnished the scientific reputations of both men. Shortly thereafter, Congress eliminated the entire Department of Paleontology and drastically cut the US Geological Survey budget. In the process, Marsh lost his position and power, and indeed, much of his income.

The feud died down somewhat with the removal of Marsh from a position of power. Without resources, neither man could mount the expeditions of earlier years. Cope ultimately sold much of his collection to the American Museum of Natural History and became more financially stable. He died prematurely at age fifty-six in his study at his museum with his last conscious sight one of "*giant bones, piled on every side of his cot*" (Osborn 1930, p. 141). Marsh followed just a few years later after suffering serious financial hardships. In his will, Cope left his brain to the Academy of Science in Philadelphia. This act spurred an apocryphal rumor that upon his deathbed, he challenged Marsh to do the same so that their rivalry could ultimately be settled. While this is a great story, it appears untrue.

So who won? While Marsh discovered many more new fossils, a hundred years later Cope's scientific legacy and reputation is more secure. Marsh has remained a somewhat murky figure, but Cope seems less tarnished. In part, this may be because Cope was likeable. By all accounts, he was more charismatic and well regarded than Marsh. But Cope's prodigious scientific output and his evident love of science were also responsible. What is clear is the high personal and professional price both men paid for their bitter rivalry; to this day the Bone Wars are more memorable than any of their remarkable accomplishments.

unknown cat of some sort (Fig. 2.3). Presenting his findings at the American Philosophical Society meetings in Philadelphia in March of 1797, Jefferson said, "*Let us only say then, what we may safely say, that he was more than three times as large as the lion: that he stood as pre-eminently at the head of the column of clawed animals as the mammoth stood at that of the elephant, rhinoceros and hippopotamus: and that he may have been as formidable an antagonist to the mammoth as the lion to the elephant*" (Jefferson 1799, p. 251).

Jefferson named this creature *Megalonyx* ("great claw"). Along with other ideas of his day, that of extinction was a novel and implausible concept; he was convinced that these animals were to be found

somewhere within the unexplored reaches of the vast American continent. He bolstered his argument with vivid accounts of frightened frontiersman that had been terrified by large unknown animals that roared, as well as rock drawings resembling lions that had been found out west. He went on to say, "*In the present interior of our continent there is surely space and range enough for elephants and lions, if in that climate they could subsist; and for mammoths and megalonyxes who may subsist there. Our entire ignorance of the immense country to the West and North-West, and of its contents, does not authorize us so say what is does not contain*" (Jefferson 1799, p. 252).

Regrettably for Jefferson, he soon discovered he was wrong on both counts. A few months before his talk, he encountered an article written by the eminent scientist Georges Cuvier in Paris describing a newly found fossil from Paraguay. It closely resembled his. The engravings of the almost complete skeleton depicted a creature with huge claws, but it was clearly related to living sloths. Accordingly, Jefferson added an addendum in which he acknowledged the find simply as *"also of the clawed kind"* (1799, p. 259). However, he remained hopeful that Lewis and Clark might encounter his beast in their travels. That, alas, did not come to pass.

Jefferson was credited with the discovery of the giant ground sloth, and in 1822 the French scientist Anselme Desmarest named it after him (*Megalonyx jeffersonii*). From more complete fossils, we now know it was a massive animal about 3 m long and weighed over 600 kg. Its three well-developed claws were probably used to strip leaves or manipulate branches (Fig. 2.3). The original bones are archived in the Thomas Jefferson Fossil Collection at the Academy of Natural Sciences in Philadelphia.

Fossils and Fossil Types

So what exactly *is* a fossil? The simplest explanation is that they are the preserved remains (or traces) of something that was once alive. They can range in size from microscopic bacteria preserved in rock to parts of giant trees or sauropod dinosaurs that weigh many tons and take a helicopter to remove. Fossils are *the* basic unit of paleoecological study and inference. They are crucially important. Without fossils, we would know little about the history of life on Earth. Because they are the actual remains of an organism that once lived, they provide the only direct historical evidence we have.

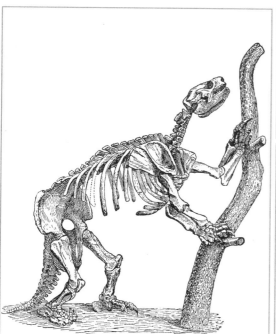

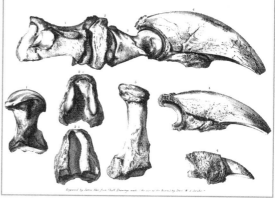

Fig. 2.3. *Megalonyx jeffersonii,* or Jefferson's ground sloth. *Left*: An illustration of how this animal may have used its large claws to feed on the leaves of trees. Note the long muscular tail and plantigrade stance that allowed a semistanding posture. *Top right*: Thomas Jefferson's measurements of various *Megalonyx* bones and the comparison to published accounts of the African lion. Note the substantial differences in the size, which led Jefferson to conclude the bones belonged to an enormous lion. *Bottom right*: Drawings of the strange bones Jefferson received from West Virginia. Caspar Wistar, a physician and anatomist, provided these detailed illustrations and descriptions of Jefferson's specimen in the same issue of the *Transactions of the American Philosophical Society*. His examination did not support the idea that these belonged to a large lion. *Clockwise from left*: Wistar 1799; Jefferson 1799; Wistar 1799.

While we can glean considerable information from indirect proxies (e.g., the analysis of genetic sequences from extant organisms), interpretation of these is always somewhat subjective. They can be misleading sometimes when recreating evolutionary events. The recent controversy over the origin of placental mammals is a classic case of how indirect evidence, in this case from molecular studies, can lead to vastly different conclusions than the fossil record (see for example: Bininda-Emonds et al. 2007; Benton and Donoghue 2007; Wible et al. 2007; dos Reis et al. 2012). Estimates based on molecular studies of extant mammals place the origin of placental mammalian orders at 100-85 million years ago, much earlier than the fossil record would indicate. If true, this means there is a 20-50 million year "gap" in the fossil record where modern forms should, but have not been, recovered. Many paleontologists argue this as highly implausible, especially considering the many well-resolved Cretaceous mammal fossil localities. Instead, they critique the methodologies used in the molecular studies, especially the underlying assumption that evolution proceeds at a constant pace. The ultimate resolution of the divergence in time of placental mammals is likely to require both the fossil record and methodologies from molecular biology. Stay tuned.

Fossils can be found in two major forms: body fossils, which record morphology and thus tell us what organisms looked like, and trace fossils (or ichnofossils), which provide information about how an organism lived (Fig. 2.4a). Body fossils include bones, shells, molds, or impressions (Figs. 2.4b and 2.4c). Sometimes we are lucky and we get the entire organism—mammoths stuck in permafrost or insects embedded in amber. But, generally, it's just a part of the organism. Trace fossils consist of footprints or walkways, burrows, root traces, or even the byproducts of biological processes, such as digestive waste (known in polite circles as "coprolites"; Fig. 2.4d).

Fig. 2.4. Types of fossils. **a**, Miocene root trace from Espanola, New Mexico. **b**, Pleistocene Mastodon tooth (locality unknown). **c**, Endocast of nautiloid from the Upper Ordovician of northern Kentucky. **d**, Massive woodrat paleomidden collected in 2009 from Death Valley National Park, California. Panels a, b, d: by the author; c: by Mark A. Wilson, courtesy of Wikipedia Commons.

Because trace fossils can be difficult to attribute to a particular organism, an entire different taxonomy is assigned to them. This results in the peculiar situation where a walkway of a particular mammal or dinosaur can have a different scientific name than the bones of the same animal.

The fossils that Mary Anning collected along the cliff shores of Dorset were primarily body fossils, although she also sometimes recovered trace fossils. Indeed, Mary Anning was perhaps one of the first to realize what coprolites were. She occasionally found "bezoar stones" (strange conical objects named after stones sometimes found in the intestine) in or near the specimens she collected. The renowned English geologist William Buckland from Oxford was a relatively frequent visitor to Mary's shop. On one such occasion, she suggested to him that the fossilized bezoar stones she sometimes found associated with ichthyosaur fossils might actually be fossilized dung. She had noticed that when bezoar stones were broken open they often contained parts of fish and other bones. Buckland, who subsequently had the dubious pleasure of coining the word "coprolite," acknowledged Mary's help when he presented his findings, unlike many of the other scholars of his day (Emling 2009).

A Very Brief Review of the Fossil Record

While paleontologists no longer debate about how to categorize a fossil, there remains a lively debate about the validity of the oldest ones (Schopf 1993; Brasier et al. 2002; Dalton 2002; Schopf 2006; Marshall et al. 2011). This controversy stems from several reasons: the very oldest specimens are microscopic, making it extremely difficult to unambiguously determine whether they were once alive, and about 90% of Precambrian rocks have been reworked by geologic processes (Garrels and Mackenzie 1971), making it diminishingly improbable that fossils can even be recovered (Schopf 2006). While there isn't much one can do about the scarcity of Archean rock, rigid criteria have been established to help differentiate between microfossils and abiotic "look-alikes" (Buick 1990). In general, these set up benchmarks to determine if specimens are biogenic (usually taken to mean having morphological structures that are consistent with living things and/ or evidence of biochemical products) and whether

the putative fossils are clearly indigenous to rocks of known provenance and well-defined Archean age.

To date, the majority of ancient fossils—from Archean formations in Western Australia—are generally agreed to be around 3.5 to 3.4 billion years old (Schopf 2006; Wacey et al. 2011; Fig. 2.5). These microorganisms lived shortly after the formation of liquid water and cessation of cosmic bombardment. The Earth was a vastly different place at this time; there was virtually no land, volcanic activity was rampant, a day was only about ten hours long, and the oceans were hot—around 40-50°C. The moon was much closer to the Earth and would have been huge in the nighttime sky. Because little oxygen existed in the atmosphere, these early organisms were most likely sulfur-based (Wacey et al. 2011).

Quite a lot of Earth history passed between those early microorganisms and the evolution of the subject of this book, mammals. The first true mammals made their appearance in the late Triassic (~210 Ma). Over the ensuing 3.2 billion years, the Earth slowed, solar radiation increased, life profoundly altered the composition of the atmosphere, land masses formed and moved around the globe, ice ages came and went, meteorites slammed into the Earth, and volcanoes erupted (Fig. 2.6, Table 2.1). The first microorganisms were eventually joined by much larger ones; the range in body size increased by approximately sixteen orders of magnitude (Payne et al. 2009). Interestingly, these increases in size occurred in a stepwise fashion, with jumps of about six orders of magnitude in the mid-Paleoproterozoic (~1.9 Ga) and again during the late Neoproterozoic/early Paleozoic (600-450 Ma). They also co-occurred with major innovations in organismal complexity—first the evolution of the eukaryotic cell and later multicellularity—and in association with substantial increases in atmospheric oxygen concentration (Bambach et al. 2007; Payne et al. 2009). Since the early Paleozoic, even considering the evolution of giant sauropods and baleen whales, the size of living things has increased by only a few more orders of magnitude.

As organisms became biologically more complex, they developed different modes of life—active hunting versus passive, for example (Bambach 1983). The expansion of ecological niches is recorded in the

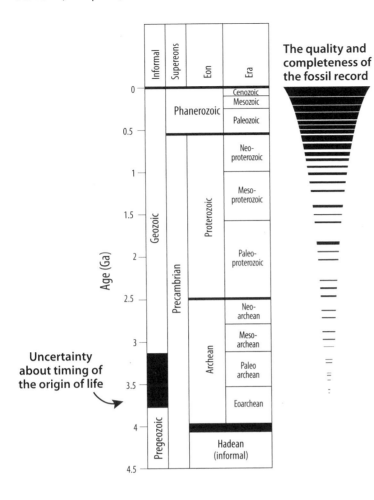

Fig. 2.5. The Geozoic supereon. A few years ago, a group of paleontologists realized that there was no longer a term that described the period of Earth time where life had existed (Kowalewski et al. 2011). The term Phanerozoic (visible life) was no longer useful since it was apparent that life had originated and diversified long before the Cambrian (e.g., Schopf 2006). Thus, they proposed the creation of a chronostratigraphic unit, the Geozoic, as an informal supereon to denote the time of life's existence on Earth; the Pregeozoic refers to the time before life existed on the planet. The terms come from the Greek *geo* (earth) and *zoic* (life). The lower and upper boundaries of the Geozoic were defined by the first and last appearance of life, respectively. Redrawn from Kowalewski et al. 2011.

increased diversity of trace fossils from the Cambrian (some 541 Ma). And this too is when the progenitors of almost every modern phylum of metazoan evolved (Fig. 2.7). The first early vertebrates arose in the oceans shortly thereafter in the Ordovician, which also saw the very beginnings of the colonization of land. Certainly by sometime in the Silurian (440 Ma), land plants had colonized terrestrial environments. The first reptiles appeared in the Pennsylvanian and were spectacularly diverse by the Permian, as were insects and other invertebrates. However, it would be wrong to conclude that the evolution of life on Earth has been progressive. Complex life is not necessarily any more successful than unicellular life. Indeed, the first forms of unicellular organisms were so well adapted that something akin to these forms continues to persist today. Some argue that microscopic life has been the most successful over the history of the Earth; they

certainly represent most of the standing biodiversity of our planet (Horner-Devine et al. 2004).

Intertwined with the history of life is the geological history of the planet. Over the past 4.6 billion years, there have been major changes in continental shape and location, in the composition of the atmosphere and oceans, and in Earth's climate. These and other events influenced the trajectory of evolution; in addition to restricting the taxonomic and geographic distribution of species, they have had a detectable and lasting impact on origination rates and overall diversity (Krug et al. 2009). Paleontologists commonly recognize five mass extinctions, when a significant and diverse array of life around the globe went extinct in a relatively short time frame. The cause for most of these is still hotly debated, but possibilities include both extraterrestrial and terrestrial factors such as impacts, glaciation, or volcanism; changes in the ocean circulation and/or chemistry; and plate

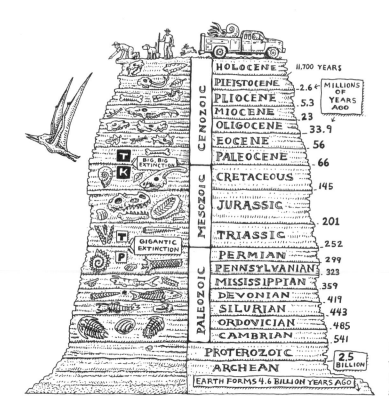

Fig 2.6. Depiction of Earth history over time. Copyrighted by Ray Troll 2019. Reprinted with permission from the artist.

tectonics (see Raup and Sepkoski 1982; Ward 2007; Bailer-Jones 2009; Munnecke et al. 2010).

The oldest resolvable event is the end Ordovician (~445 Ma), which resulted in the extinction of perhaps 65% of all species on the planet. It particularly affected brachiopods and bryozoans; two-thirds of all families perished. This was followed by the late Devonian (~365 Ma), which may have affected 70%-72% of species but was most severe in marine realms. The closest life on our planet has come to total annihilation was during the end Permian event (~251 Ma), which appears to have wiped out 90%-95% of all species (Erwin 2006). Some have argued that only a few thousand species survived (Erwin 2006). There were two further mass extinctions in the Mesozoic: the end Triassic (~210 Ma), which affected about 60%-65% of species, and the one most are familiar with, the end Cretaceous (65 Ma), which eliminated around 60% of species, including all the non-avian dinosaurs. Only for this last event is the evidence clear enough to unambiguously assign a cause (Alvarez et al. 1980; Schulte et al. 2010). Understanding the consequences and causes of previous

mass extinctions is hampered by the diminishing rock record as one goes back further in time, the higher probability of the reworking of the Earth surface, and a lowered resolution of the fossil record.

The Early History of Mammals

Synapsids—the early ancestors of mammals—arose in the late Carboniferous around 315 Ma. This group was distinguished from other amniotes by the presence of a temporal fenestra—a bilaterally opening in the skull behind the orbit of the eye, which may have been associated with more robust jaw architecture. Both synapsids and their more advanced mammal-like form, the therapsids, which arose later, probably laid parchment-shelled or leathery eggs, similar to that of modern monotremes. They had a more erect posture, may have had hair and appear to have had differentiated and ecologically specialized teeth (Benton 2005; Kemp 2006; Ji et al. 2006; Angielczch et al. 2013; Zhou et al. 2013). Therapsids became quite diverse in both ecology and morphology during the Permian, dominating terrestrial ecosystems. The therapsid species that

Table 2.1. An abbreviated history of the Earth, with the approximate timing of various evolutionary innovations and geologic events. The five major mass extinctions are not shown.

Eras		Periods		Epochs	Aquatic Life	Terrestrial Life
Cenozoic (65) Age of Mammals	Paleogene | Neogene	Quaternary (2.6)	Glacial cycles (~20) Continent drift continues	Holocene (.0117)	All modern groups present	Humans colonize the Americas
				Pleistocene (2.6)		*Homo* evolves; large forms of mammals present
				Pliocene (5.3)		Arctic ice cap forms
		Tertiary (66)		Miocene (23)		Widespread grasslands develop
				Oligocene (33.9)		Adaptive radiation of birds
				Eocene (56)		Adaptive radiation of mammals and herbaceous angiosperms
				Paleocene		
Mesozoic (251.9) Age of Reptiles		Cretaceous (145)	North America and Northern Europe are still attached, as are Australia and Antarctica; shallow seas recede and transgress		Modern bony fishes; extinction of ammonites, plesiosaurs, etc.	Ends with the extinction of non-avian dinosaurs; rise of woody angiosperms, snakes
		Jurassic (201)	Africa and South America begin to drift apart at the end; North America and South America separate; global transgression of shallow seas		Plesiosaurs and ichthyosaurs are abundant	Dinosaurs dominant; mammals, lizards, and angiosperms appear at beginning; insects abundant
		Triassic (251.9)	Pangaea splits into Laurasia and Gondwana		First plesiosaurs; ammonites abundant; rise of bony fishes	Adaptive radiation of reptiles; therapsids, turtles, crocodiles, and dinosaurs arise
Paleozoic (541) Ancient Life		Permian (299)	Periodic glaciation; arid climate; ends with formation of Pangaea		Ends with worst mass extinction in history	Reptiles abundant; cycads, conifers, ginkgos
		Pennsylvanian (323)	Warm humid climate; Pennsylvanian and Mississippian together make up the Carboniferous, or Age of Amphibians		Ammonites, bony fishes	First reptiles
		Mississippian (358.9)			Adaptive radiation of sharks	Forests of lycopsids, seed ferns; amphibians abundant; land snails
		Devonian (419) Age of Fishes	Extensive inland seas		Cartilaginous and bony fishes, ammonites, and nautiloids	Ferns, lycopsids, first gymnosperms, insects, and amphibians
		Silurian (443.8)	Mild climate; inland seas; Australia near equator		Nautiloids and other molluscs; jawed fish at end	More organisms colonize land
		Ordovician (485.4)	Mild climate/glaciation; inland seas; most landmasses located in southern or equatorial latitudes		Trilobites abundant; jawless vertebrates	Fungi and bryophytes; millipedes?
		Cambrian (541)	Periodic glaciation at start; single continent of Gondwana		Crustaceans, molluscs, annelids sponges, echinoderms, etc.	Lichens, mosses, perhaps some vascular plants?

(Phanerozoic Eon appears vertically along the left margin alongside the Mesozoic and Paleozoic eras.)

(continued)

Table 2.1. (continued)

Eras	Periods		Epochs	Aquatic Life	Terrestrial Life
Proterozoic Eon (2,500)	Neo-, Meso-, Paleo-		Oldest eukaryotics ~2.0–1.8 Ga; evolution of multicellularity, sex, metazoans, etc.; Snowball/Slushball Earth		
Archean Eon (4,600)	Neo-, Meso-, Paleo-, Eo-		Earth formed; earliest life evolves around 3.6 to 3.8 Ga		

Source: Dates are from the International Commission on Stratigraphy's International Chronostratigraphic Chart, version 2018/07 (https://stratigraphy.org/files/ChronostratChart2018-07.pdf).

Note: Starting dates in millions of years ago (Ma) are shown in parentheses. Table is not to scale; keep in mind that approximately seven-eighths of Earth history is represented by the Proterozoic and Archean Eons.

survived the catastrophic end-Permian extinction ultimately gave rise to mammaliaforms around 225 Ma (Kemp 2006).

The oldest described mammal fossil is that of a shrew-like form from the early Jurassic (Luo et al. 2001). It lived around 195 Ma in what is now China and weighed around 2 g (about the size of a standard paper clip). These primitive mammals were likely monotremes and laid eggs; only five species of this clade are extant today (four species of echidna and the duck-billed platypus). Metatherians (extant marsupials and their relatives) originated by the late Jurassic or early Cretaceous and radiated across Laurasia by the late Cretaceous (Williamson et al. 2014). While somewhat limited in size (the largest species was only ~180g), they remained taxonomically and morphologically diverse until the early Cenozoic (Cifelli 2004; Williamson et al. 2014), when Eutherian (placental) mammals replaced them on many continents. The divergence of Metatheria and Eutheria likely occurred around 168-178 Ma (dos Reis et al. 2012).

While mammals got larger over the Mesozoic—the largest therians were about 10-15 kg (Hu et al. 2005; Smith et al. 2016a)—they still remained a fairly insignificant part of the terrestrial biosphere. Certainly a variety of feeding strategies had evolved by the middle of the Mesozoic, but the restricted range of body mass severely constrained the potential roles of mammals in paleocommunities (Lillegraven et al. 1979; Crompton 1980; Kielan-Jaworowska et al. 2004). Most mammals were probably insectivores, omnivores or scavengers (Lillegraven et al. 1979). Mammalian herbivory, as we

think about it today, probably did not evolve until the Cenozoic; various physiological constraints make it difficult to digest plant materials for animals less than 5 kg (Lillegraven et al. 1979; Van Soest 1994). A few of the larger mammals may have preyed on the smaller-bodied dinosaurs (Hu et al. 2005), but this cannot have been that common given the large disparity in body size between the taxa (Fig. 2.7).

In contrast to the limited role of mammals, dinosaurs evolved gigantic sizes and diversified into a wide variety of forms and lifestyles (Benson et al. 2014). Why dinosaurs were so diverse and mammals so restricted in their ecologies may have been a matter of timing: dinosaurs had a 20 million year head start. Although mammals may have begun to outcompete the smaller-body-sized dinosaurs in the late Mesozoic (Fig. 2.7), it was really only with the extinction of the dinosaur clade at the Cretaceous-Paleogene (K-Pg) boundary around 66 Ma that mammals were able to fully radiate. This they did quickly; by 41 Ma mammals had diversified into the full range of ecological roles and body sizes they occupy today (Alroy 1999a; Smith et al. 2010a; Fig. 2.8).

Since the Cenozoic, Eutherian mammals have dominated the terrestrial biosphere, both morphologically and ecologically. The largest mammals—members of the elephant family such as *Deinotherium* and the Perissodactyla *Indricotherium*, each weighing around 18 tons—evolved by the late Oligocene or early Miocene (Smith et al. 2010a). While ecological release from dinosaur competition was clearly important in allowing the evolution of larger body size in mammals, the maximum size attained over the Cenozoic appears to be strongly

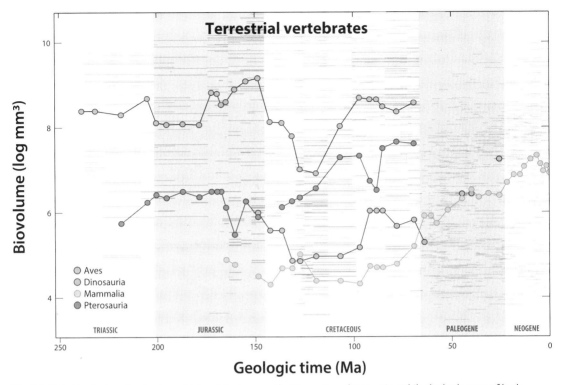

Fig. 2.7. Global body size of major terrestrial vertebrates over the Mesozoic and Cenozoic. While the body mass of both dinosaurs and pterosaurs display a passive trajectory, the mass of mammals and birds were active, or directed, evolutionary trends; notice the increases prior to the Cretaceous-Paleogene boundary (K-Pg). Redrawn from Smith et al. 2016a.

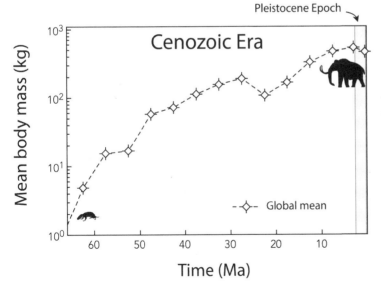

Fig. 2.8. Global mammal body size over the past 66 million years. After their origination in the early Mesozoic, mammals were restricted to small-bodied forms, although they did eventually come to occupy a variety of terrestrial and aquatic niches. It was only after the extinction of non-avian dinosaurs about 66 Ma that there was a rapid radiation of body mass. Redrawn from Smith et al. 2010a. Animal silhouettes from PhyloPic (http://phylopic.org) under Creative Commons license.

constrained by temperature (Smith et al. 2010a; Saarinen et al. 2014). Indeed, environmental temperature has had a significant role in constraining mammal body size across *both* space and time (Bergmann 1847; Mayr 1956; Smith et al. 1995; Smith and Betancourt 1998, 2003, 2006; Ashton et al. 2000; Millien et al. 2006; Smith et al. 2010a); this is discussed further in chapters 6 and 10. Interestingly, the trends in body size over the Cenozoic are replicated on all continents, both at the class and ordinal level (Smith et al. 2010a; Smith and Lyons 2011), with mammalian orders reaching their maximum size at the same times on the different continents despite dissimilar taxonomic compositions (Saarinen et al. 2014).

FURTHER READING

Alvarez, W. 1997. *T. rex and the Crater of Doom.* Princeton, NJ: Princeton University Press.

Benton, M.J. 2003. *When Life Nearly Died: The Greatest Mass Extinction of All Time.* London: Thames and Hudson.

Benton, M.J. 2005. *Vertebrate Paleontology.* Oxford: Blackwell Science.

Eldredge, N. 2014. *Extinction and Evolution: What Fossils Reveal about the History of Life.* Ontario: Firefly Books.

Emling, S. 2009. *The Fossil Hunter: Dinosaurs, Evolution, and the Woman Whose Discoveries Changed the World.* Basingstoke, UK: Palgrave Macmillan.

Erwin, D.H. 2006. *Extinction: How Life Nearly Ended 250 Million Years Ago.* Princeton, NJ: Princeton University Press.

Raup, D.M. 1991. *Extinction: Bad Genes or Bad Luck?* New York: W.W. Norton.

3 Taphonomy

Putting the Dead to Work

At Midnight in the museum's hall
The fossils gathered for a ball
There were no drums or saxophones
But just the clatter of their bones,
A rolling, rattling, carefree circus
Of mammoth polkas and mazurkas.
Pterodactyls and brontosauruses
Sang ghostly prehistoric choruses.
Amid the mastodonic wassail
I caught the eye of one small fossil.
Cheer up, sad world, he said, and winked—
It's kind of fun to be extinct.
—Ogden Nash, "Carnival of Animals," *New Yorker*,
   January 7, 1950, p. 26

Fossilization is rare. To become a fossil (Fig. 3.1), an organism must be buried by sediments fairly rapidly, or it decomposes and/or is eaten by hungry scavengers or microbes or destroyed by physical processes. Moreover, the organism generally has to have hard parts that can be preserved in the first place. And it has to die in—or be carried rapidly to—an appropriate site with a high rate of sedimentation. Furthermore, even when an organism does become a fossil, the probability that we will encounter it is fairly low (Fig. 3.2a). Over time, fossils are often reworked or destroyed by geological forces (Fig. 3.2b). Many are buried under sediments and not exposed, or in more recent times,

they are run over by overzealous tractor operators at construction sites. Further, a paleontologist has to be there to find and identify a fossil as something of interest when, and if, it ultimately sees the light of day.

There are always some exceptions to these general principles. For example, mammoths sometimes had the misfortune to take a bad step and fall into an ice crevice or pothole where they froze or were buried quickly (Fig. 3.3). This seems to have happened with some regularity in the Pleistocene. With our rapidly changing climate, these animals are reappearing at Far North latitudes: a grim scientific bonanza (Ryder 1974; Guthrie 1990; Boeskorov et al. 2007; Pabst et al. 2009; Kosintsev et al. 2010; Thali et al. 2011; Fisher et al. 2012; Papageorgopoulou et al. 2015; Boeskorov et al. 2016; Grigoriev et al. 2017; Boeskorov et al. 2018). The freeze-dried finds of mammoths, horses, steppe bison, shrub ox, and even lions have allowed a much more fine-grained examination of the ecology of ancient animals; often the hair, flesh, and even gut contents are intact (e.g., Ryder 1974; Mol et al. 2001; Kosintsev et al. 2010; Fig. 3.4).

But, in general, all we typically get is a part of something that was once alive. The study of the natural processes influencing fossilization and the biases in the fossil record is a very active area of research (Behrensmeyer and Hill 1980; Shipman 1981; Behrensmeyer and Kidwell 1985; Behrensmeyer et al.

Fig. 3.1. Vertebrate Paleontology Laboratory research collection at Texas Memorial Museum in Austin, Texas. Photo by author.

2000; Terry 2004). A comprehensive understanding of how something became a fossil is crucially important to understanding its life history and ecology and interpreting the remains. This subdiscipline of paleontology is called *taphonomy*.

What Is Taphonomy?

The term "taphonomy" comes from the Greek *taphos* (τάφος) meaning "burial," and *nomos* (νόμος) meaning "law," thus, the word means "laws of burial." It is a discipline within paleontology that seeks to understand the biases in the formation and preservation of fossils, which can influence our interpretation of the fossil record (Shipman 1981; Carroll 1988; Behrensmeyer and Kidwell 1985; Behrensmeyer et al. 2000). Thus, it includes postmortem surface processes such

as scavenging, microbial degradation, exposure, or transport; subsurface processes such as burial compaction, bioturbation, erosion, and fossilization; and even postcollection processes such as collecting bias and decomposition that can occur during storage (Fig. 3.2a). While the term was coined in 1940 by Russian scientist Ivan Efremov to denote the formal study of the transition of organisms and their parts or products from the biosphere to lithosphere, Efremov certainly wasn't the first to study these processes. For example, by comparing the properties of living and dead bivalves, Leonardo da Vinci concluded that fossils he found in the nearby Italian mountains hadn't gotten there by a biblical flood but rather had lived and died in situ at a time when these areas were underwater (Baucon 2010).

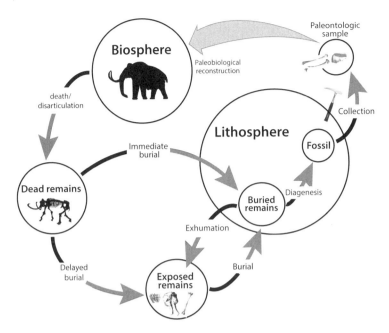

Fig. 3.2a. Taphonomic cycle. Paleo-ecologists try to reconstruct the life history and ecology of animals from their dead remains. However, at each stage of the fossilization process, numerous factors can lead to distortion of the signal (indicated by the transition from black to grey arrows). The trick is to figure out what these biases are and to compensate as much as we can for them. Animal silhouettes from PhyloPic (http://phylopic.org) under Creative Commons license. Other fossil images by author.

Fig. 3.2b. Photograph of the inside of Hall's Cave in 2014, looking out. Deposition at this site largely results from fluvial processes; note in the foreground the bones of recently dead animals, which have washed into the cave. An over six-foot-tall geoarchaeologist (Mike Waters) is shown for scale. Courtesy of Thomas Stafford, Jr.

More recent taphonomic studies have demonstrated that swifter river channels tend to accumulate bones from a regional rather than local level, but oxbows (sharp U-shaped curves in a river channel) or meander bends usually represent a more localized assemblage; these distinct modes of bone accumulation can be recognized in fluvial deposits throughout the Phanerozoic (Behrensmeyer 1988). A particularly interesting example of paleontological taphonomy, or paleoforensics, was in determining the probable death of the Taung Child. This fossil of a young *Australo-pithecus africanus* was discovered in 1924 by workers in a quarry in South Africa. It was described as a new species of hominin by Raymond Dart (1925), who assumed that the child had lived and died in the cave where it was found. Based on this context, he proposed that early hominins lived in open grasslands like those near the cave and, moreover, based on the other bones found, had hunted other mammals.

However, recent studies of the surface damage on the skull yielded another story: the gouges and punctures in the orbital, frontal, temporal, parietal,

Fig. 3.3. An in situ "ice mummy." Shown is a six-to-nine-month juvenile woolly mammoth baby nicknamed Dima. She was found in exposed permafrost near the Kirgiljach River in northeast Siberia. Her remains dated to 37,000 B.C. Photo by A.V. Lozhkin for the US National Oceanic & Atmospheric Administration, 1977, courtesy Wikimedia Commons.

Fig. 3.4. Not all animals found in permafrost are giants. Shown here is a mummified arctic ground squirrel (*Spermophilus parryii*), which was unlucky enough to have been frozen in permafrost some 20,000 years ago. "Squirrel Mummy" by Ryan Somma, CC BY-SA 2.0, courtesy of Wikimedia Commons.

and occipital regions suggested the child had been captured and eaten by a large raptor (Berger and Clarke 1995; Berger and McGraw 2007). This interpretation was possible because of detailed taphonomic studies conducted on the types of damage caused by modern raptor predation, including the crown eagle (Berger and McGraw 2007). Raptors that large can prey on mammals weighing as much as 30 kg, which is well within the mass of the Taung Child (Ferguson-Lees and Christie 2001; Berger and McGraw 2007). A more likely story is that most of the bones within the cave were the remains of carnivore foraging, and that these early hominins were probably prey and not (yet!) predators (Berger and Clarke 1995; Berger and McGraw 2007). Moreover, the child was probably transported from some distance, which also cast doubt on the interpretation that these early *Australopithecus africanus* lived in the open grasslands near the cave (Berger and Clarke 1995; Berger and McGraw 2007).

Modern taphonomy studies have expanded the focus on postmortem biases in preservation to include the processes that lead to accumulation of biological remains, the microbial and biogeochemical influences on tissue preservation, and especially, through the use of live-dead studies, the spatial and temporal resolution, documentary quality, and ecological fidelity of the fossil record (Behrensmeyer and Hill 1980; Behrensmeyer and Kidwell 1985; Kidwell and Flessa 1995; Behrensmeyer et al. 2000). I elaborate more on these issues with particular emphasis on how they pertain to mammals.

Types of Fossils

As discussed in chapter 2, fossils can be found in two major forms: body fossils, which record morphology and thus tell us what organisms looked like, and trace fossils (or ichnofossils), which provide information about how an organism lived (see Fig. 2.4). Body fossils include bones, shells, molds, or impressions while trace fossils consist of footprints or walkways, burrows, root traces, or even the byproducts of biological processes such as digestion. Thus, a fossil does not necessarily have to be "fossilized" to be called a fossil, it just needs to be preserved in some way. Although some researchers restrict the term to specimens older than some minimum age, often, and somewhat arbitrarily, more than 10,000 years old (Shipman 1981; Carroll 1988).

The fossils we find today are usually those parts of an organism that were partially mineralized when it was alive. For vertebrates, this usually means the bones and teeth; soft tissue preservation is only rarely found in the fossil record (see "Fossil Lagerstätten" later in this chapter). However, we do sometimes encounter impressions of soft parts, such as skin (Fig. 3.5). Work on fossil integument from *Tyrannosaurus* and other tyrannosaurids has revealed that these large-bodied theropods had lost the

Fig. 3.5. Sometimes skin impressions—or even hair—are left intact on specimens. Here, fur was preserved on a woolly mammoth (*Mammuthus columbi*) frozen in permafrost. This specimen is on display at the Naturhistorisches Museum in Basel, Switzerland. Image by Jonathan R. Hendricks, CC BY-SA 4.0.

feathers common earlier in the lineage (Bell et al. 2017). Rather than looking like overgrown chickens, which is somewhat less terrifying, these tyrannosaurids had smooth, scaly skin. The mechanisms behind this secondary loss of feathers are unclear, although it may be related to the evolution of giant size in these dinosaurs (Bell et al. 2017). Also a recent study found "exquisitely preserved" skin traces of the smallest known ichnogenus theropod, *Minisauripus,* in a trackway excavation near Jinju City, Korea (Kim et al. 2019). The footprints of this diminutive theropod were only an inch long but contained tiny scale traces of between 0.3 mm and 0.5 mm in diameter in a perfect array. The authors postulated that unusual sedimentological conditions had allowed the preservation of these footprints without smearing the skin texture patterns (Kim et al. 2019).

Process of Fossil Formation

In regard to mammiferous remains, a single glance at the historical table published in the Supplement to Lyell's Manual, will bring home the truth, how accidental and rare is their preservation . . .
—Charles Darwin, "On the Imperfection of the Geological Record," *On the Origin of Species by Means of Natural Selection*, 1859.

The processes that influence the likelihood that a fossil is created, as well as the documentary quality of the geologic record, can be roughly divided into two phases. The first, often called *biostratinomy,* refers to the processes that take place after something dies but before its final burial (Weigelt 1927; Lawrence 1968). The second phase is called *diagenesis;* this includes the events that take place after the final burial of the organism (Müller 1951). As you might expect, different factors influence the rate or likelihood of both processes.

Biostratinomy

Most animals do not die of old age. Instead most die of starvation, injuries, or (especially) predation. Studies of mortality in mammals highlight the importance of predation and, moreover, demonstrate that fewer than 1% ever even approach senescence (Murie 1944; Caughley 1966). Thus, mammals tend to be eaten by other mammals, which accelerates the postmortem processes of disarticulation, disintegration, and ultimately decay—the three Ds of biostratinomy (Shipman 1981). As flesh and tendons are either eaten or decay, they no longer hold together the skeleton. This means that the pieces of a mammal begin to be separated across the landscape; dispersal is exacerbated by natural events, such as rain or floods, or biological agents, such as scavengers (Behrensmeyer and Kidwell 1985: Behrensmeyer et al. 2000).

Unless a mammal's parts are buried fairly rapidly, they will be destroyed by these physical and/or biological processes (Shipman 1981). As Darwin noted in his writings on the geological record, "*Wherever sediment did not accumulate on the bed of the sea, or where it did not accumulate at a sufficient rate to protect organic bodies from decay, no remains could be preserved*" (Darwin 1859, p. 300). Typical depositional environments are lakes, river bends or deltas, ocean basins, or even potholes; all places where water slows down and sediments—or would-be fossils—are knocked out of suspension, deposited, and buried (Shipman 1981; Behrensmeyer 1988). Indeed, the fossil record of gastropods and molluscs is excellent because these animals lived in ideal depositional environments where they were often and rapidly buried postmortem (Valentine 1989). In contrast, the number of fossils of dinosaurs is relatively meager because these animals often lived in habitats where rapid burial was not as likely. The fossil abundance of terrestrial mammals is somewhat intermediate between these extremes, likely in part because of their higher species diversity and numerical abundance, which increase the chances of preservation in the record.

Transport to an appropriate depositional site is another filter on preservation. Parts of the organism may be transported different distances; the teeth and bones of vertebrates, for example, have different buoyancies and thus tend to be transported different distances depending on water flow (Behrensmeyer and Hill 1980; Behrensmeyer 1988; Behrensmeyer and Hook 1992). Size or shape sorting can occur because of differences in density or buoyancy. And, of course, the process of transport itself also often leads to breakage or even the complete destruction of bones (Behrensmeyer and Hill 1980; Behrensmeyer 1988). Thus, in most instances the biostratinomic processes of disarticulation, disintegration, decay, and inadequate or delayed burial lead to the total loss of the original remains. However, if a few elements of a mammal do happen to be buried under the appropriate conditions fast enough, preservation *may* occur. This is where diagenesis takes over.

Diagenesis

Diagenesis is the postburial modification of fossil elements; it can lead to destruction or to varying levels of preservation depending on whether materials are mineralized, dissolved, compacted, or deformed (Shipman 1981; Behrensmeyer and Kidwell 1985: Behrensmeyer et al. 2000). The most common mode of preservation is *permineralization*, where after burial, the pores and cavities within the original organic material are filled with mineral-rich groundwater (Fig. 3.6a). These minerals—often calcites, phosphates, silicas, and iron—crystalize and fill the central cavity of the cell, preserving the original cell structure. Permineralized fossils tend to be much heavier than the original element because of the weight of the added minerals. Because vertebrate bones house marrow, blood vessels, and nerves, they tend to be very porous, which favors this process. Consequently, most preserved dinosaur and mammal bones are permineralized.

Sometimes, *recrystallization* occurs. This is when the internal physical structure of the would-be fossil is lost because of changes in the microstructure of minerals; while there is a change in the crystal structure, there is not a change in the mineral chemistry. Recrystallization tends to be common in calcareous fossils; many are composed of the mineral

aragonite, which over time tends to recrystallize into calcite, a more stable form of $CaCO_3$. So, although the original materials remain in the fossil, the internal structure is blurred or lost completely.

The minerals within a fossil can also be altered. The molecule-by-molecule substitution of another mineral of different composition for the original material is referred to as *replacement*. Here, despite the loss of the original organic material, the fine details of cell or shell structures are generally preserved. Petrified wood is formed by replacement; over time, silica replaces the original wood, literally turning the remains to stone (Fig. 3.6b). Other minerals that are often exchanged for the original organic material include pyrite or phosphates.

If organisms are rapidly buried under low-oxygen (i.e., anaerobic) conditions, they can sometimes be carbonized. *Carbonization* occurs when the hydrogen, oxygen, and nitrogen in the original organic material is driven off, leaving a thin two-dimensional film of carbon behind. While fine details can be preserved, some distortion occurs from the compression of the three-dimensional organism to two dimensions. Carbonization is relatively common in plants, such as ferns in fine-grained sediments or shales (Fig. 3.6c); sometimes soft-bodied animals such as jellyfish or worms are also preserved in this way. The most famous examples are the fossils of the Cambrian-age Burgess Shale.

Of course, sometimes a mold or cast is all that remains of the ancient organism. Molds are impressions, either internal or external, of the original animal. The level of detail preserved depends on how the mold or cast was formed. Sometimes an external mold is filled with sediment that later lithifies, essentially creating a cast of the animal.

Unaltered Fossilization or "True-Form" Fossils

The most exciting fossil discoveries are those where the organism underwent minimum alteration (Fig. 3.6d). Under highly unusual conditions, organisms may undergo fast postmortem burial in a biologically inert environment. This limits diagenesis and tends to preserve fine details of the morphology and sometimes even details of the life

Fig. 3.6. Types of fossil preservation. **a**, A permineralized fossil of a *Tyrannosaurus rex* skull. **b**, Petrified wood from Arizona. Petrification can occur when a constant flow of water filters through remains, leaving deposited minerals within the dead cells. Thus, the original organic material is replaced with minerals while retaining the original structure. **c**, Fern fossil. Ferns and other plants are often preserved through the process of carbonization. This occurs when the remains of the organism are crushed beneath the weight of overlying sediments. Because of generated heat and compression, hydrogen, nitrogen, and oxygen are off-gassed, leaving behind a carbon film of the former living thing. **d**, *Knightia* (a species of freshwater herring) from the Green River Formation in Wyoming. This fossil is from the Eocene (52 Ma). The Green River Formation is particularly known for its superbly preserved fossil fish. **e**, Bones of the dodo (*Raphus cucullatus*), an extinct flightless bird endemic to the island of Mauritius, east of Madagascar in the Indian Ocean. Near-time fossils such as this dodo are generally not permineralized and, thus, are sometimes called subfossils. Panel a: *T. rex* skull, American Museum of Natural History No. 5027, photo by A.E. Anderson, 1910; Panels b, c, d, e: Photos by author.

history or ecology. Such preservation can occur in a number of ways. For example, in 2011 a wonderfully preserved three-dimensional ankylosaur (*Borealopelta markmitchelli*) was discovered during overburden removal in a bituminous sand mining operation in northeastern Alberta, Canada. The dinosaur appears to have been washed out to sea, ending up upside down on the sea floor, and then completely encased in a very dense siderite concretion; all of this happened very quickly postmortem preventing decay or scavenging (Brown et al. 2017). The end result was exceptional preservation of both the hard and soft parts. Almost the entire skin of *Borealopelta* was preserved, including the osteoderms (or bony armor), the keratin sheaths of the horns, and even the stomach contents from the animal's last meal (Brown et al. 2017). Mass spectroscopy and other analyses indicated the presence of benzothiazole,

which is diagnostic for phaeomelanin, suggesting that the dinosaur had reddish-brown coloration.

Amber, or fossilized resin, is another source of exceptional preservation (Fig. 3.7). Insects may be entombed if they become trapped in the sticky resin produced by conifer trees, which were widespread in the Mesozoic. If more resin fell on the insect and was polymerized, it could be fossilized and persist for many tens of millions of years. For example, in the popular novel (and blockbuster movie) *Jurassic Park*, dinosaurs were genetically engineered in a lab using dinosaur DNA from the blood meals of mosquitos that had been trapped in amber during the Mesozoic (Crichton 1990). While making for a fun movie, in reality it is farfetched; amber resins are not airtight and so such DNA is highly degraded and unlikely to produce much useful genetic information for cloning (Stankiewicz et al. 1998; Shapiro 2015). Amber is

Fig. 3.7. Amber, or fossilized tree resin, can lead to unusually good 3-D preservation. Pictured here is an approximately 50-million-year-old beetle (Coleoptera) trapped inside amber from the Baltic. "Baltic Amber Coleoptera Brentidae Apion 3" by Anders Leth Damgaard, CC BY-NC-ND 3.0, at www.amber-inclusions.dk.

Fig. 3.8. Blue Babe. A famous steppe bison dating to around 36,000 years ago on display at the University of Alaska Museum of the North in Fairbanks. It was found in permafrost in 1979 by gold miners in Alaska and donated to the museum. It has claw and tooth marks on the rear of its body that likely came from an American cave lion. Steppe bison were found throughout interior Alaska during the last glacial period. The skin has a bluish color, which was likely caused by phosphorus in tissues reacting with the iron in the surrounding soil. This produced a mineral coating of vivianite, which when exposed to air, turns a brilliant blue. This unusual color led to the specimen being nicknamed Blue Babe after the famous fictional companion of Paul Bunyan, a giant lumberjack of American and Canadian folklore. He was known for his superhuman strength and his trusty oversized companion, Babe the Blue Ox. "Blue Babe @ Museum of the North" by Bernt Rostad, CC BY 2.0, Flickr.

mostly found in Cretaceous or younger rocks, much of it from Baltic Sea region of Europe.

Organisms can also be mummified through desiccation or flash freezing (Aufderheide 2003). Desiccation is extremely rare. It results when animals die in an environment with little moisture, such as a desert or dry cave. Perhaps the most famous example of dinosaur desiccation is the Trachodon mummy, which is housed at the American Museum of Natural History (Osborn 1911, 1912). This hadrosaurid dinosaur was discovered in 1908 near Lusk, Wyoming; it was the first specimen found with a nearly full skeleton and much of its original skin. The skin was not only still tightly attached to the bones but also drawn into the body cavity, suggesting the animal had been dehydrated prior to a burial that led to its unusual and excellent preservation (Osborn 1912).

In a similar fashion, animals can be mummified under extremely cold, dry conditions (Aufderheide 2003). As permafrost or glaciers in northern latitudes or high elevations continue to melt under anthropogenic warming, they are yielding many examples of exquisitely preserved Ice Age mummies who stumbled, fell, or were otherwise trapped in ice crevices in the distant past (Fig. 3.8). These animals include adult and baby mammoths, steppe bison, horses, wolves, cave lions, ground squirrels, and even ancient humans (e.g., Ryder 1974; Guthrie 1990; Boeskorov et al. 2007; Pabst et al. 2009; Kosintsev et al. 2010; Thali et al. 2011; Fisher et al. 2012; Papageorgopoulou et al. 2015; Boeskorov et al. 2016; Grigoriev et al. 2017; Boeskorov et al. 2018). Not only do these remains generally contain soft tissues, but also the tissues retain enough chemical integrity to make possible histological, isotopic, and ancient DNA studies (Aufderheide 2003). One of the most recent spectacular finds was the discovery of the Yukagir bison, an extinct steppe bison (*Bison priscus*), which emerged from permafrost in northern Yakutia, Eastern Siberia, Russia. This nearly 11,000-year-old animal was almost entirely complete with much of his fur, all of his internal organs, and his stomach contents intact; analysis of the latter revealed that he was a selective grazer feeding on grasses and sedges, with some dwarf birch and legumes (Serduk et al. 2014; Boeskorov et al. 2016).

In 2007, a baby woolly mammoth (*Mammuthus primigenius*) nicknamed Lyuba (Fig. 3.9) was found along the Yuribey River in northwest Sibèria, after being washed out of permafrost upstream (Fisher et al. 2012). After a series of misadventures, which included her carcass being stolen and exhibited outside a local store where feral dogs chewed off her tail and part of her right ear lobe, she was ultimately brought to the Shemanovskiy Museum and Exhibition Center in Salekhard (Fisher et al. 2012). Studies using growth rings in her tusk and $\delta^{15}N$ analysis (see chapter 6) indicated the mammoth was about one month old at the time of her death, and moreover, she had probably drowned in muddy water around 43,500 years ago (Kosintsev et al. 2010; Fisher et al. 2012; Papageorgopoulou et al. 2015). Lyuba was so beautifully preserved that soft tissues such as hair, muscle, lungs, stomach, and liver could be examined; she still had milk in her stomach from her last meal (Kosintsev et al. 2010; Fisher et al. 2012; Papageorgopoulou et al. 2015). Also in Siberia, two baby cave lion cubs were found in 2015. Complete with fur and whiskers, these tiny fluffy mummies apparently had not yet nursed before they died, perhaps buried by a landslide (Liesowska 2017).

As captivating as these flash-frozen mammals are, they can only provide a glimpse into the most recent geological past. Indeed, the advance, retreat, and re-advances of the ice sheets over the past 2.6 million years (Dyke and Prest 1987) mean that it is highly unlikely that we will recover true form fossils much older than about 100,000 years, that is, the last glacial cycle.

In the heart of a major metapolitical area—Los Angeles, California—sits one of the most important late Pleistocene faunal sites, and one with unusually good preservation (Fig. 3.10). The Rancho La Brea Tar

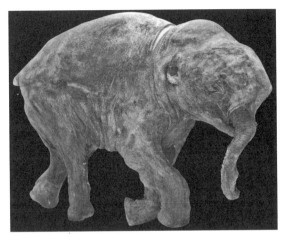

Fig. 3.9. Picture of a frozen baby mammoth (*Mammuthus primigenius*) found in 2007 in the Yamal Peninsula, in a remote area of Siberia. Lyuba, as she is called, was one month old when she died around 42,000 years ago. CT scanning suggested that she choked to death after being trapped in mud. Her internal organs were so well preserved that milk and pollen were found inside. "Lyuba at the Shemanovskii Regional Museum" by James St. John, CC BY 2.0, courtesy of Wikipedia.

Pits houses the largest collection of late Pleistocene asphaltic fossils in the world and is the type locality for the late Pleistocene Rancholabrean North American Land Mammal Age, which extends from about 240 ka to 11 ka (Savage 1951; Woodard and Marcus 1973; Harris and Jefferson 1985; Stock and Harris 1992). Despite the name "tar pits," the animals that perished here did not actually die in pits but rather were trapped in asphalt seeps (Woodard and Marcus 1973). These were places where crude oil was pushed up from deeper sediments and subsequently saturated the surface sediments through fissures in the Earth's crust; as the lighter fraction of the oil evaporated, it left behind heavy tar where animals were episodically trapped (Woodard and Marcus 1973; Stock and Harris 1992). The tar pits we see today are the result of human excavation at the site and not reflective of the original taphonomy (Woodard and Marcus 1973).

While there is a rich accumulation of bones at Rancho La Brea, predators are by far the most common (Stock and Harris 1992). For example, more than half the recovered bones are from canids, and most of these are from the extinct dire wolf, *Canis dirus* (Stock and Harris 1992; Coltrain et al. 2004).

About a third of the bones are felids, with the extinct saber-tooth cat, *Smilodon fatalis*, particularly well represented (Stock and Harris 1992; Coltrain et al. 2004). This abundance of predators leads scientists to speculate that the site was a carnivore trap (Fig. 3.11)—a site where predators were attracted by trapped herbivores and then became trapped

Fig. 3.10. A bubble forming in the La Brea tar pits. Photograph by Daniel Schwen, CC BY-SA 2.5, courtesy of Wikimedia Commons.

Fig. 3.11. Carnivore trap. A depiction of saber-tooth cats (*Smilodon*) and dire wolves (*Canis dirus*) fighting over a mammoth (*Mammuthus columbi*) carcass in the Rancho La Brea Tar Pits. This image is meant to explain why the fossil assemblage at La Brea is so heavily composed of carnivores. As large herbivores became trapped in the tarry substrate, their struggles attracted sequential waves of predators, who became entrapped in turn. Painting by Robert Bruce Horsfall in 1911 and reprinted from the frontispiece of William Berryman Scott, 1913, *A History of Land Mammals in the Western Hemisphere* (New York: Macmillan).

themselves (Harris and Jefferson 1985; Stock and Harris 1992; Van Valkenburgh and Hertel 1993).

Naturally occurring asphalt or bitumen (tar, or *brea* in Spanish) has high preservation potential (McMenamin et al. 1982). Thus, the bones and organic material within are well preserved but not the soft tissue surrounding the bones; furthermore, most skeletons are largely disarticulated (McMenamin et al. 1982; Stock and Harris 1992). Such excellent preservation has allowed genetic and dietary analyses through the extraction of ancient DNA and stable isotope analysis (e.g., Janczewski et al. 1992; Coltrain et al. 2004). While it is the best known, Rancho La Brea is far from the only locality where animals were trapped in hydrocarbon-saturated sediments. For example, Tanque Loma in the Santa Elena Peninsula in southwestern Ecuador is a late Pleistocene tar seep that also contains a rich mammal fauna (Lindsey and Lopez 2015).

Fossil Lagerstätten

On very rare occasions, we encounter a fossil locality with exceptional preservation of an entire community, sometimes including those organisms made up entirely of soft parts. These are known as *fossil Lagerstätten*, from the German *Lager* meaning "storage" or "lair" and *Stätte* meaning "place" (Seilacher et al. 1985). These assemblages likely result from extremely rapid or catastrophic burial in an anoxic environment that minimized exposure to scavengers, bacteria, and decomposition. Such extraordinary sedimentological conditions with minimum diagenesis can lead to preservation of nonmineralized structures, including plants and soft-bodied organisms, and of articulated skeletons (Seilacher et al. 1985). The known Lagerstätten (Table 3.1) span much of the late Proterozoic and the entire Phanerozoic geological record, from just prior to the Cambrian to the Pleistocene. Moreover, the rapid burial generally means the assemblage is contemporaneous, with minimal time-averaging (Kidwell and Flessa 1995).

Fossil Lagerstätten provide an unusual and important glimpse into the past by preserving soft-bodied taxa and morphologies that we would otherwise not know about. And some of them are very strange indeed. While Adolf Seilacher proposed two kinds—(1) Konzentrat-Lagerstätten, sites with anomalously high abundance of fossil materials, and (2) Konservat-Lagerstätten, sites with unusually well-preserved fossils—the term more typically is applied to the latter.

The Burgess Shale is arguably the most well known of Lagerstätten; perhaps because of Stephen Jay Gould's 1989 award-winning book *Wonderful Life: The Burgess Shale and the Nature of History*. In it, Gould mused about the nature of contingency in the evolutionary history of life. Using the extraordinarily well-preserved fossils of the Burgess Shale, he pointed out that the variety of anatomical body plans was greater than found today but that most of these organisms left no modern descendants. This led to the proposition that if we were to "rewind the tape of life" and let it play again, we might find a vastly different world. Although some of his observations have since been disputed (Briggs 2015), he captured the imagination of paleontologists and the public alike. Gould's arguments were persuasive largely because the Burgess Shale contains highly unusual forms, such as *Opabinia regalis*, a stem arthropod with five compound eyes, a backward-facing mouth under the head, and a proboscis that may have been used for foraging (Whittington 1975).

The Burgess Shale is a mid-Cambrian (~508 Ma) fauna from the Canadian Rockies in British Columbia, Canada. The fossils are deposited in slightly calcareous dark mudstones at the base of a cliff; this location may have protected the organisms from tectonic compression (Briggs et al. 1995). The organisms underwent carbonization and so are preserved as black carbon films (Briggs et al. 1995). In some cases, this process was enhanced by phosphatization and/or pyritization, leading to finer resolution of three-dimensional structures. The excellent preservation of the Burgess Shale may have been facilitated by the unusual chemistry of the oceans at the time, which had low sulfate concentrations and high alkalinity (Gaines et al. 2012).

Faunal remains include free-swimming benthic and sessile organisms with high taxonomic and morphological diversity (Fig. 3.12; Briggs 2015). These forms would not appear in the fossil record; perhaps only 14% of the genera and 2% of the individual fossils would have been preserved in a "normal" Cambrian fossil bed (Morris 1986). This was probably not an unusual assemblage, but rather it reflects what Cambrian faunas actually looked like if more of the species were preserved. As you might expect based on its importance, the Burgess Shale has been the subject of many studies (see Briggs et al. 1995; Briggs 2015, and references therein).

Other notable examples of fossil Lagerstätten include the Carboniferous Mazon Creek fauna from Pennsylvanian Illinois (Table 3.1) and the Lower Cretaceous Jehol Group of northeastern China. The Mazon Creek Lagerstätten contains an exceptional diversity of soft-bodied organisms from both freshwater and marine environments, which were preserved in ironstone concretions (Selden and Nudds 2012). The Jehol Group includes articulated skeletons as well as an abundance of flowers, insects, and even twigs with leaves and flowers still attached; it also has preserved vertebrate integument, including mammal fur, reptile scales, and dinosaur feathers (Zhonghe et al. 2003).

Fidelity of Fossil Assemblages

The interpretation of fossil assemblages is complicated by taphonomic biases that dampen or distort the biological signal. A number of physical and biological agents as well as the intrinsic traits of organisms all color our interpretation of patterns of evolution as revealed by the fossil record (see Kidwell and Flessa 1995; Behrensmeyer et al. 2000). Moreover, these processes can act differentially on the biochemical, anatomical, spatial, temporal, or compositional fidelity of the fossil record (Behrensmeyer et al. 2000). For example, noncontemporaneous material can be preserved together if burial is delayed; this blurs the resolution of ecological communities since these organisms may not have interacted (Kidwell and Flessa 1995; Behrensmeyer et al. 2000; Behrensmeyer 2007). Known as time-averaging, this process is a major concern in the fossil record (Walker and Bambach 1971; Kidwell and Flessa 1995; Behrensmeyer et al. 2000). Live-dead studies of modern and fossil vertebrate communities from the same habitat generally demonstrate that the number of species represented in the fossil record is inflated by two to ten times (Behrensmeyer 1993).

Spatial fidelity assumes that organisms are not transported out of their life position or life habitat,

Table 3.1. Exceptional fossil localities with notes on their significance. These include both Konzentrat-Lagerstätten, sites with anomalously high abundance of fossil materials, and Konservat-Lagerstätten, sites with exceptional preservation, especially of soft-bodied forms or tissues.

Geological stage and related localities	Beginning of geological stage[a]	Significance	Region
Precambrian			
Bitter Springs	>850 Ma	Chert beds with >30 species of microfossils, including cyanobacteria, algae, fungi, and bacteria	South Australia
Doushantuo Formation	635–551 Ma	Minutely preserved phosphatic microfossils, including algae, seaweeds, sponges, acritarchs, and others	Guizhou Province, China
Mistaken Point	565 Ma	Preserved on individual bedding planes with fine volcanic ash; diverse and obscure forms; many with large, frond-like leafy forms	Avalon Peninsula, Newfoundland, Canada
Ediacara Hills	550? Ma	Diverse assemblage of soft-bodied fossils; mostly preserved on undersides of quartzite and sandstone slabs	South Australia
Cambrian			
Qingjiang biota	518 Ma	Newly described Lagerstätten with soft-bodied taxa, including cnidarians and many undescribed species	Hubei Province, China
Sirius Passet	518 Ma	Fauna not fully described, but includes arthropods, sponges, and other enigmatic forms	Greenland
Chengjiang biota; Maotianshan Shales	515 Ma	Very diverse assemblage of soft-bodied forms (>185 species to date), including many found in the Burgess Shale, trilobites and other enigmatic forms	Yunnan Province, China
Emu Bay Shale	514–509 Ma	Preservation similar as in Burgess Shale; >50 species of trilobites, as well as endemic arthropods, worms, and other forms	South Australia
Kaili Formation	513–506 Ma	Highly diverse assemblage with 110 genera; trilobites, eocrinoids and many soft-bodied organisms	Guizhou Province, south-west China
Blackberry Hill	~510–500 Ma	Some of the first land animals; 3-D casts of soft-bodied forms, trace fossils	Central Wisconsin, United States
Burgess Shale	508 Ma	Unparalleled preservation of soft-bodied forms; many arthropods and sponges	British Columbia, Canada
Wheeler Shale	507 Ma	Many Agnostida and trilobites; soft-bodied forms	Western Utah, United States
Kinnekulle Orsten and Alum Shale	500 Ma	Exceptional 3-D preservation of soft-bodied organisms and larval stages	Sweden
Ordovician			
Fezouata Formation	~485 Ma	Both mineralized and soft-bodied forms; bryozoan and graptolites, echinoderms and many similar to Burgess Shale	Draa Valley, Morocco
Beecher's Trilobite Bed	445 Ma	Exceptionally well-preserved trilobites with ventral anatomy and soft parts	New York, United States
Walcott-Rust Quarry	457–454 Ma	Exceptional preservation of >18 species of trilobites with appendages; other soft-bodied and mineralized fauna	New York, United States

Table 3.1. (continued)

Geological stage and related localities	Beginning of geological stage[a]	Significance	Region
Soom Shale	450? Ma	Wonderful preservation of characteristic Ordovician micro and macro fauna as well as soft-bodied forms, cephalopods, trilobites, eurypterids, etc.	South Africa
Silurian			
Wenlock Series	420 Ma	Fine-grained volcanic ash preserving soft-bodied 3-D animals, including polychaete worms, sponges, starfish, and graptolites	Herefordshire, England, United Kingdom
Devonian			
Rhynie chert	410 Ma	Plant, fungi, algae, lichen, and animal remains overlaid with volcanic ash; high resolution of internal structures including stomata and lignin; early representation of the colonization of land	Scotland, United Kingdom
Hunsrück Slate	408–400 Ma	>260 animal species from a shallow water habitat; includes brachiopods, sea cucumbers, crinoids, fish, corals, trilobites, sponges, corals, and cephalopods	Rheinland-Pfalz, Germany
Gogo Formation	380 Ma	Reef community with 3-D soft-tissue preservation	Western Australia
Miguasha National Park	370 Ma	Estuary environment with high diversity of fossil fish, worms, eurypterids, and water-to-land transition fossils	Québec, Canada
Canowindra	360 Ma	Diverse fish fauna in single bedding plane	New South Wales, Australia
Carboniferous			
Bear Gulch Limestone	320 Ma	Shallow marine or estuary; diverse fossil fish assemblage	Montana, United States
Joggins Fossil Cliffs	310 Ma	Tropical rainforest ecosystem with tetrapods, ferns, early amphibians and many fish and arthropods; trackways	Nova Scotia, Canada
Linton Diamond Coal Mine	312 Ma	~40 vertebrate genera represented, including fish, coelacanths, sharks, and amphibians	Ohio, United States
Mazon Creek	310 Ma	Mostly terrestrial fossils, finely preserved plants, some insects, centipedes, arachnids, and small amphibians found in nodules of iron carbonate	Illinois, United States
Montceau-les-Mines	300 Ma	Freshwater environment with rich and diverse flora and fauna; includes lycopsids, ferns, pteridosperms, bivalves, annelids, crustaceans, insects, and tetrapods	France
Hamilton Quarry	300–298 Ma	Estuary environment; preserved diverse group of marine, freshwater, volant, and terrestrial plants and animals	Kansas, United States
Permian			
Mangrullo Formation	~286–273 Ma	Petrified wood, many vertebrate and invertebrate remains; abundant fossils of mesosaurs, including gut contents and coprolites; oldest known amniote embryos	Paraná Basin, Uruguay

(continued)

Table 3.1. (continued)

Geological stage and related localities	Beginning of geological stage[a]	Significance	Region
Triassic			
Madygen Formation	230 Ma	Terrestrial lake and river deposit; includes many vertebrate fossils and insects	Kyrgyzstan
Ghost Ranch	205 Ma	Many Triassic dinosaurs, including basal forms; >1,000 *Coelophysis* (the New Mexico state fossil)	New Mexico, United States
Jurassic			
Holzmaden/Posidonia Shale	180 Ma	Exceptionally well-preserved and close to complete skeletons of marine fish and reptiles; includes ichthyosaurs and plesiosaurs	Württemberg, Germany
Mesa Chelonia, part of Qigu Formation	164.6 Ma	Large accumulation of freshwater turtles	Shanshan County, China
La Voulte-sur-Rhône	160 Ma	Marine habitat with fossilized cephalopods, as well as fish, crustaceans, and other pyritized fossils	Ardèche, France
Karabastau Formation	155.7 Ma	Pterosaurs, earliest salamanders, as well as invertebrates	Kazakhstan
Solnhofen Limestone	155 Ma	Shallow lagoon environment with jellyfish, invertebrates and vertebrate animals, land plants, protists, as well as the earliest known bird, *Archaeopteryx*	Bavaria, Germany
Canjuers Limestone	145 Ma	Lagoon environment containing a diverse assemblage of fossil plants, invertebrates and vertebrates	France
Cretaceous			
Las Hoyas	~129–126 Ma	Inland lacustrine environment with many beautifully preserved fish, arthropods, molluscs, annelids, and unique dinosaurs	Cuenca, Spain
Yixian Formation	~129–122 Ma	Extensive forest environment with flowering plants, horsetails, ginkgoes, conifers, cycads, ostrocods, clam shrimp, insects, dinosaurs, and pterosaurs; major radiation of birds and mammals	Liaoning, China
Xiagou Formation	~125–113 Ma	High diversity of birds, including the earliest true modern bird	Gansu, China
Crato Formation	~117 Ma	Limestone accretions with nodules encompassing fossil organisms; contains fish with stomach contents, pterosaurs, reptiles, amphibians, plants, and insects	Northeast Brazil
Haqel, Hadjula, and Al-Namoura	~95 Ma	Many free-living fossil polychaetes	Lebanon
Santana Formation	115–108 Ma	Lacustrine to subtidal shallow marine environment with many fish, arthropods, insects, turtles, snakes, dinosaurs, pterosaurs, and bird feathers	Brazil
Smoky Hill Chalk	87–82 Ma	Many marine taxa, including plesiosaurs, large bony fish, mosasaurs, pterosaurs, and turtles	Kansas and Nebraska, United States

Table 3.1. (continued)

Geological stage and related localities	Beginning of geological stage[a]	Significance	Region
Ingersoll Shale	85 Ma	Contains theropod feathers	Eastern Alabama, United States
Auca Mahuevo	83.5–79.5 Ma	Rookery of excavated nest structures with dinosaur eggs containing embryonic remains	Patagonia, Argentina
Zhucheng	66 Ma	Many dinosaur fossils, including fossil eggs dating to just before the K-Pg extinction	Shandong, China
Tanis	66 Ma	Part of Hell Creek Formation, captures the moment by moment details of the Chicxulub meteor event	Southwestern North Dakota, United States
Eocene			
Fur Formation	55 Ma	Tropical or subtropical site with abundant fossils of fish, insects, reptiles, as well as plants and birds	Fur, Denmark
London Clay Formation	56–49 Ma	Tropical or subtropical forest; many plant fossils with seeds and fruits	England, United Kingdom
McAbee Fossil Beds	52.9 ± 0.83 Ma	Old lake bed with high diversity of plant, insect, and fish fossils, including leaves, shoots, seeds, and flowers	British Columbia, Canada
Green River Formation	53.5–48.5 Ma	Multiple fossil beds containing near-continuous 5-6 million year record of fine-grained and detailed fossils, including complete insects, many plants, and particularly fish	Colorado/Utah/Wyoming, United States
Klondike Mountain Formation	53.5–48.5 Ma	Mesic forest with very diverse and well-preserved plant, fish, and insect fossils	Northeast central Washington, United States
Monte Bolca	49 Ma	>250 species of fish, some with finely detailed organs, skin, and color; jellyfish, poly-chaetes, crustaceans, bird feathers, and tortoise scutes	Verona, Italy
Messel Pit/Messel Oil Shale	49 Ma	Extensive preservation of skeletons, sometimes even with fur and feathers; early primate fossils; turtles during mating; >10,000 fish specimens; insects; many mammals including horses, mice, hedgehogs, aardvarks, and bats; birds, reptiles, and amphibians	Hessen, Germany
Quercy Phosphorites Formation	25–45 Ma	Mammals, unusual mummified amphibians, squamate with articulated skeletons	Southwestern France
Oligocene-Miocene			
Dominican amber	~25 Ma	Derives from the extinct tropical tree *Hymenaea protera*; generally transparent with many fossil inclusions	Dominican Republic
Riversleigh	25–15 Ma	Capture changes from humid lowland rainforest to dry eucalypt forest/woodland; found in noncompressed freshwater limestone, so 3-D structure preserved; fossils include birds, reptiles, and monotreme and marsupial mammals	Queensland, Australia

(continued)

Table 3.1. (continued)

Geological stage and related localities	Beginning of geological stage[a]	Significance	Region
Miocene			
Clarkia fossil beds	~15 Ma	Best known for fossil leaves, which maintain color; DNA may be preserved	Idaho, United States
Barstow Formation	19-13.4 Ma	Abundant vertebrate fossils with bones, teeth, and footprints	Mojave Desert, near Barstow, California, United States
Ashfall Fossil Beds	11.83 Ma	Volcanic ashfall-entombed animals at a waterhole; many birds, horses, camels, rhinos with stomach contents, and embryos	Northeastern Nebraska, United States
Pleistocene			
Mammoth Site	26 ka	Entrapment in a sinkhole; one of the largest concentrations of mammoth remains as well as other vertebrates	South Dakota, United States
Rancho La Brea Tar Pits	40-12 ka	Many vertebrates, particularly predators such as dire wolves, saber-tooth cats, short-faced bears	California, United States
Waco Mammoth National Monument	68-53 ka	A large concentration of mammoth remains from several rapid but episodic events, possibly flash floods	Texas, United States
El Breal de Orocual	3-2.5 ka	Asphalt pit that trapped many mammals such as sabertooths, toxodontids, glyptodontids, rodents, llama, etc.	Monagas, Venezuela
Mene de Inciarte	25.5-28 ka	An asphalt pit that preserves mammal fossils from a savanna/forest community	Zulia, Venezuela

[a] Based on the 2018 International Commission on Stratigraphy's International Chronostratigraphic Chart (http://stratigraphy.org).

Fig. 3.12. The Middle Cambrian–aged Burgess Shale Lagerstätte. This is one of the most famous localities of exceptional preservation of soft-bodied forms in the world; many of the fossils have had their appendages and internal organs preserved. *Left*: Charles Doolittle Walcott (standing in side profile, leaning on knee) excavating the Burgess Shale in 1915. This site is located near the town of Field in Yoho National Park, southeastern British Columbia, western Canada. *Right*: Diorama reconstruction of the Burgess Shale on public exhibit at the Nebraska State Museum of Natural History in Lincoln, Nebraska. "Charles Doolittle Walcott Excavating Burgess Shale" by unknown photographer, 1915, courtesy of Wikimedia Commons; "Diorama of the Burgess Shale Biota" by James St. John, CC BY 2.0, Flickr.

Box 3.1
Putting the Dead to Work

Not everyone enjoys studying smelly, slowly decomposing mammals. But to Dr. Kay Behrensmeyer, watching the gradual disintegration of a wildebeest up close is fascinating. In fact, she has been going to South Africa for more than forty years to do just that. Kay is a pioneer in the field of taphonomy—a subdiscipline within paleontology that seeks to understand the biases in the formation and preservation of fossils that influence our interpretation of the geological record.

Although dying is inevitable, becoming a fossil is not. Actually, it's pretty hard. To become a fossil, you have to be buried fairly rapidly, or you decompose or are eaten by hungry opportunistic scavengers, or destroyed by wind, water, or other factors. Moreover, you also have to be made out of the right stuff—generally that means hard parts that can be preserved in the first place. And you have to manage to die in just the perfect site where preservation is possible. Moreover, even if you do somehow manage to become a fossil, the probability that you actually get found at some point by an intrepid paleontologist is like winning the lottery. Over time, most fossils are reworked or destroyed by geological forces. Many are buried under sediments and never exposed, or in more recent times they are crushed by inattentive tractor operators at construction sites. And then of course, even if you survive all that and are exposed, somebody needs to come along and find you. Not an easy path to immortality.

The study of all of these biases or "laws of burial" is called taphonomy. Remarkably—despite the strong gender bias that persists in paleontology—many of the pioneers in modern taphonomy are women. Of these, Dr. Kay Behrensmeyer, who is the curator of vertebrate paleontology in the Department of Paleobiology at the Smithsonian Institution's National Museum of Natural History, has had a profound influence on our understanding of the taphonomy of vertebrate fossils and, in particular, how sedimentary processes influence the destruction or preservation of bones.

Box fig 3.1 Kay Behrensmeyer. Photo by author.

Her influence is clear in her remarkable scholarly record of publication and mentorship (she has an h-index of 72 and more than 20,000 citations). Indeed, *Discover* magazine named Kay as one of the fifty most important woman scientists in 2002.

Kay is known for her work in sedimentary taphonomy—understanding the biases and processes of how bones are transported in river systems, become buried, and are fossilized. As she said in a recent acceptance speech for a prestigious award, she grew up beside the Mississippi River in western Illinois and, thus, both modern and ancient rivers have very much shaped her professional life.

Kay's work on taphonomy began as a graduate student at Harvard University. Among other projects, she had the opportunity to work with Richard Leakey in Africa and became interested in the paleoecological context of human evolution. One of her frequently cited experiments is a long-term study of modern vertebrate remains in Amboseli National Park in Kenya, which she began with a colleague in 1975. Over the years, they have watched animals disintegrate; the carcasses and bones of animals in the park are surveyed every five to ten years, along with a census of the live animals. The Amboseli Bone Taphonomy Project has addressed the issue of just how well mammal death assemblages reflect the abundance and composition of the living population. Are the same species present in the same abundance? In the course of this now forty-year study they defined weathering stages and determined how long

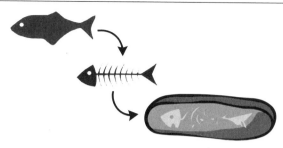

Box Fig 3.2 Going from a live fish to a fossil involves a lot of steps. The study of these steps is part of taphonomy. Figure drawn by Jenna Mccullough.

it took a bone to reach each stage of decomposition. As Kay said in a recent interview, *"There's so much focus now on recycling and trying to make it happen, but nature's been making it happen forever."*

At Amboseli, Kay and her colleagues documented a shift in the composition and preservation of bones from the 1970s to the 2000s. Interestingly, this mirrored a shift from predation to scavenging activities by the carnivores in the park. As lions decreased in abundance—and hyenas began to become more dominant—there were fewer bones in the environment, and the ones left were larger. No longer was there an abundance of large and small bones, nor did the site support the same diversity of predators. The "survivability" of a bone was influenced not just by its size and structure but also by its density and grease content. Thus, the carnivores on the landscape significantly influenced what fossils were likely to form.

So, the next time you encounter a dead animal on a hike or nature walk, consider what information it can provide as it slowly disintegrates into the environment. After all, the dead have much to tell us.

yet this is common for vertebrates (Behrensmeyer et al. 2000). And anatomical fidelity can be compromised if different parts of the organism vary in their mineralization potential or even the likelihood of physical deformation (Behrensmeyer et al. 2000). Thus, characterizing what the taphonomic filters are, and how they influence a particular fossil assemblage, is essential for interpreting paleoecological patterns.

Preservation Potential: Biological Filters

The intrinsic traits an organism possesses are the first-order constraints on the likelihood of its preservation. These traits include its morphology, body size, behavior, and especially, biochemistry (Shipman 1981; Kidwell and Flessa 1995; Behrensmeyer et al. 2000). Organisms with "hard parts," that is bones, teeth, and shells, are much more likely to make it into the fossil record than those composed of soft tissue (Shipman 1981; Kidwell and Flessa 1995). Thus, molluscs; corals; foraminifera and some sponges composed of calcite ($CaCO_3$); other sponges in general; diatoms or plant parts composed of silica (SiO_2); and vertebrate bones, teeth, and fish scales made of hydroxyapatite ($Ca_5(PO_4)_3OH$) all have good preservation potential. But having bones is not enough; birds, with their small, hollow, and fragile bones, do not often survive as intact fossils (Chiappe 1995). Nor is the fossil record of bats very complete (Carroll 1988; Plotnick et al. 2016).

While soft-bodied life forms can represent 30%–100% of the species diversity in some marine or terrestrial communities, they do not usually fossilize (Lawrence 1968; Kidwell and Flessa 1995; Behrensmeyer et al. 2000). Thus, taxa such as fungi, worms, jellyfish, and even most cephalopods are not usually found in the fossil record, yielding a biased view of the structure and function of ecological communities (Kidwell and Flessa 1995; but see the discussion under "Fossil Lagerstätten" earlier for exceptions). In one classic study, Lawrence (1968) examined both recent and Cenozoic marine communities from the Atlantic coastal plain off North Carolina. He found that most species were soft-bodied with little preservation potential. Indeed, a comparative analysis between these revealed that only 25% of the species would leave a fossil record (Lawrence 1968). Without Lagerstätten, we would have little information about these nondurable soft-bodied forms.

Life history traits also matter. It is generally assumed that larger, abundant, and more widely distributed animals have higher preservation potential (Cummins et al. 1986; Behrensmeyer and Chapman 1993), although some of these factors (e.g., abundance and body size) are negatively correlated with each other in some clades such as mammals (Peters 1983). In a study of the differential preservation potential of modern mammals, Plotnick et al. (2016)

demonstrated that smaller animals and those with larger geographic ranges were better represented (Fig. 3.13). This suggests that high abundance and large geographic range influence the likelihood of fossilization (Plotnick et al. 2016)—factors that are not always correlated. Indeed, among contemporary threatened or endangered mammals that tend to be large-bodied, less than 9% were represented in the fossil record compared to 20% of other modern mammals. This led to the sobering conclusion that many species probably go extinct without leaving a tangible record (Plotnick et al. 2016). Other traits can also influence preservation potential. For example, difference in mobility can lead to catastrophic mass burial assemblages with limited representation of very mobile species (Behrensmeyer et al. 2000). In general, sedentary, herbivorous filter feeders are more likely to be represented than mobile, carnivorous forms because of the former's greater and localized abundance.

Biological agents in the environment also influence the documentary quality of the fossil record. Scavengers and microbes may recycle or rework dead material (Behrensmeyer et al. 2000). Scavengers, in particular, may churn sediments, thereby increasing exposure to oxygen and, indirectly, also increasing decay. Such biological agents may differently influence the preservation of various tissue types (Behrensmeyer et al. 2000). Indeed, predators and scavengers preferentially target particular skeletal elements depending on their nutritional value. For example, a long-term taphonomic study in Amboseli National Park demonstrated that the proximal and distal ends of limb bones, which contain high levels of grease, are particularly attractive to carnivores (Faith and Behrensmeyer 2006). Moreover, as the dominant predators in the community changed—from a lion- to hyena-dominated predator assemblage—the types and sizes of bones on the landscape changed as well (Faith and Behrensmeyer 2006).

Preservation Potential: Environmental Factors

Where an animal lives and dies also influences the likelihood it will become a fossil. For example, among mammals, bats are not well represented in the fossil record (Plotnick et al. 2016). While their fragile skeletons certainly contribute to the lack of preservation, bats also often live in moist tropical habitats, which are particularly poor depositional environments. For terrestrial mammals, bones of animals that die in cool and dry habitats ("erosive" environments) tend to have higher durability prior to burial (Behrensmeyer et al. 2000). Yet, barring unusual circumstances, for vertebrates to make it into the fossil record, their postmortem remains then must be transported into an appropriate accumulation site, such as a fluvial system

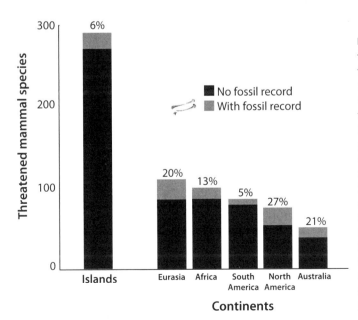

Fig. 3.13. Statistical studies of biases in the fossil record are a powerful way to understand filters and biases in our interpretation and reconstruction of ancient life. Here, the number and percentage of mammals listed in threatened categories by the International Union for Conservation of Nature (IUCN) are plotted by continent and/or island. Those currently threatened mammals without a fossil record are shown in black, and those with a fossil record are in gray. Strikingly, overall only a small and biased fraction of threatened species (< 9%) has a fossil record; in contrast, about 20% of nonthreatened mammal species have a fossil record (not shown). Moreover, there is a strong taphonomic bias of which mammals form fossils that is related to body size and the size of the geographic range. Redrawn from Plotnick et al. 2016.

(i.e., river channel, floodplain), coastal environment (i.e., beach, delta, or swamp), lake, bog, or ocean. In most instances, vertebrate postmortem remains do not travel far; "out of habitat" hydraulic transport of bones is fairly restricted in the fossil record (Behrensmeyer 1988; Kidwell and Flessa 1995; Behrensmeyer et al. 2000), suggesting this is not a major source of bias.

A number of other environmental factors also play a role in preservation (Behrensmeyer et al. 2000). These include the likelihood of immediate burial, biogeochemistry (e.g., oxygen, acidity), and the temperature and energy of the environment. The rate of burial is paramount; if the postmortem remains lie exposed on the surface, they are more likely to be modified or destroyed. However, once buried, other diagenetic conditions also matter. For example, low oxygen and nondynamic environments favor preservation because of the reduced number of scavengers and microbes. Low-oxygen environments also favor mineral precipitation that can aid preservation (Behrensmeyer et al. 2000). Similarly, lower temperatures result in less microbial activity and decay of the remains. Acidic environments also tend to destroy mineralized skeletons (e.g., $CaCO_3$ and $Ca_5(PO_4)_3(OH)$), although under certain conditions they may also retard microbial degradation, enhanc-

ing preservation (Behrensmeyer et al. 2000). Peat bogs have yielded many fossils of late Quaternary mammals that are exceptionally well preserved (Table 3.1). Here, decomposition can be arrested by an unusual combination of low temperature, acidity, and anaerobic conditions. The most famous of the animals preserved in this way is *Megaloceros giganteus,* or the Irish elk, known for its enormous and distinctive antlers (Fig. 3.14 and Fig. 3.15).

The energy of the environment has long been thought to be a major contributor to the disarticulation of complex skeletons. Wave action, wind, or other turbulent conditions were thought to weather fossil elements (Shipman 1981). Recent work, however, suggests that the fragmentation of animal hard parts may have more to do with the biogenic processes of scavenging and predation rather than the distance or speed of hydraulic transport (Behrensmeyer et al. 2000). Still, it is common for the different individual bones of vertebrates, which vary in their hydrodynamic properties, to be transported different distances after death. Finding a complete skeleton of a vertebrate is very uncommon and generally arises if the animal died as a result of a catastrophic event, such as a flood or ash flow. Indeed, most vertebrate specimens on display at

Fig. 3.14. The Irish elk (*Megaloceros giganteus*) is an extinct species of giant deer that ranged across much of Eurasia during the Pleistocene. It was very large—about 2.1 m tall at the shoulders—and was noteworthy for its huge antlers, which could reach 3.65 m from tip to tip and weighed as much as 40 kg. Considerable scientific speculation has arisen about the evolutionary drivers behind these massive antlers. *Left*: Skeleton of a male Irish elk. *Right*: Painting of male and female Irish elk as they might have appeared in the Pleistocene. "Irish Elk Side" by Franco Atirador, CC BY-SA 2.5, courtesy of Wikimedia Commons; Reproduction of a painting by John Henry Smith (active 1852–1893), licensed under the CC BY 4.0, courtesy Wellcome Collection.

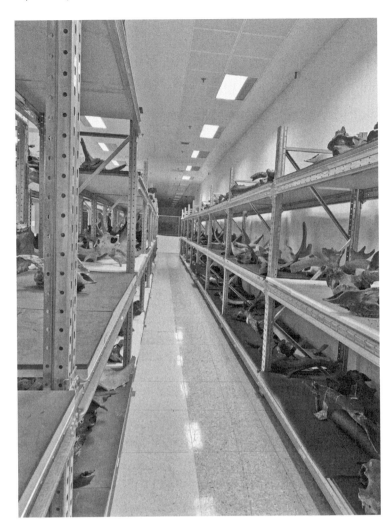

Fig. 3.15. Racks of Irish elk skeletons. Because many specimens have been found in bogs in Ireland, the National Museum in Dublin has the largest collection of bones. Photo by author.

museums contain many fabricated parts or else tend to be cobbled together from multiple individuals.

Moreover, rocks vary in their preservation potential and in the types of fossils they hold. Igneous rocks generally do not preserve remains, although ashfalls can entomb organisms under some circumstances (Shipman 1981). The process of metamorphism usually destroys fossils that might have been in the original sedimentary rock, although sometimes they survive. For these reasons, most fossils are found in sedimentary rocks (Shipman 1981). However, not all sedimentary rocks are likely to house fossils; intrinsic characteristics of the rock, such as particle size and composition, matter as does where the rock forms and ends up (Behrensmeyer et al. 2000). Fossils are rare in conglomerates, for example,

which are too coarse to preserve larger bones. Similarly, while bones are not often preserved in sandstones, trace fossils such as footprints are. In contrast, fossils are common in siltstones, mudstones, and limestones—all rocks that solidify relatively quickly and tend to seal out oxygen. Rocks tend to survive longest in stable environments, such as continental rift margins (Behrensmeyer et al. 2000).

Megabiases

Sometimes broadscale changes in taphonomy influence the fossil record; these are referred to as megabiases (Behrensmeyer et al. 2000). A bias is a skewing of information, which can either arise from natural taphonomic processes or from analytical methods employed; megabiases are those that are particularly large-scale in scope (Behrensmeyer et al. 2000). These can include the evolution of new life forms or innovations that increase (or decrease) preservation potential, changes in ocean chemistry or global climate, and/or geography (Behrensmeyer et al. 2000). Such biases can have a large influence on paleoecological inferences. For example, Sepkoski (1978, 1979, 1984) compiled data on orders and families of marine metazoans to examine broadscale diversity patterns over the geologic record (Fig. 3.16). He found a striking pattern of increasing diversity over time, continuing into the present. However, his original data were not corrected for a number of megabiases, including variations in the volume of sedimentary rock over time, which could bias the record of benthic continental-shelf organisms. Later work corrected for this, and other taphonomic biases, resulting in a somewhat dampened diversity curve (i.e., Alroy et al. 2001, 2008).

Other large-scale factors influencing the documentary quality of the fossil record include those relating to the basic geological processes: (1) the effects of rock recycling, which means that there are not as many old rocks or fossils as young ones; (2) the state of fossil preservation decreasing with greater age and those relating to scientific or human biases/error; (3) "the pull of the recent," the idea that our intensive sampling of extant organisms extends the stratigraphic ranges of even poorly sampled species; and (4) the "monograph effect," the idea that the extensive scientific study of a clade results in higher diversity because we know more about it. While the bias

associated with geology is generally dealt with statistically, that associated with humans can be altered by changing our collection methods. For example, early on, fossil hunters would collect only the large, highly visible remains of mammals and dinosaurs and leave behind smaller bones; this has been replaced by fine-scaled screening of fossil localities to obtain a more comprehensive sampling.

How much a taphonomic bias influences our perceptions depends on the scale and type of scientific question (Kidwell and Flessa 1995). In some instances, paleontologists assume that biases or filters are second-order effects with minimal influence on the process under study, or else that biases are randomly distributed with respect to the question under study (Behrensmeyer et al. 2000). And this may be true in many cases. Others may try to compensate for known effects by normalizing data. For example, one can examine data trends as a function of sampling intensity using techniques like rarefaction, which are common in ecological studies, or by varying the size of the temporal or spatial bins (Behrensmeyer et al. 2000). Some analytical metrics are less influenced by potential taphonomic biases: the median is less sensitive than the mean, patterns of change are more robust than absolute numbers, and so on.

Live-dead studies have been effectively used to investigate the fidelity of taxonomic and age-class compositions as well as the degree of time-averaging of death assemblages (Kidwell and Flessa 1995). Particularly for vertebrates, death assemblages

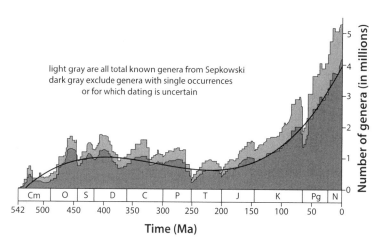

Fig. 3.16. Changes in Sepkoski's Phanerozoic diversity curve. This plot represents the accumulation of new generic diversity over the Phanerozoic based on fossil marine organisms. Increased diversity can result from a reduction in extinction rate, an increase in origination rate, or both. Note the shifts that occur when data are restricted to only well-defined genera (i.e., those with multiple occurrences and well-defined dating). The line is a third-order polynomial fit only to these well-defined genera. Modified from Sepkoski 1981, original licensed under CC BY-SA 3.0.

demonstrate high fidelity to the original community (Kidwell and Flessa 1995; Terry 2010). For example, a number of such studies conducted in Africa demonstrate that the abundance of mammal bones accurately reflects the representation of those species in the habitat; this is especially true for larger-bodied animals (Behrensmeyer and Dechant Boaz 1980; Behrensmeyer and Chapman 1993; Tappen 1994; Behrensmeyer et al. 2000). Thus, even processes such as postmortem transport do not appreciably homogenize the death assemblage. Similarly, late Quaternary live-dead studies of long-term predator accumulations from owl pellets show high fidelity to the modern local small mammal community (Terry 2010; Fig. 3.17). Here, taxonomic richness and evenness were in good agreement despite potential selectivity on the part of the raptors; it was only when the temporal window was enlarged to include centuries that the death assemblage became enriched relative to the modern (Terry 2010). Moreover, the analysis of

death assemblages can yield information about historical population fluctuations of species on landscapes. After demonstrating high fidelity of the Yellowstone death assemblage to the living community in both species richness and community structure, Miller (2011) used dated ungulate bones naturally left by predators in Yellowstone National Park to recreate species abundance and diversity. He was able to demonstrate dramatic ecological changes in mammal populations following both the 1988 wildfires and 1995 wolf reintroduction.

How to Become a Fossil

The bottom line from the discussion so far is that the best way to become a fossil depends on what type of organism you are. For vertebrates, the combined effect of taphonomic processes and simple mathematical probability means that fossilization will favor species that have abundant hard body parts; are geographically widespread, abundant, and

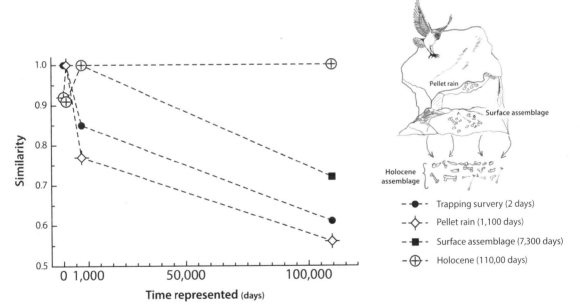

Fig. 3.17. Taphonomic similarity between modern and subfossil assemblages over time. In this study by Rebecca Terry, she compared the mammal species composition at a site over time by live-trapping animals near the cave site, dissecting owl pellets produced by raptors living in the cave, sampling the surface accumulation of bones with the cave, and sampling the subfossil remains. Each of these represented a different amount of time, ranging from a few days from the live-trapping to over 110,000 days as determined by radiocarbon-dating the fossil remains. She found that the agreement between modern live surveys (filled circles) and modern pellet rain samples (open diamonds) was consistently high and, moreover, that the time-averaged surficial death assemblage showed only slightly higher levels of estimated species richness than seen in the modern trapping surveys or modern pellet rain data. The late Holocene fossil strata, however, were different than other collections, likely owing to the increased species diversity. Drawn from data presented in table 3 of Terry 2010.

large-bodied; are found in cooler environments; and have a longer stratigraphic duration. Conversely, it is very unusual for small, soft-bodied, rare, geographically restricted, and stratigraphically ephemeral animals to fossilize.

So, How Reliable Is the Fossil Record?

Despite the myriad potential taphonomic biases, it turns out that the fossil record, even for vertebrates, is actually pretty damn good (Valentine 1989; Kidwell and Flessa 1995; Benton 1998; Terry 2010). Indeed, our qualitative understanding of evolution has not changed drastically over the past one hundred years, even as we accumulate much more data (Maxwell and Benton 1990; Benton 1998). The quality of the fossil record has been tested in a number of ways, including using collector curves where sampling effort is compared against yield and by comparing phylogenetic and stratigraphic evidence (Benton 1998). In this latter case, the assumption is that morphological and molecular phylogenies are constructed independently of the fossil record; thus, if the order and timing of splitting events corresponds with the fossil record, it suggests that the record is fairly complete. There is good correspondence with vertebrates and other taxa (Benton 1998). Thus, our understanding of the overall evolutionary pattern of diversification, as well as the timing and relative magnitudes of major

extinctions events, appears to be fairly good for both the terrestrial and marine realms (Maxwell and Benton 1990; Sepkoski 1993; Benton 1998). Although there is a low probability of any individual organism becoming a fossil, when you consider how many billions of organisms have lived on the planet over Earth's history, it is not surprising that the geologic record is robust.

FURTHER READING

Behrensmeyer, A.K., and A.P. Hill. 1980. *Fossils in the Making: Vertebrate Taphonomy and Paleoecology.* Chicago: University of Chicago Press.

Benton, M.J. 1998. The quality of the fossil record of the vertebrates. In *The Adequacy of the Fossil Record*, edited by S.K. Donovan and C.R.C. Paul, 269–300. New York: John Wiley and Sons.

Briggs, D.E.G., et al. 1995. *Fossils of the Burgess Shale.* Washington, DC: Smithsonian Institution Press.

Carroll, R.L. 1988. *Vertebrate Paleontology and Evolution.* New York: W.H. Freeman.

Efremov, I.A. 1940. Taphonomy: a new branch of paleontology. *Pan-American Geology* 74:81-93.

Kidwell, S.M., and K.W. Flessa. 1995. The quality of the fossil record: populations, species, and communities. *Annual Review of Ecology and Systematics* 26:269-299.

Shapiro, B. 2015. *How to Clone a Mammoth.* Princeton, NJ: Princeton University Press.

Shipman, P. 1981. *Life History of a Fossil: An Introduction to Taphonomy and Paleoecology.* Cambridge, MA: Harvard University Press.

4   Determining Age and Context

A mammal fossil is no more than a pretty curiosity unless it has context. The removal of fossils from their environments without proper documentation—however well meaning it may be—drastically reduces their scientific value. This is because a fossil is more than the record of an individual of a particular mammalian species; the combination of fossil and context can tell you how it lived, how it died, and how it was preserved. Moreover, by comparing with other taxa within the rock stratum, you can also learn something about the past environment. Thus, for a fossil to be useful, you need an understanding of both the environmental setting in which it was found and its geological age.

There are two basic approaches to dating fossils. The first is through the use of relative aging: figuring out how old a sample is in relation to other fossils or rock units. This involves using the layering of rock strata, index fossils, or combinations of fossils to put geologic events in chronological order without specifying an exact date. For many years, this was the only means for determining how old something was. The second technique is absolute dating, which uses radiometric dating methods, tree rings, or amino acid dating to obtain a precise age. Radiometric dating is the gold standard, but it is dependent on the fossil having the right composition of minerals and, moreover, can be very expensive.

How Does Relative Aging Work?

The oldest, and still most often used, method of aging fossils is relative dating. By working out the chronology of a fossil in a series of ordered horizontal rock layers, or strata, it is possible to identify events that occurred before and after it was deposited. The study of rock layers is called stratigraphy, and when the focus is on using the composition of fossils within rocks to correlate horizons or assign dates, it is called biostratigraphy. The International Commission on Stratigraphy (http://stratigraphy.org) is the official body determining divisions of time and space for the history of the Earth. Although it can get complicated, the basic principles underlying stratigraphy are pretty straightforward (Fig. 4.1). The first is the *law of superposition*, the idea that if undisturbed, rock layers get progressively younger as you move up a stratigraphic column (Brookfield 2008). As you may recall from chapter 2, this idea and several other basic principles were first developed by Nicolas Steno based on his geological rambles over the countryside of Italy. For example, his observations suggested that the force of gravity caused marine sediments to form distinct horizontal layers over time; this process is now called the *principle of original horizontality*. If undisturbed by geologic processes or if not deposited on an inclined surface, sediment settles into layers with the oldest rock

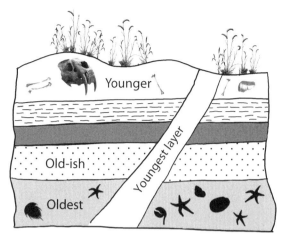

Fig. 4.1. Illustration encompassing the three major principles in stratigraphy. The law of superposition is the principle that rock layers, unless disturbed, get progressively younger as you move up a stratigraphic column; here, the oldest layers are on the bottom of the column. Next, the principle of original horizontality is the idea that the force of gravity causes sediments to form distinct horizontal layers over time. Finally, the third major principle, that of cross-cutting relationships, is the idea that a geologic feature (e.g., a fault, fracture, or igneous intrusion) that "cuts" or disrupts another feature must be younger than the stratum it is disturbing (Brookfield 2008). Organism silhouettes from PhyloPic (http://phylopic.org) under Creative Commons license; grass drawings from USDA-NRCS PLANTS Database 1950; *Smilodon* skull modified from photo by Wallace63, CC BY3.0; other images drawn or taken by author and/or courtesy of the public domain.

found on the bottom and the younger ones on top (Brookfield 2008). Of course, interpretation is often complicated by the dynamic nature of the Earth, which means that stratigraphic sections can be faulted or tilted. That leads into the third major principle: *cross-cutting relationships*. This is the idea that a geologic feature (e.g., a fault, fracture, or igneous intrusion) that "cuts" or disrupts another feature must be younger than the stratum it is disturbing (Brookfield 2008).

These approaches have been used to demonstrate that the lithology of the Grand Canyon, Zion National Park, and Bryce Canyon are all part of a continuous stratigraphic sequence often referred to as the Grand Staircase (Keyes 1924; Hintze 1988; Doelling et al. 2000). The Grand Canyon lies at the lowest level with the oldest visible horizons dating to the Precambrian,

while those found at the top of the canyon are Permian or early Triassic (Fig. 4.2). Through a series of topographic benches and cliffs, the surface steps up progressively as you travel north. Thus, the oldest visible rocks at the bottom of the sequence at Zion National Park are part of the Kaibab Formation (Permian), which stratigraphically is near the top of the older Grand Canyon sequence. Moving further north, Navajo sandstone (Jurassic), which lies in the middle of the sequence at Zion, is close to the bench at Bryce Canyon; the Tertiary-aged Wasatch Formation is found at the top (Fig. 4.2). The entire sequence encompasses more than 600 million years of geologic time (Doelling et al. 2000), providing arguably the most highly resolved and continuous sedimentary records on Earth. (Learn more at the National Park Service's Grand Staircase webpage: www.nps.gov /brca/learn/nature/grandstaircase.htm).

While stratigraphy continues to provide an important framework to correlate the relative position of rock formations across space, it is of more limited use when trying to assign dates. Here is where fossils come in. The *principle of faunal succession* is analogous to the law of superposition, but it is based on the contents of sedimentary rock strata rather than the organization of the layers themselves. Because sedimentary rocks very often contain fossilized remains of plants and animals, it follows that over geologic time as species and genera originate and go extinct, the composition of fossils in different rock layers varies in a fairly consistent order. Moreover, because the "lifespan" of a typical species is about 2–8 million years, depending on the clade (Foote and Raup 1996), each fossil reflects a particular time period in Earth history (Fig. 4.3). For example, finding a trilobite in a horizon immediately allows you to date it to the Paleozoic, since these interesting and diverse arthropods went extinct at the end of the Permian. Thus, we can reasonably assume that the first fossil assemblage in Fig. 4.3 dates to sometime in the Paleozoic; use of other index fossils in this stratigraphic sequence could further refine the chronology.

This leads to the idea of a biostratigraphic unit, or biozone, which is defined based on the type of characteristic fossils it contains. Despite the incongruous mix of extinct and extant taxa found in a lot of old

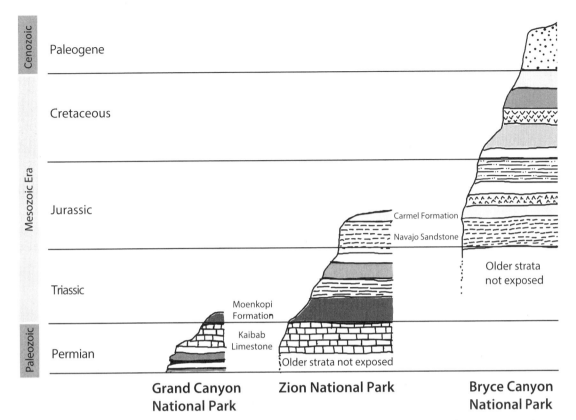

Fig. 4.2. The stratigraphy of the Grand Staircase in the western United States, a continuous stratigraphic sequence ranging from Grand Canyon National Park in Arizona to Bryce Canyon National Park in Utah. Note the sequences shared between the top of the Grand Canyon and the bottom of the exposed layers at Zion National Park (Moenkopi Formation and Kaibab Limestone), and between the top of Zion National Park and the bottom of the exposed layers at Bryce Canyon National Park (Navajo Sandstone and Carmel Formation).

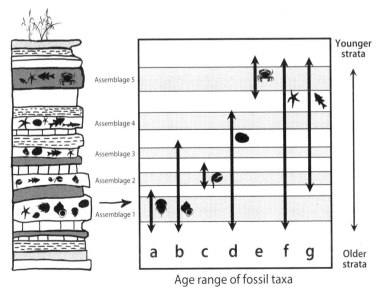

Fig. 4.3. A stratigraphic column with five distinct fossil assemblages, or biozones, interspersed with nonfossiliferous layers. The panel to the right illustrates the age range of each fossil taxon. Note that each biozone contains a characteristic group of fossils. Some taxa, such as f, range throughout the entire column, while others (a, c, and e) are limited to particular horizons. Fossils with a limited temporal extent are good candidates for index fossils (see p. ***). Also note that even wide-ranging taxa are not always present in each layer. Organism silhouettes from PhyloPic (http://phylopic.org) under Creative Commons license.

Hollywood movies (i.e., the coexistence of dinosaurs and humans in *One Million Years B.C.* or *Journey to the Center of the Earth*), there are actually quite regular patterns of faunal composition over the history of life on our planet. For example, unless the stratigraphic unit is disturbed, a fossil hominin tooth would never be found in the same layer as that of a *T. rex* femur. So the notion that an angry *T. rex* ever chased a human is a fictional invention of Hollywood . . . but a mammoth might have done so if provoked sufficiently by an ancient human. We often find mammoth teeth or bones associated with human fossil remains.

Not all of the fossils found in a biozone are informative. For example, fossil taxon f in Fig. 4.3 is not particularly useful because it has such a long geologic duration. Thus, the choice of fossil to use in characterizing biozones turns out to be important.

Index Fossils

The most useful fossils for biostratigraphic analysis are called index fossils. Looking at Fig. 4.3 again, the candidate index fossil taxa include species a, c, and e; note that these are all limited to a specific rock strata, which makes them at least potentially informative. But, index fossils need to have some other fairly specific ecological and life history characteristics as well (Brookfield 2008). First, they must be geographically widespread so that they can correlate horizons across space; plants and animal that are endemic to a particular area are of limited utility unless you are studying that specific time and place. Second, a candidate index fossil must be sufficiently abundant in the geologic record that their absence from a stratum is meaningful. This implies, of course, that they easily form fossils, which tends to exclude any plant or animal without hard parts. Third, index fossils should be morphologically distinct and easy to identify in rock strata. Finally, they must be rapidly evolving, which implies they will have relatively short geologic durations. This last requirement is vitally important. A taxon that otherwise fits the bill but demonstrates little morphological change in the geologic record will be found so broadly across horizons as to be of little use; temporal resolution is key for dating. Fortunately there are many fossils that fit these requirements. These change, of course, over different geologic periods, but typically they include

species of fusulinids, graptolites, trilobites, conodonts, ammonoids, and planktonic microfossils (Fig. 4.4; see also "List of index fossils" on Wikipedia at https://en .wikipedia.org/wiki/List_of_index_fossils).

Biostratigraphy is a powerful approach: if a fossil is found in the same stratigraphic layer as an index fossil, it follows that the two taxa coexisted (Fig. 4.3). And if an index fossil is found in widely separated geographic regions, it suggests that those strata were deposited at the same time. Different types of biozones have been identified, which represent different lengths of time (or biochrons; Fig. 4.5). Thus, biostratigraphy can allow the determination of relative ages of unknown fossils and the correlation of rock layers across large discontinuous areas. It was also foundational for the development of the geologic time scale in widespread use today (see http://stratigraphy.org /chart for the latest version). But how did these ideas come about?

William Smith and the "Map That Changed the World"

These days it is ridiculously easy to find a map replete with detailed geographic features for virtually any locality on the planet with just a quick search on the web. Or, if you prefer, you can go to a library and find entire rooms devoted to maps of all kinds. We know so much about the surface and structure of the Earth that it is hard to conceive of a time before such widespread knowledge was readily available. But it was only about two hundred years ago that the first detailed geologic maps were created by the Scottish-born American William Maclure (1809, 1818) and the British surveyor and engineer William Smith (1816-19). In particular, Smith's 1815 highly detailed and large-scale map of England, Wales, and Scotland was an amazing achievement that revolutionized geology (Winchester 2001; Sharpe 2016). This map detailed the local sequences of rock strata at a resolution of five miles per inch; overall, the map measured some six by eight and a half feet. To obtain a 3-D interpretation of surfaces, Smith used color-coding to indicate various rock layers or strata; an idea he got from studying an agricultural map of soil types (Fig. 4.6 and Fig. 4.7).

To put this into context: At this time in England, coal was emerging as an important energy source,

Fig. 4.4. Examples of some common index fossils. **a**, Brachipods. Shown is a common brachiopod, *Mucrospirifer mucronatus*, dorsal valve view, from the Silica Shale Formation in Paulding County, Ohio, USA. Specimen is Middle Devonian. **b**, Fusulinids. Shown is a fusulinid-bearing limestone from the Iola Formation, Elk County, southeastern Kansas, USA. This dates to the Upper Pennsylvanian. The photo is 13.7 cm across at its widest. Fusulinids were a widespread group of unicellular, marine benthic organisms with large, elongated, rice-like, microgranular, calcareous skeletons (sometimes called "tests"). **c**, Ammonites. Shown is *Parkinsonia* sp., a genus of fast-moving nektonic carnivores that lived in the Middle Jurassic. **d**, Trilobites. One of the most successful groups of arthropods in Earth history, there were ten orders and perhaps as many as 50,000 species of trilobites during various periods of the Paleozoic. Moreover, trilobites occupied a wide range of ecological niches and ranged in size from 3 mm to 1 m. The example shown is *Olenoides superbus* from the Burgess Shale in Canada. This species dates from the Cambrian period and grew up to 10 cm long. Panel a: Photo by Wilson44691, CC BY-3.0; b: Photo by James St. John, CC BY 2.0; c: Photo by Kaldari, CC0; d: Photo by Daderot, CC0 1.0. All courtesy of Wikimedia Commons.

and companies were intent on exploring potential sources. Because Smith wasn't wealthy, he had found work as a land surveyor and engineer (Torrens 2016). By this time in history, some general physical principles were widely accepted by geologists, including the idea that older rock layers were found below younger ones, and that fossils represented the remains of once-living organisms. In his capacity as a surveyor, Smith was sent across the country to inspect and survey coal mines; he also conducted extensive surveys of surrounding landscapes to determine if canals could be built to transport the materials across England (Fig. 4.6; Torrens 2016; Sharpe 2016). In the process of this work, which often involved going down mine shafts or inspecting cross-sectional cutouts of excavations, Smith noted a striking pattern of similarities in the type and color of rock layers and, especially, in the types of fossils they

contained. For example, he noticed a distinct shift in the composition of fossils between two rock layers near Bath, England; while the lower layer was full of plant fossils, the upper level contained seashells. We now know these layers represent the boundary between the Carboniferous and Permian periods (~300 Ma), a time when Earth's climate changed drastically from warm and wet to cooler and drier. Smith's observations led him to the idea that rock strata were laid down in a unique and predictable fashion and, moreover, that the fossils they contained could be used to correlate rock layers across distances. That is, he concluded that rocks containing the same fossil taxa had to have been deposited during the geologic duration of the taxon. Hence the idea of the principle of faunal succession was born.

These principles allowed Smith to draw his detailed geologic map. And so, on August 1, 1815, he

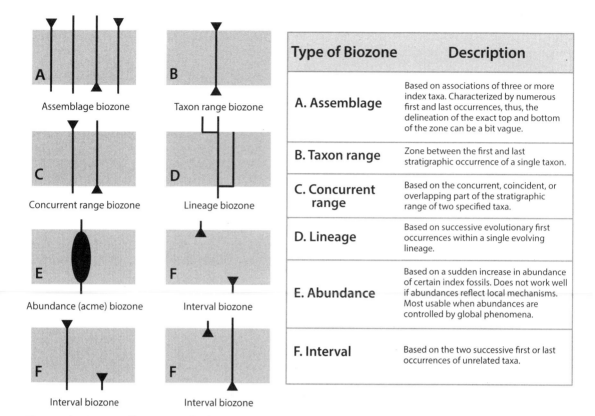

Type of Biozone	Description
A. Assemblage	Based on associations of three or more index taxa. Characterized by numerous first and last occurrences, thus, the delineation of the exact top and bottom of the zone can be a bit vague.
B. Taxon range	Zone between the first and last stratigraphic occurrence of a single taxon.
C. Concurrent range	Based on the concurrent, coincident, or overlapping part of the stratigraphic range of two specified taxa.
D. Lineage	Based on successive evolutionary first occurrences within a single evolving lineage.
E. Abundance	Based on a sudden increase in abundance of certain index fossils. Does not work well if abundances reflect local mechanisms. Most usable when abundances are controlled by global phenomena.
F. Interval	Based on the two successive first or last occurrences of unrelated taxa.

Fig. 4.5. The six types of biozones used in biostratigraphy. These are sometimes called Oppel zones after the German paleontologist Carl A. Oppel, who introduced the concept. Each rectangle represents bodies of sedimentary rocks and each line is the geologic duration of a particular taxon. The arrows indicate the taxon's first or last appearance, when known. Redrawn from Vera Torres 1994 (CC BY-SA 4.0).

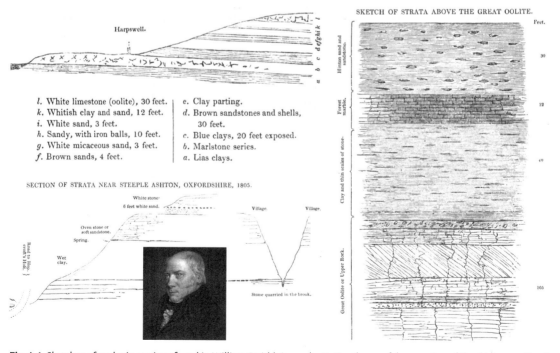

l. White limestone (oolite), 30 feet.
k. Whitish clay and sand, 12 feet.
i. White sand, 3 feet.
h. Sandy, with iron balls, 10 feet.
g. White micaceous sand, 3 feet.
f. Brown sands, 4 feet.

e. Clay parting.
d. Brown sandstones and shells, 30 feet.
c. Blue clays, 20 feet exposed.
b. Marlstone series.
a. Lias clays.

Fig. 4.6. Sketches of geologic sections found in William Smith's journals. Notice the careful measuring of the thickness of each layer of strata. *Inset:* Photo of Smith, aged 69. Images from Phillips 1844.

Fig. 4.7. William Smith's 1815 map of England, Wales, and part of Scotland. The original was in hand-tinted color. Photo by author of Smith's original 1816–1819 work.

published *A Delineation of the Strata of England and Wales with a Part of Scotland*, which he dedicated to the great biogeographer and his long-time supporter Sir Joseph Banks (Phillips 1844). Sadly, his map did not sell well. There are a number of theories why, including the oft-repeated story that a competitor redid his maps and sold them more cheaply, which may not be true (Winchester 2001; Torrens 2016; Sharpe 2016). Regardless, the poor sales of his map, combined with a number of unfortunate or ill-timed business ventures that went sour, led to a stint in a debtors' prison in 1819 (Torrens 2016). This was a common fate at the time for middle- and lower-class people unable to pay their debts. And science was certainly not a well-paid profession; most scientists of the day were wealthy nobleman who had no need of financial support. John Phillips, Smith's nephew, rather indirectly alluded to these unfortunate circumstances when he wrote in his "purposely softened" biography about his famous uncle, *"Science, indeed, is a mistress whose golden smiles are not often lavished on poor and enthusiastic suitors"* (Phillips 1844, pp. 77-78). However, there was ultimately a happy ending for Smith: his incarceration was brief, only three months (Torrens 2016), and after a further period of financial struggles, Smith finally received belated recognition for his work. In 1831, he was awarded the Wollaston Medal by the Geological Society, who acknowledged him the "Father of English Geology." More importantly for his financial health, in 1832 he received a lifetime pension from King William IV of England.

The Geologic Time Scale

Smith's work laid the groundwork for biostratigraphy, which allowed the relative dating and correlation of strata worldwide. Moreover, biostratigraphy led to the creation of a global chronosequence of fossils and ultimately to a geologic time scale for the Earth. An early attempt by the German geologist Abraham Werner divided the Earth's rock layers into four time periods: Primary, Secondary, Tertiary, and Quaternary. While this division did not last long, the latter two names stuck and remain as the major divisions of the Cenozoic era. Much of this early work on the geologic time scale was conducted by British geologists, who colorfully named many of the geologic periods after parts of England and Wales (Rudwick 2008). For example, the Cambrian, Ordovician, Silurian, and Devonian are all named after ancient Welsh tribes or English counties. Indeed, it was Smith's nephew, John Phillips, who in 1841 named the major eras in use today (Rudwick 2008). His time scale was based on turnover in the characteristic fossils present, many of which were earlier collected by his uncle. Thus, the Paleozoic, or ancient life; Mesozoic, or middle life; and Cenozoic, or new life, eras were established.

As mentioned earlier, the geologic time scale is maintained by the International Commission on Stratigraphy, the official body tasked with establishing divisions of time and space over the history of the Earth. It is a chronological dating system based on geologic strata and turnover in fossils. Because it was largely constructed in the nineteenth century before absolute dating was developed, the geochronologic units have very different durations (although tweaks are continually ongoing). For example, the Pliocene epoch lasted from around 5.3 to 2.58 Ma, while the Miocene ranges from about 23 to 5.3 Ma; similar inconsistences in temporal duration exist for eras, periods, and epochs. The Cenozoic era is defined as about 66 Ma to present, while the Paleozoic stretches over almost 300 million years (from ~541 to 252 Ma), about five times as long. However, because divisions in the geologic time scale are fundamentally based on differences in fossil composition, the temporal units tend to tightly correlate with major shifts in life on Earth—in particular, mass extinction events.

The three recognized eons (i.e., Archean, Proterozoic, and Phanerozoic) are each divided into eras, which are then divided into periods, and finally epochs and ages (see Fig. 2.6 in chapter 2 and http://stratigraphy.org/chart). These units are often modified using the terms "lower," "middle," or "upper" when talking about the rocks or "early," "mid," or "late" when talking about time.

Paleomagnetism

Paleomagnetism can be used as an indirect way of dating fossils (Tauxe 2010). By figuring out where in the sequence of magnetic events a fossil is situated, you can bracket its probable age. This works because the Earth has a magnetic field. Early naturalists and explorers exploited this by using lodestones, naturally occurring magnetic rocks, for navigation. Lodestones are created when magnetite or titanomagnetite iron ore is struck with lightning (Wasilewski and Kletetschka 1999). Lodestones were the first compasses; when hung on a bit of string so they can turn freely, they orient in a magnetic north/south direction (Wasilewski and Kletetschka 1999).

The paleomagnetic field of the Earth results from a *magnetohydrodynamic dynamo* (Opdyke and Channell 1996; Butler 2004). In brief, this means that the magnetic field is induced and maintained by the convection of liquid iron and nickel in the Earth's outer core. As the Earth rotates, these electric currents produce a magnetic field (Opdyke and Channell 1996; Jeanloz and Romanowicz 1997; Butler 2004; Tauxe 2010). This leads to an interesting phenomenon where we have *both* magnetic and geographic North and South Poles (Fig. 4.8). The geographic North and South Poles are defined by the axis of rotation of our planet. But, as any backpacker

or hiker knows, a compass works by pointing to the *magnetic* north, not the geographic north. While the magnetic north and south poles are in the same general region of the geographic poles, the magnetic core is actually inclined about 10 degrees from the Earth's axis of rotation (Butler 2004; Tauxe 2010).

To make it even more complicated, the Earth's magnetic field is not constant over time (Tauxe 2010). The position of the north geomagnetic pole drifts over time, by about 0.05 degrees per year (Fig. 4.8; Butler 2004; NCEI 2019). This doesn't occur in a predictable manner; rather, polar drift best models a random walk (Butler 2004). And over longer time scales, the orientation or polarity of the Earth's magnetic field actually switches (Tauxe 2010). That is, rather than drifting in the same general area, the poles rapidly flip, with the result that the magnetic north pole is near the geographic South Pole, and the magnetic south pole is near the geographic North Pole

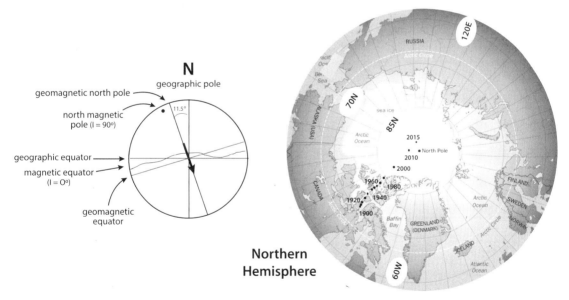

Fig. 4.8. While the geographic North and South Poles are located at the spin axis of the planet, the geomagnetic poles reflect the location of the Earth's magnetic field, which is not aligned with the rotation axis. These geomagnetic poles do not change much over time and are currently tilted about 10 degrees relative to the axis of rotation (Butler 1992). The north and south magnetic poles are where the inclination of the Earth's magnetic field points vertically downward (i.e., is equal to 90 degrees). These poles move over time because of fluctuations within the Earth's core as well as solar insolation and solar wind. *Right*: Locations of the north magnetic pole from 1900 to 2015. Note the considerable wandering, or polar drift, over the past century. The pace of drift has increased over time, and currently the north magnetic pole is moving toward Siberia at about 34 km per year (https://www.ncei.noaa.gov/news/world-magnetic-model-out-cycle-release). Interestingly, the north and south magnetic poles are not antipodal, that is, a straight line going through the center of the Earth does not bisect both. If the Earth's magnetic field was a simple dipole, this would occur. In contrast, the geomagnetic north and south poles are defined as the antipodal points that reflect a line bisecting the Earth's core.

(Fig. 4.9). Field reversals have occurred many times in Earth history (Opdyke 1985; Opdyke and Channell 1996; Butler 2004). It is suspected that changes in convection in the Earth's core lead to such changes in the magnetic field (Opdyke and Channell 1996; Butler 2004; Tauxe 2010). When the magnetic north pole is oriented toward the geographic North Pole (i.e., it is in the Northern Hemisphere, as today), this is called normal polarity; the opposite is called reverse polarity. For example, if you had used a compass to orient yourself to due north 1.5 Ma, you would be pointing toward Australia and not the Arctic because the polarity of the Earth's geomagnetic field was reversed in the middle Pleistocene.

While Alfred Wegener proposed in 1915 that the Earth's crust was composed of plates that moved over time, just how dynamic the Earth's lithosphere was, was not recognized until after World War II. The war led to many advances in seismic imaging and sonar and to the discovery of curious magnetic

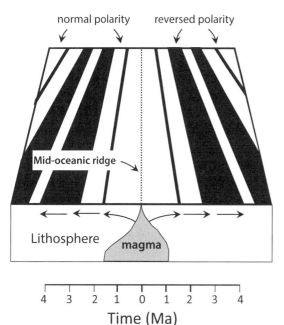

Fig. 4.9. Illustration of paleomagnetic "stripes" on the seafloor. These are the result of geomagnetic field reversals. Because seafloor is created in mid-oceanic trenches, the patterns of reversals are symmetrical on either side of the ridge. These observations, which were not possible before the advent of sonar and seismic imaging, led to the acceptance of the idea of plate tectonics after nearly fifty years of debate (Irving 1988; Butler 2004; Opdyke and Channell 1996).

anomalies: symmetrical, parallel stripes of rock with similar magnetization bracketing mid-ocean ridges (Fig. 4.9; Tauxe 2010). It wasn't too long before scientists recognized that these were due to changes in the Earth's magnetic field over time. Indeed, the eventual acceptance of the theory of plate tectonics in the 1960s was dependent on these paleomagnetic studies, which provided quantitative information about the movement of the lithosphere over geologic time (Opdyke 1985; Irving 1988; Opdyke and Channell 1996; Butler 2004; Tauxe 2010).

So, what does the Earth's magnetic field have to do with the dating of fossils? It turns out that the small bits of magnetic materials within rocks acquire and retain a remanent magnetization that records the direction of the Earth's geomagnetic field at the time they were formed. Thus, unless reheated or disturbed, rocks can serve as compass points frozen in time. The direction of the remnant magnetic polarity recorded in a stratigraphic sequence allows subdivision into *chrons*, units characterized by their magnetic polarity. In the same way that stratigraphy allows the ordering of rock layers across wide geographic areas to determine approximate ages, geologists use the pattern of reversals to figure out the timing of events and/or bracket the age of fossils. This has been facilitated by the use of absolute radiometric dating of some of the chrons to determine when magnetic reversals occurred in the past. Using these techniques, geologists have devised the geomagnetic polarity time scale, which is used to help calibrate the geologic history of the Earth.

"Absolute" Dating

Absolute dating involves using the physical or chemical structure of a fossil to obtain a calendar age. While these can provide a precise age for a fossil, the term "absolute" is a bit of a misnomer. The typical examples useful for paleoecology include radiometric dating methods, tree rings, or amino acid configurations. Which of these, if any, are useful for a particular fossil depends on factors such as preservation, mineral and elemental content, and the likely age of the specimen.

Radiometric Dating

The gold standard for figuring out the age of a fossil is through the use of radioactive elements. Radioactivity was discovered by Henri Becquerel in 1896 and investigated extensively by Marie and Pierre Curie (Mould 1998; Goldsmith 2005). Their work led to the development of radiometric dating methods by the early and mid 1900s. Many elements have isotopes that are radioactive, often because they have too many neutrons to be balanced by the number of protons in the nucleus. Radioactive elements turn out to provide a sort of atomic clock: by exploiting the fact that they are unstable and break down spontaneously, one can determine an actual date of deposition and age of a fossil or rock.

Radioactive elements decay from a "parent" to a "daughter" at a rate unique to that element; the time it takes for half the original material to decay is known as the half-life and it is invariant (Fig. 4.10). The half-life for carbon 14, for example, is 5,730 years; other elements have much longer or shorter half-lives. By measuring the ratio of parent product to daughter product with a mass spectrometer, you can precisely determine the amount of time

that has elapsed since the "radiometric clock" started ticking (i.e., the age of the sample).

The start of the radiometric clock is influenced by temperature. When rocks are melted by Earth processes, the daughter isotopes that have accumulated within are lost; this resets the radiometric clock to zero. The temperature at which this occurs is known as the closure temperature; below this, the mineral and/or rock can be considered a closed isotopic system. Different minerals have very different closure temperatures (Faure and Mensing 2004). For example, while zircon can have a closure temperature in excess of 900°C, hornblende is around 500°C, and biotite is around 300°C. Minerals with particularly high closures can maintain their original crystallization age even if redeposited within a newer rock if it forms at lower temperatures, complicating interpretation (Kelley 2002; Faure and Mensing 2004).

Which isotopes are useful for a particular study (Fig. 4.11) depends on the elemental composition of the sample or rock, the closure temperature, and the time frame of interest. For example, potassium-argon, argon-argon, the uranium series, and, for late Quaternary fossils, radiocarbon dating, have all been used to obtain absolute ages for fossils, including stromatolites that are over 3.4 billion years old. Radiometric dating is particularly useful for dating inorganic materials, especially igneous rocks. When molten rock cools, radioactive atoms are trapped inside and decay at a predictable rate. Because fossils may not contain sufficient quantities of some of the most useful radiometric elements (e.g., uranium and potassium), sometimes scientists date the layers of igneous rock or volcanic ash above and below the target fossil and "bracket" the age of the fossil.

For radiometric dating to yield accurate results, three critical assumptions must be true. First, it must be possible to accurately measure the starting amount of the daughter element (Fig.4.10). Second, the sample must have existed in a closed environment relative to the isotope system. That is, the entire amount of the daughter element must be attributable to radioactive decay. Finally, the decay rate of the parent element must have been constant over time. In practice, these assumptions are not always met, and departures can skew results and interpretation.

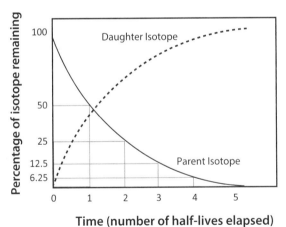

Fig. 4.10. Radiometric dating. Radioactive isotopes are unstable and, thus, spontaneously decay from parent isotopes to daughter isotopes. When the process of releasing energy and matter from the nucleus leads to a new element, it is known as transmutation. For example, uranium 238 undergoes decay to produce thorium and gamma rays; that is, $^{238}_{92}U \rightarrow {}^{230}_{90}Th + {}^{4}_{2}He +$ gamma rays. The time it takes for half the parent isotope to decay is a fixed rate for each element and known as the half-life. In this example, it would take 75,380 years for half of the original uranium 238 to decay to thorium; in 150,760 years, only a quarter would remain. The increase in the daughter isotope mirrors the decrease in the parent isotope

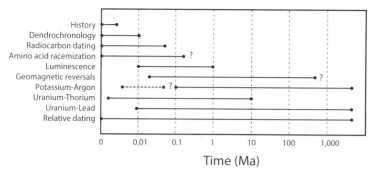

Fig. 4.11. The time frame for some common dating methods used in paleontology and geology. Some of these (history, dendrochronology) are not discussed in this text, but they can provide dating for Holocene samples (Walker 2005). The time frame for some of these methods has expanded as new analytical techniques and more sophisticated machines have come online.

Below I discuss in more detail a few of the more commonly employed radiometric isotopes in paleoecology. This is not meant to be an exhaustive list. Quite a few other radiometric systems are employed in particular contexts (e.g., rubidium-strontium, chlorine-36, samarium-neodymium, etc.).

Radiocarbon Dating (^{14}C to ^{14}N)

For late Quaternary fossils, which may still retain appreciable quantities of collagen, radiocarbon is the best direct or absolute method of dating. Since Willard Libby developed it in the late 1940s (Libby 1946; Arnold and Libby 1949; Libby 1965), radiocarbon dating has been extensively employed in paleoecological, archaeological, and anthropological studies (Walker 2005); it even has its own scientific journal, *Radiocarbon* (http://radiocarbon.webhost.uits.arizona.edu). By far (98.9%) the most abundant isotope of carbon on Earth is ^{12}C, which has 6 protons and neutrons. Some carbon, however, exists in the form of ^{13}C and ^{14}C, about 1.11% and 1.0×10^{-10}%, respectively. While the ^{13}C isotope is relatively stable and extremely useful for stable isotope analysis (chapter 7), the isotope ^{14}C is radioactive and spontaneously decays to ^{14}N. The half-life from the parent isotope (^{14}C) to the daughter isotope (^{14}N) is 5,730 years.

So how can this be used to date fossils? It turns out that carbon dioxide in the atmosphere is composed of both $^{12}CO_2$ and $^{14}CO_2$. As plants incorporate carbon dioxide during the process of photosynthesis, they take in both forms. Animals eat plants, thus the ^{14}C is incorporated into their tissues in equilibrium with their environment. However, when an animal or plant dies, it stops exchanging ^{14}C with its environment. The amount it contains in its tissues and bones decreases as the radioactive carbon begins to decay; this can be accurately measured using a mass spectrometer (chapter 6) and used to date the bone or tissue. For example, imagine that a paleontologist finds a bone in a horizon, sends it off for analysis, and learns that the abundance of ^{14}C and ^{14}N is approximately equal. This suggests an age of 5,730 years (i.e., one half-life) for the fossil. If analytical results had revealed that there was three times less ^{14}C than ^{14}N (i.e., only 25% of the original material was left), it would have suggested an age of 11,460 years (i.e., two half-lives; Fig. 4.10). Radiocarbon dating is useful for dating materials up to about 50,000 years (Walker 2005); after that, the amount of ^{14}C remaining is generally too small for accurate measurement even on the most powerful machines.

One caveat to this method is that radiocarbon years do not neatly match up to calendar years (Libby 1965). Thus, calibration curves relating radiocarbon estimates with other dating proxies are employed to convert radiocarbon years to calendar years (e.g., see CalPal Online at www.calpal-online.de). The difficulty in relating calendar and radiocarbon ages is that the $^{14}C/^{12}C$ ratio in the atmosphere, and hence the biosphere, has not been consistent over Earth history. This occurs for a number of reasons, including fractionation (chapter 7), fluctuations in cosmic ray production (influenced by solar intensity), the marine reservoir effect (oceans have two sources of radiocarbon, the atmosphere and the deep ocean), and anthropogenic causes such as the Suess effect (a change in the atmospheric concentrations of carbon isotopes because of the admixture of fossil fuel-derived carbon dioxide, which is depleted in the ^{13}C and ^{14}C carbon isotopes) (Suess 1955; Libby 1965; Keeling 1979; Tans et al. 1979; Farquhar et al.

1989; Bowman 1995; Usoskin 2017). Indeed, the testing of nuclear weapons in the 1950s and 1960s left a legacy in terms of artificially increasing the amount of radiocarbon in the atmosphere (Vogel et al. 2002; Hua and Barbetti 2004; Levin and Kromer 2004; Uno et al. 2013); this is called the bomb effect. It allows samples younger than 1950 to be accurately dated to within 0.3-2 years (Geyh 2001; Uno et al. 2013). Studies have used the bomb effect to date human remains as well as a variety of plant and animal remains, including elephants and hippos (Spalding et al. 2005; Uno et al. 2013); this has useful for human and wildlife forensics as well as modern and historical ecology. In particular, the ability to use bomb curves to calibrate overlapping serial samples of tree rings or tusks allows the development of longer-term ecological records.

It is hard to overstate the importance of radiocarbon dating to late Quaternary studies, particularly those dealing with the terminal Pleistocene and Holocene megafauna extinctions, where a chronology is important for resolving causation (e.g., Hester 1960; Martin and Klein 1984; Haynes 1993; Steadman 1995; Koch and Barnosky 2006; Haynes 2009; Turvey 2009; Bourgeon et al. 2017; Smith et al. 2018) For example, radiocarbon dating of three hundred extinct elephants and mammoth from the New and Old World demonstrated a striking difference in the extinction dynamics (Martin and Stuart 1995). While the extinction was gradual in Eurasia, it was abrupt in the Americas. In our own work with late Quaternary mammals, the radiocarbon dating of materials within an excavation has allowed the development of an "age model," or a chronology of deposition of fossil materials. Time series analyses are dependent on accurate calibration, which can be obtained from radiocarbon if sufficient organics are present in the fossils. Because we can correlate depth with age, we can compare the morphology, abundance, and turnover of mammal fossils from different levels with known climatic, environmental, and biotic changes (Smith et al. 2015).

Potassium-Argon Dating (^{40}K to ^{40}Ar)

While radiocarbon is extremely useful for fossil materials that are less than 50,000 years old, absolute dating of deep-time fossils requires the use of isotopes with much slower decay rates (i.e., longer half-lives;

Fig. 4.11). Thus, potassium-argon or rubidium-strontium are often employed when interested in time spans from 4 Ga to 100 ka (McDougall and Harrison 1999). While these isotopes may be rare in fossils, they are much more abundant in rocks and minerals and so can be measured in the rocks surrounding or straddling a fossil if it is carefully removed—another reason why a fossil removed from context is of little value!

Potassium (^{39}K) is an alkali metal found in many materials (particularly silicas) within the Earth's crust. It is the eighth most abundant element and makes up about 2.4% of the crust by mass (McDougall and Harrison 1999). Potassium has three isotopes, including the radioactive form, ^{40}K. Over time, about 11% of the radioactive ^{40}K spontaneously decays to ^{40}Ar, with a half-life of 1.248 billion years (McDougall and Harrison 1999; Kelley 2002); the remainder decays to ^{40}Ca and is not of relevance here. Because argon is a noble gas, it is essentially chemically inert and does not bind to other atoms within the lattice. Moreover, it is a rare trace element, meaning that virtually any argon found within a rock comes from the decay of potassium (Kelley 2002). If a potassium-containing rock is melted, the argon is released into the atmosphere; as it cools and recrystallizes, the rock becomes impermeable to gasses (McDougall and Harrison 1999). Thus, the ^{40}K-^{40}Ar clock starts when the igneous rock is initially formed, and measurement of the isotope ratios tells you how long the clock has been ticking. Most minerals containing K-Ar have very high closure temperatures, which means that this system can be used for dating very old substrates since the probability they have been reworked since formation is low. K-Ar dating was one of the earliest isotope dating methods used (Aldrich and Nier 1948); it is particularly useful for dating igneous or volcanic rocks that form over strata containing the fossils of interest. Dating of this layer provides a minimum age of deposition.

As with other radiometric techniques, there are caveats. Most important is the assumption of a closed system, which assumes no argon has escaped since rock formation, nor has any atmospheric argon been incorporated into the rock (McDougall and Harrison 1999; Kelley 2002). Thus, all argon present is assumed to have arisen from in situ decay of radioactive potassium. Whether this assumption is valid may depend on the type of rock; quartz, metamorphic rocks, and those associated with hydrothermal systems are more likely to contain excess argon (Kelley 2002). Moreover, the measurement of potassium and argon isotopes must be conducted on separate aliquots requiring additional material. Thus, heterogeneous samples can be an issue (McDougall and Harrison 1999; Kelley 2002). Violation of these assumptions can lead to erroneous ages to be assigned to the samples (McDougall and Harrison 1999).

A more sophisticated variant of potassium-argon dating relies on the measurement of relative abundances rather than the absolute determination of isotopic abundance. Here, the ratio of ^{40}Ar-^{39}Ar is used, measured from a single aliquot. This allows greater internal precision and also allows the analysis of small and heterogenous samples (Merrihue and Turner 1966; Alexander et al. 1978; Deino et al. 1998; McDougall and Harrison 1999; Kelley 2002). It is now preferentially employed in many geological applications (Kelley 2002).

Potassium-argon dating has been used to determine important Earth transitions, including mass extinctions, volcanic eruptions, human evolution, and even moon rocks (Kelley 2002). For example, the first rocks collected on the moon by the Apollo 11 mission were dated using K-Ar (and later by Ar-Ar), returning ages in the range of 3-4 Ga (Turner 1970). Studies relating the development of African grasslands to large-scale mammal turnover and the evolution of the genus *Homo* have only been possible through the radiometric dating of volcanic strata (Behrensmeyer et al. 1997; Bobe and Behrensmeyer 2004). A particularly fascinating use of K-Ar dating was on pumice deposited by the catastrophic eruption of Mount Vesuvius. The pyroclastic surges produced by this event destroyed several Roman cities including Herculaneum and Pompeii; Herculaneum was buried by 23 m of tephra. Dating of the pumice suggested the eruption occurred some 1,925 (± 94) years ago (i.e., around 75 CE; Renne et al. 1997), which was in close agreement with the date recorded by the Roman writer Pliny the Younger (Roberts 2007), who was one of the few surviving eyewitnesses. This study was particularly noteworthy because it demonstrated the potential utility of K-Ar for dating events and artifacts in historical time.

Uranium-Lead Dating (^{238}U and ^{235}U Series)

Uranium-lead (U-Pb) is one of the oldest and well-used dating methods, going back to the early twentieth century (Barrell 1917). It relies on the decay of multiple parent isotopes through separate decay chains or radioactive cascades as uranium does not decay directly to a stable state but undergoes a sequential series of alpha and beta decay until a stable isotope of lead is reached (Schoene 2014). This radiometric dating provides two basic clocks, which each start with a different parent isotope of uranium. The isotope ^{238}U ultimately decays to ^{206}Pb after the loss of six alpha and eight beta particles, with a half-life of 4.47 billion years. Similarly, ^{235}U decays to ^{207}Pb after the loss of seven alpha and four beta particles, with a half-life of 704 million years. Using the ratios of intermediate daughters can also be useful (Schoene 2014). For example, uranium-thorium (U-Th) is useful for dating materials up to about a million years old. This system relies on the ratio of the two alpha emitters, ^{238}U and ^{230}Th; while the former has a half-life of 4.47 billion years, thorium's half-life is only 75,380 years.

While terrestrial vertebrate fossils can contain significant amounts of uranium—from 1 to 1,000 ppm (van der Plicht et al. 1989; Balter et al. 2008; Greene et al. 2018)—almost all of this is acquired postmortem during diagenesis; that is, the chemical and physical changes that occur during the fossilization of the animal. Fresh bones usually contain less than 0.1 ppm (Schwarcz 1982; van der Plicht et al. 1989). Thus, the chronological potential of U-Pb dating depends on careful characterization of the history of uranium uptake and the sedimentary environment where fossilization occurs (van der Plicht et al. 1989; Rae and Ivanovich 1986). Bones are recrystallized during the decay of the organics within the original bone. If uranium uptake is geologically abrupt, the system can be assumed to be closed and dates obtained can be relatively accurate (Greene et al. 2018). However, if the system is open, and uranium exchange occurs between the bones and/or teeth and the environment, dates will be tend to be much younger than the actual deposition of the fossil itself (Rae and Ivanovich 1986; van der Plicht et al. 1989; Balter et al. 2008).

One way to quantify the open or closed nature of the system is through the use of the dual decay paths of uranium (Schoene 2014). Using the expected values of ^{206}Pb-^{238}U and ^{207}Pb-^{235}U over time, a plot can be constructed relating the two, called a concordia diagram (Wetherill 1956). If measured samples fall on the line, it suggests that the isotopic system has been closed since the time of formation. If values do not fall on the line, they are considered discordant and strongly suggest open-system behavior, complicating the validity of the dates (Schoene 2014). Considerable work has gone into understanding the uptake and fixation of uranium by fossil bone and how knowledge of the distribution of uranium and the local deposition and sedimentary environment can be used to develop more accurate dating (Rae and Ivanovich 1986; Balter et al. 2008; Greene et al. 2018). But, the difficulties inherent in interpretation of the results on fossil materials mean that the surrounding or bracketing rocks are often used preferentially for U-Pb dating.

Thus, U-Pb dating is often conducted on zircon (Fig. 4.12). Zircon is a ubiquitous mineral in all types of rocks within the Earth's crust. It is also particularly amenable to radiometric dating because of its durability and chemical inertness (Schoene 2014). Moreover, while the crystal structure of the mineral contains trace quantities of uranium and thorium, it does not readily incorporate lead. Thus, virtually all of the lead present in a zircon mineral comes from the radiometric decay of uranium (but see Schoene 2014 for caveats). Zircon also has a very high closure temperature (around 900°C), which means it often maintains its internal radiometric clock even when redeposited.

Uranium-lead dating has been used to date the formation of the solar system, the surface of the moon, and meteorites; calibrate the geologic time scale; and better pinpoint major tectonic events, including the rise of atmospheric oxygen (e.g., Patterson 1956; Behrensmeyer et al. 1997; Bekker et al. 2004; Condon et al. 2005; Rasbury and Cole 2009; Gradstein et al. 2012; Guex et al. 2012; Amelin and Ireland 2013; Schoene 2014). For example, although no crustal rocks have survived from the formation of the Earth, the high closure temperature of zircon means that some have been found dating to this time in metamorphosed sediments in Western Australia. The oldest zircon dated from Western Australia is from 4.404 Ga, just after the formation of the Earth (Wilde et al. 2001).

Of particular interest here, U-Pb has allowed the dating of important events in the evolution of life (Schopf 1993; Lamb et al. 2009; Knoll and Nowak 2017). For example, U-Pb dating of zircons extracted from volcanic ash beds in China has been used to precisely date animal fossil beds from the Neoproterozoic (Condon et al. 2005) and noteworthy Mesozoic mammals (Luo et al. 2001; Qiang et al. 2006). But, more generally, radiometric dating has allowed the timing of crucial evolutionary transitions, such as the development of eukaryotes and then metazoans, the evolution of sex, the colonization of land and evolution of plants, and the evolution of the major clades of life (Butterfield 2000; Gensel 2008; Lamb et al. 2009; Knoll and Nowak 2017).

Amino Acid Geochronology

Amino acids are the building blocks of proteins and thus ubiquitous within living things. With the exception of glycine, which has the simplest structure of all amino acids, they are optically active with asymmetric carbon atoms. "Optically active" refers to the plane of polarization of light though the material. While amino acids could potentially have either a D (dextro, or clockwise) or L (laevo, or

Fig. 4.12. Zircon is a mineral often found in all types of rocks within the Earth's crust. It is particularly useful for radiometric dating because of its durability and chemical inertness. Photo by author.

counterclockwise) configuration, most living things keep their amino acids in the thermodynamically unstable L configuration (Bada 1985; Kaufman and Manley 1998). Once organisms die, control over this configuration ceases and there is a shift toward chemical equilibrium (Bada 1985). Thus, the ratio of D to L changes from a value near 0 in the living organism to an equilibrium value of closer to 1 in the dead organism (i.e., where D and L configurations are found in equal amounts). This chemical reaction is called racemization: the interconversion of amino acid enantiomers (Bada 1985; Kaufman and Manley 1998). And measuring the ratio of D to L tells you how long ago something died. The different amino acids have different racemization rates, preservation potentials, and analytical precision (Bada et al. 1979; Bada 1985). For example, aspartic acid is useful for Holocene or late Pleistocene fossils (and yields more reproducible results), whereas isoleucine racemizes almost ten times slower and, thus, is useful for much older bones (Bada et al. 1979; Bada 1985).

High performance liquid chromatography is typically used to analyze the extent of racemization in amino acids within carbonate-containing fossils (Bada 1985; Kaufman and Manley 1998). This technique is useful for many types of organic fossils, including bones, teeth, molluscs, and forams, and can yield dates from time scales of decades to millions of years, depending on environmental temperature and the amino acid employed (Bada 1985; Kaufman 2000). For fossil bones, racemization

half-lives are around 10^4-10^5, much longer than radiocarbon (Bada 1985). Moreover, it is relatively inexpensive, with current sample costs running around seventy-five dollars.

However, there are pitfalls. In particular, amino acid racemization rates are dependent on both time and temperature (Bada 1985; Kaufman and Manley 1998). Thus, it may be necessary to determine the temperature history of a bone before this technique can be applied. Moreover, depending on the depositional environment, it is possible for secondary amino acids to be incorporated into fossil bone (Bada 1985). Various calibration techniques have been proposed to deal with these contingencies (Bada 1985); however, there remains some variation in analytic results among laboratories (Powell et al. 2013).

FURTHER READING

Bowman, S. 1995. *Radiocarbon Dating*. London: British Museum Press.

Brookfield, M.E. 2008. *Principles of Stratigraphy*. New York: John Wiley and Sons.

Faure, G. and T.M. Mensing. 2004. *Isotopes: Principles and Applications*. 3rd ed. New York: John Wiley and Sons.

Gradstein, F.M. et al. 2012. *The Geologic Time Scale 2012*. Waltham, MA: Elsevier.

Tauxe, L. 2010. *Essentials of Paleomagnetism*. Berkeley: University of California Press.

Walker, M. 2005. *Quaternary Dating Methods*. New York: Wiley.

Winchester, S. 2001. *The Map That Changed the World: William Smith and the Birth of Modern Geology*. New York: HarperCollins.

Part Two
Characterizing the Ecology of Fossil Organisms

On Being the Right Size

For every type of animal there is a most convenient size, and a large change in size inevitably carries with it a change of form.
—J.B.S. Haldane, "On Being the Right Size," *Possible Worlds and Other Essays*, 1928

Body mass is one of the most important axes of biological diversity. The size of an animal (or plant, or protist, or other organism) strongly influences all biological rates and processes, including the essential activities of metabolism, reproduction, and growth (Peters 1983; Calder 1984; Schmidt-Nielson 1984). Indeed, body size turns out to be crucially important in virtually all aspects of physiology, life history, and behavior; it also mediates the way an animal interacts with other organisms and the abiotic environment. In short, the size of an animal is arguably the most important thing to know about it. It should come as no surprise that scientists and philosophers from Aristotle, Galileo Galilei, Charles Darwin, J.B.S. Haldane, George Gaylord Simpson, and D'Arcy Thompson to the present have pondered what underlies the incredible diversity of the size and shape of life: why organisms tend to evolve to a certain size and what the corresponding ecological and evolutionary consequences and trade-offs are. Fortunately, body size is something we are able to characterize fairly easily using the fossil record.

How Do We Measure "Size"?

The most direct way of estimating size—assuming your animal is alive and manageable—is to put it on a scale and weigh it. This isn't always possible of course, and so a number of alternative methods have been used in both paleontology and biology. For

example, one approach is to use Archimedes' principle, which is often done with very large vertebrates, especially the very largest sauropods (Alexander 1971; Larramendi 2016). You may recall from somewhere in your academic past that density (d) equals mass (m) divided by volume (v). If we want to determine mass, then, all we need is volume and density. Density is straightforward: while bone has an average density of 1.75 g/cm^3, most other components within a body hover around 1.0 g/cm^3, which of course is also the canonical density of water. For example, blood, muscle, and fat have values of 1.04, 1.06, and 0.91 cm^3, respectively (Morales et al. 1945). Thus, the basic idea behind the volumetric method is to build an accurate scale model and estimate its volume, which as the ancient Greeks noted is easily determined by measuring the amount of water displaced (Fig. 5.1). Since most of an animal is made up of water (70%), and virtually all of the remainder (~24%) by soft parts such as skin, viscera, and blood (Morales et al. 1945), we can reasonably assume the density of a dinosaur or extinct mammal was 1.0 g/ml; this is the integrated density of modern reptiles and mammals. Note this technique is dependent on having a good quality, calibrated scale model of an organism and not a Disney version with distorted proportions and big round soulful eyes. Determining the scale of the model is not difficult if you know the approximate length your animal was/is. For example,

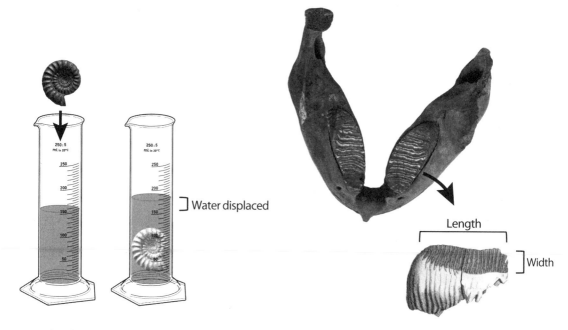

A. Volumetric Measurement   ## B. Morphological Measurement

Fig. 5.1. Two commonly employed methods for estimating the mass of extinct animals. Archimedes' principle uses water displacement to compute biovolume (A), which is then converted to mass given a known density. Alternatively, many skeletal and dental features scale predictably with mass (B). These include molars (especially the M1) and many postcranial bones (e.g., femur length and shaft, etc.). The size of other animals, such as bivalves and gastropods, can be estimated in a similar fashion from surface area. Image of ammonite by Kaldari, CC0, courtesy of Wikimedia Commons; mammoth jaw from Leunis and Ludwig 1891; mammoth tooth courtesy of the public domain.

if an actual mammoth was about 6 m long and your model is 15 cm long, then the scale is (6 m/0.15 m) or 40/1 (40:1).

Turning water displacement into mass is straightforward. Once you have the volume of your mammal or dinosaur, you multiply the water displacement (in ml or cm^3) by the scale of the animal cubed. This last step is important: since your model is scaled linearly, but you are working with volumes (i.e., cubic dimensions), you have to cube your estimate. Thus, if the scale is 40:1, you multiply your estimated volume by 40^3, or 64,000. For example, if you had a mammal model that displaced 200 ml of water, you would estimate the biovolume of the actual critter at 200 ml × 64,000 = 12,800,000 ml or 1.28×10^7 ml. (Keeping your units consistent is important, but fortunately 1 ml = 1 cm^3). To convert biovolume to body mass, which is generally more useful, you multiple the density of the animal by the volume (i.e., d = m/v, thus m = dv). Because the density of an animal is about

1.0 g/cm^3 (and 1 kg = 1,000 g), your conclusion is that this animal weighed 1.28×10^7 g, or 1,280 kg. Thus, overall, estimated mass (g) = density (g/cm^3) × water volume displaced (cm^3) × scale of model.

While it may be fun to create a scale model and dunk it in water, this is a great deal of work for a single body-mass estimate. Fortunately, there are much easier ways for most fossils. Pragmatically, the best way to estimate mass is to measure some linear dimension of the fossil, and using regression equations based on extant, closely related animals, translate that into body mass (Table 5.1). This works for virtually any fossil that has an ecological analog in the modern age (e.g., Niklas 1984; Damuth and MacFadden 1990; Gingerich 1990; Payne et al. 2009; Smith et al. 2010a, 2016a; Heim et al. 2015; Larramendi 2016). For some animals or plants, especially those without a modern analog, mass or biovolume is approximated by the geometric shape that best represents the organism. Body mass for extinct trilobites, for example, can be estimated using an

Table 5.1. Examples of regression equations useful for translating cranial and postcranial measurements of fossil mammals into estimates of body mass. Many proxies exist for long bones and various cranial measures. A few of the most common are shown here.

Order	Variable	Slope (a)	Intercept (b)	Units
Artiodactyla	Lower M1 length	3.263	1.337	cm, kg
	Lower M2 length	3.20	1.13	cm, kg
	Humerus length	3.395	−2.513	cm, kg
Carnivora, overall	Lower M1 length	2.97	−2.27	mm, kg
	Skull length	3.13	−5.59	mm, kg
Family Canidae	Lower M1 length	1.82	−1.22	mm, kg
Family Felidae	Lower M1 length	3.05	−2.15	mm, kg
Family Ursidae	Lower M1 length	0.49	1.26	mm, kg
Marsupials	Lower M1 length	3.562	1.749	cm, kg
	Lower M2 length	3.480	1.505	cm, kg
Perissodactyla	Lower M1 length	3.187	1.264	cm, kg
	Humerus length	2.714	−1.357	cm, kg
	Tibia length	2.253	−0.824	cm, kg
Primates	Femur length	0.331	1.912	mm, kg
Proboscidea	Humerus length	0.99	0.66	cm, kg
	Femur length	0.83	1.10	cm, kg
	Femur minimum circumference	1.12	−0.09	cm, kg
Rodentia	Lower M1 length	3.31	0.611	mm, g
All mammals	Humerus length	2.675	−4.558	mm, kg

Source: All equations from appendices in Damuth and MacFadden 1990 with the exception of Proboscidea, which derive from Jukar et al. 2018.

Note: Regression equations take the form:

Log (body mass estimate in grams or kilograms) = $a \times$ log (variable in mm or cm) + b.

Clade-specific values are derived by measuring elements on extant mammals; for these equations to be robust, taxa must span about an order of magnitude in body mass. Note that base 10 logs are used here and that units of measurement may vary.

ellipsoid (i.e, mass = $4/3 \times \pi \times (l/2) \times (w/2) \times (h/2)$, where l = length, w = width, and h = height; Payne et al. 2009). Similarly, the body size of trees is generally approximated by a cylinder (i.e., mass = $\pi \times r^2 h$, where r = the radius and h = height; Niklas 1984).

Vertebrate paleoecologists that have the good taste to specialize on mammals often use teeth or long bones to estimate the body mass of extinct species. Teeth are especially useful, not just because they are readily preserved in geological excavations, which they are, but also because they are highly diagnostic and can help identify the species or genus (see chapter 6 for more on teeth). Moreover, mammals have heterodont dental morphology, which means that variations in the shape and relative size of premolars, molars, and incisors also provide useful information about diet. It turns out that there is generally a robust predictive relationship between

the length (or area) of the first molar (M1) and body mass (Table 5.1). These allometries are based on relationships constructed using extant taxa and, as you might expect, vary between different mammalian orders. It thus becomes possible to examine patterns of body-size evolution and diet across time based solely on recovered fossil teeth (Box 5.1; also see chapter 2). These techniques have been used successfully in many studies and are the basis of compilations of large-scale data of animal body size over time (Sepkoski 1981, Jablonski and Raup 1995; Alroy 1998; Smith et al. 2010a, 2016a), as well as that of body size of virtually all life over the Geozoic (Heim et al. 2015).

Two important parameters are often computed to evaluate the precision and quality of the estimates used to obtain body mass estimates. Recall from basic statistics that a regression is a line that minimizes the sum of the squared deviations of prediction, which are often called the sum of squares error. In developing an allometry to predict body size from some dental or postcranial element, it is likely that you have experimental error in both the *x*- and *y*-axis. Thus, it is obviously useful to be able to assess just how well your regression does at estimating body mass. Most often used are the percent pre-dicted error (%PE) and the percent standard error (%SEE) of the estimate (Smith 1984). The %PE is computed as:

$$\%PE = \left(\frac{(Observed\ value - predicted\ value)}{Predicted\ value} \right) \times 100.$$

It provides an assessment of how well the regression does in predicting the actual mass of the animal.

Box 5.1
Estimating the Size of an Extinct Titanothere

BOX FIG 5.1 *Left:* Courtesy of Wikimedia Commons. *Right:* Photo by author.

The drawing here is of an extinct ungulate, *Brontops* sp., which was found during the Eocene of North America. This large-bodied mammal was a member of the now extinct family Titanotheriidae (Brontotheriidae), which is related to modern horses and rhinos (order Perissodactyla, the odd-toed ungulates). Interestingly, *Brontops* fossils have been found with fractured ribs, which, as you can see, artist Robert Bruce Horsfall in 1913 artistically interpreted as indicating males battling for hierarchical dominance. Given the paucity of other large-bodied mammals at this time, it is not implausible. The photo to the right of the drawing is of the author holding a right lower first molar (M1) of a *Brontops*. This particular specimen came from an Eocene formation in the White River Badlands of South Dakota.

Suppose you wanted to estimate the body mass of the animal from this molar. Using digital calipers or a digital microscope, you obtain a length of 4.8 cm for the M1. This can be transformed into a body mass estimate using the slope and intercept values for Perissodactyla from Damuth and MacFadden (1990), found in Table 5.1:

Log body mass (kg) = 3.187 × (log 4.8 cm) + 1.264

or,

Log body mass (kg) = 3.44

Since this is Log Body Mass, you need to convert to just body mass (i.e., take the inverse). Do this by raising both sides by 10. That is, $10^{Log(Body\ Mass)} = 10^{3.44}$ which yields an estimate of 2,754.23 kg (2.75 tons); note that $10^{Log} =$ 'cancels'. Typical estimates in the literature for the mass of *Brontops* range from 2 to 3 tons, so this value is in the right ballpark. The computation of %PE and %SEE is an exercise left to the reader.

Moreover, comparing the %PE of regressions constructed on different dental or postcranial material provides a sort of comparative index useful in determining the relative predictive accuracy of different fossil elements or equations developed at different hierarchical levels (e.g., order versus family regressions).

The %SEE also yields information about the predictive accuracy of the regression (Smith 1984). It is computed as:

$$\%SEE = (10^{(2 + \log(\textit{standard error of the estimate})}) - 100.$$

Assuming a normal distribution, if the computed %SEE were 90, this can be interpreted to mean that 68% of the actual mass values should fall within ±90% of the predicted mass estimate.

The Range of Size in Living Organisms

The range of body sizes among living things spans close to an amazing twenty-two orders of magnitude (Fig. 5.2). This includes *Mycoplasma*, a genus of anaerobic bacteria that weigh around 10^{-13} g, to the giant sequoia at 9×10^9 g (or 9×10^{12} mm³); the largest animal on Earth has been the blue whale (*Balaenoptera musculus*) at near 10^8 g (180–200 tons) (Peters 1983; Schmidt-Nielson 1984; Payne et al. 2009; Smith and Lyons 2011, 2013; Smith et al. 2016a). Interestingly, much of the increase in the body size of living things over Earth history was achieved in two jumps: the first during the Paleoproterozoic era (~1.9 Ga) and the second during the late Neoproterozoic (~0.6 Ga) (Payne et al. 2009). Each of these events resulted in an expansion of body size morphospace by upwards of six to seven orders of magnitude; they were also associated with major evolutionary transitions in organismal complexity. The first jump was associated with the origin of the eukaryotic cell, and the second, eukaryotic multicellularity (Payne et al. 2009). While life continued to evolve larger size, other increases were much more gradual (Payne et al. 2009; Heim et al. 2015). The drivers behind the jumps are unclear, although it is speculated that major shifts in the oxygen content of the atmosphere at these times may have facilitated evolutionary innovations (Payne et al. 2009).

Over evolutionary time, mammals have spanned about eight orders of magnitude in body size (Fig. 5.3), although most of this range was achieved

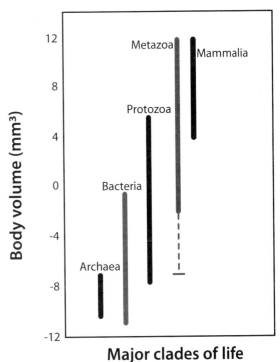

Fig. 5.2. The approximate body size of living things. Here, we show the range of size achieved over evolutionary history by each domain (archaea, bacteria, protists, and metazoans or multicelluar animals) in $\log_{10}$ biovolume (mm³). The body-size range of mammals is also indicated. The dotted line extending below Metazoa connects the smallest animal (an ostracode) in the Heim et al. (2015) dataset with the minimum size of animals, *Myxosoma chuatsi*, a tiny parasitic cnidarian. After two initial jumps in body size during the Proterozoic (Payne et al. 2009), the range occupied by each major clade has remained relatively constant. Moreover, with the exception of Archaea, which appear to have been restricted to only four orders of magnitude, most clades occupy about the same range of body size (~12–14 orders of magnitude), although their minimum and maximum body sizes vary considerably, reflecting different constraints. Redrawn from Smith et al. 2016a.

in the Cenozoic after the demise of the non-avian dinosaurs (Alroy 1998; Smith et al. 2010a). Indeed, for the first roughly 140 million years of their evolutionary history, mammals occupied a fairly restricted range of body mass and shapes (from about 2 g to a maximum of 5-10 kg), which has been attributed to the ecological dominance of the dinosaur clade (Lillegraven et al. 1979; Smith et al. 2010a, 2016a). After the extinction of the non-avian dinosaurs at the Cretaceous-Paleogene (K-Pg) boundary, there was a major and rapid radiation of morphological, ecological,

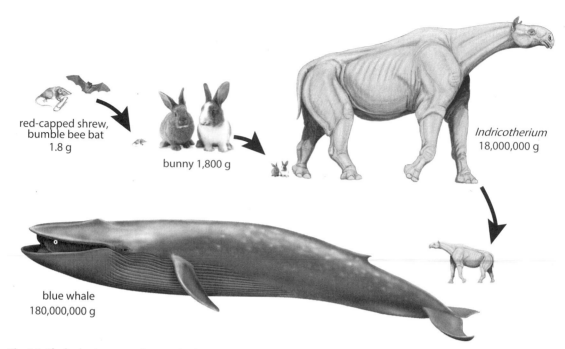

red-capped shrew,
bumble bee bat
1.8 g

bunny 1,800 g

Indricotherium
18,000,000 g

blue whale
180,000,000 g

Fig. 5.3. The body-size range of mammals, drawn approximately to scale. Over evolutionary time, mammals have spanned eight orders of magnitude in body mass, ranging from the smallest species (the red-capped shrew and bumblebee bat) at less than 2 g, to the largest terrestrial mammals (*Indricotherium* and *Deinotherium*), which weighed around 18 metric tons, to the largest animal to exist on Earth, the blue whale at over 180 metric tons (Smith et al. 2003, 2010a, 2016a). The giant sequoia has a larger biovolume than a blue whale and so is the largest organism on the planet (Payne et al. 2009). *Indricotherium* by Dmitry Bogdanov, CC BY 3.0, courtesy of Wikimedia Commons. Other images courtesy of public domain.

and phylogenetic diversity, leading to the full spectrum of body masses seen today (Alroy 1998, 1999; Smith et al. 2010a, 2016a).

Constraints on Mammal Body Size

Constraints on the maximum and minimum body size that mammals have achieved over evolutionary history are likely set by a combination of physiological and environmental factors (Thompson 1942; Gould 1966; Alexander 1971, 1982, 1989; Bourliere 1975; Smith and Lyons 2011, 2013). Certainly, interactions with the environment are mediated by body mass, as are the relative importance of forces such as gravity, temperature, or even the surface tension of water. For example, while a cat cannot walk on water or straight up a wall (although it looks that way sometimes!), some insects or lizards can. Similarly, the trunks of trees and the limbs of terrestrial animals need to be sufficiently strong to provide support against the force of gravity. But, a trunk

cannot be so thick that it interferes with efficient transport of water and nutrients, nor can legs be so wide as to interfere with efficient locomotion. Consequently, the theoretical upper body mass limit of a quadrupedal animal is about 140 tons (McMahon 1973). Building a larger animal is difficult because the limbs needed to support the mass become so wide they interfere with movement. Interestingly, the largest known terrestrial animal was the sauropod *Argentinosaurus*, who probably weighed about 100 tons, not too far from this hypothesized maximum.

Mammals have never achieved the extraordinarily large body sizes of some dinosaurs. The largest terrestrial mammals were *Indricotherium* and *Deinotherium* at around 18 metric tons, although several species of mammoth reached masses exceeding 12 tons (Fortelius and Kappelman 1993; Smith et al. 2003, 2010a) and perhaps as much as 20 tons (Larramendi 2016). The reasons for this probably have to do with energy: the maximum body size of a mammal is significantly related to the area of the island, continent, or ocean basin it occupies (Marquet and Taper 1998; Burness et al. 2001; Smith et al. 2010a). Mechanistically, this likely reflects the positive scaling of individual resources and space use with increasing body mass and the necessity to maintain sufficiently large population numbers to avoid stochastic extinction. Interestingly, ectothermic or mesothermic dinosaurs were able to achieve much larger body sizes over Earth history, but they also had much reduced mass-specific metabolic requirements relative to endotherms (Peters 1983; Calder 1984). It has also been theorized that heat dissipation, or overheating, may have limited the maximal size of mammals, but not sauropods, unless they too were endothermic (Alexander 1982, 1989). This idea too has support: in a study of deep time mammal evolution, we found that the largest terrestrial mammals were only found during cooler climates (Smith et al. 2010a).

The smallest size a mammal can attain may also be set by energic constraints (Pearson 1948; Alexander 1982; Bourliere 1975). The very smallest-bodied extant mammals, such as the red-capped shrew (*Suncus etruscus*) or the bumblebee bat (*Craseonycteris thonglongya*), weigh about 1.8 grams, the weight of a single playing card; late Paleocene and early Eocene lipotyphlans may have been a bit smaller, closer to 1.3 g (Bloch et al. 1998). Animals of such small size face extremely high mass-specific maintenance costs. Indeed, their intense energetic demands require a continual supply of high quality food resources (Pearson 1948; Peters 1983; Calder 1984). Consequently, the smallest mammals feed on high calorie foods such as nectar or insects, and further, may employ facultative ectothermy by using torpor to reduce energy demands (Pearson 1948).

Aquatic organisms face different proximate constraints, but energy may also be the ultimate factor influencing both minimum and maximum body size. In fully marine mammals, the lower limit of size is likely set by the energetic cost of thermoregulation: not for them, but for their offspring (Downhower and Blumer 1988; Smith and Lyons 2011; Gearty et al. 2018). Water has almost twenty-four times the heat conductance of air (0.58 vs. 0.024 watts per meter-Kelvin), which can make regulation of body temperature and maintaining homeostasis particularly difficult for offspring. Because the mass of neonates scale with adult mass, the minimum size of marine mammals is about 10 kg, several orders of magnitude greater than the 1.8 g minimum for terrestrial mammals (Smith and Lyons 2011).

A number of ideas have been proposed to explain constraints on the maximum size of whales, but none are as yet generally accepted (Smith and Lyons 2011, 2013). Many of the largest-bodied whales feed seasonally and migrate to warmer equatorial locations for reproduction. This has led to hypotheses that rates or limits of energy acquisition constrain maximum size, but others have pointed to trade-offs in the scaling of feeding rate versus metabolic rate, limitations of bone density and structure or heart size and circulation, and problems with heat dissipation (Downhower and Blumer 1988; Alexander 1982, 1989; Calder 1984; Smith and Lyons 2013; Gearty et al. 2018). Intriguingly, it is also possible that the largest marine mammals have yet to evolve (Uhen et al., in prep.).

Implications of Body Size

Many fundamental physiological, ecological, and evolutionary factors scale with body mass in predictable ways. If the relationship between the trait and body size is linear (or geometric), it is referred to as isometric (*iso* = same; *metric* = measure); the gut capacity of animals is an example of such a trait. However, most biological rates and processes actually scale nonlinearly with body mass (Fig. 5.4; Table 5.2), that is, with slopes less than or greater than 1. Important examples include metabolic rate, reproduction, and the cost of locomotion (Peters 1983; Calder 1984). These latter nonlinear relationships are called allometries (*allos* = different; *metron* = measure; Fig. 5.4). Although allometric scaling relationships were first quantified by Eugene Dubois in his studies of the relationship between the weight of the brain and body mass (Dubois 1897), the ideas and term were codified by and expanded upon by Julian Huxley and Georges Teissier (1936). Because body size can be reliably measured from the fossil record, allometries have been extensively explored in paleontology (e.g., Kermack and Haldane 1950; Gould 1966, 1971; Gingerich et al. 1982; Blumenberg and Lloyd 1983; Smith and Lyons 2013).

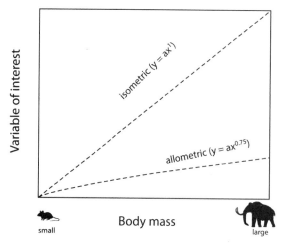

Fig. 5.4. Allometric and isometric relationships. These are often formulated as power functions: $Y = aM^b$, where Y is the variable of interest, M is body mass, b is the slope (representing how the variable of interest changes with differences in body size), and a is a taxon-specific constant (the intercept at unity body mass when M = 1). Power laws are often logarithmically transformed so that log y = log a + b log M, because the exponent then becomes the slope of a straight line. This facilitates computations and interpretations of pattern. In the example shown here, the allometric slope is 0.75, as is the case for metabolic rate. Animal silhouettes from PhyloPic (http://phylopic.org) under Creative Commons license.

All organisms require energy for the essential processes of growth, reproduction, maintenance, and survival. With environmental changes, there may be a reallocation of energy between these activities. Consequently, a knowledge of energetics is central to an understanding of the selective forces that shape an organism's physiology, natural history, and evolution. The best-known allometry is metabolic rate, which represents how energy is acquired, transformed, and used within an organism (Kleiber 1932; Peters 1983; Calder 1984).

Metabolism scales with body mass with an exponent of 0.75 (Kleiber 1932; Fig. 5.4). While this is actually true of all living things, be they protists, eukaryotes, or plants (Ernest et al. 2003), the pattern is often called "the mouse to elephant curve" because of the taxa that defined the upper and lower limits of the original regression by Kleiber. This is a robust pattern, with over 90% of the variation in metabolic rate across species explained solely by their body mass. The small amount of residual or unexplained

variation reflects unique evolutionary or biological adaptations that are specific to particular clades.

The implications of allometric scaling of metabolism are profound. Pragmatically, each gram of a mammal the size of a mouse uses about twenty times more energy than an equivalent gram of elephant or mammoth. Consequently, food acquisition, processing, and passage rates are much more rapid for small animals, leading to highly selective foraging and a reliance on high quality resources. Over the past few decades, a wide array of allometries has been demonstrated; this is especially true for mammals (e.g., Peters 1983; Calder 1984; Schmidt-Nielsen 1984; and references therein). Thus, it is possible to calculate all sorts of potentially useful biomechanical, physiological, and even ecological information from body mass. This includes such eclectic information as size of the heart or volume of blood in the body, sleeping duration, fecundity, running speed, home range, population density, or geographic range (Table 5.2).

Allometries can be particularly useful for extinct species since it is not possible to otherwise predict physiological rates or reconstruct life history. Indeed, producers of some very successful nature programs, including BBC Natural History units *Walking with Pre-Historic Beasts* or *Walking with Dinosaurs*, have employed allometric regressions to reconstruct the likely ecology and behavior of extinct animals.

Patterns in Body Mass over Time and Space

The trade-offs between energy acquisition, the body mass of animals, and challenges imposed by their environment have led to several well-documented ecogeographic gradients or "rules" (Mayr 1956; Millien et al. 2006). The existence of such ecogeographic patterns that hold across space and time reflects the strong selection imposed on animals by their environment. Environmental temperature, in particular, is an important selective force operating over both spatial and temporal dimensions (Andrewartha and Birch 1954; Birch 1957; Mayr 1956, 1963; Brown 1968; Brown and Lee 1969; Dawson 1992; Tracy 1992; Smith et al. 1995; Smith and Betancourt 1998, 2003, 2006). While animals can regulate the loss or gain of heat through behavior—circulatory mechanisms that alter blood flow, insulation (i.e.,

Table 5.2. Example of allometries relating various life history, physiological, ecological, or morphological traits with body mass in kilograms (kg). Equations take the form of a power function (i.e., $Y = a \times M^b$), except for litter size, which is a semi-log relationship.

Trait	Slope (*a*)	Intercept (*b*)	Units
Litter size	−0.30	3.03	number of offspring
Total litter mass	0.83	0.17	kg
Gestation time	0.26	65.3	day
Age at weaning	0.23	58	day
Time to maturity from birth (ungulates)	0.27	229	day
Time to maturity from birth (carnivores)	0.32	591	day
Surface area	0.73	0.08	m²
Brain mass	0.70	0.01	kg
Lung volume	1.02	56.7	ml
Feeding speed	0.70	10.7	watts
Gut capacity	1.08	0.10	kg
Metabolic rate	0.76	3.89	watts
Pulse	−0.27	3.61	number/sec
Walking speed	0.33	0.21	m/sec
Running speed	0.12	2.34	m/sec
Total sleep time	−0.15	32,500	sec
Average life-span	0.17	2,040	day
Population density (herbivores in temperate regions)	−0.61	214	number/km²
Population density (carnivores in temperate regions)	−1.16	15	number/km²
Mean size of prey (carnivore)	1.16	0.11	kg
Maximum size of prey (carnivore)	1.45	4.03	kg
Home range (herbivores)	1.00	0.03	km²
Home range (carnivores)	1.37	1.39	km²

Source: All values from Peters 1983.

fur, fat, or feathers), and evaporative mechanisms (i.e., sweating or panting)—a more direct way is to alter body size and/or shape to track the environment. Consequently, it should not be surprising that the canonical ecogeographic pattern, Bergmann's rule, involves both body size and temperature.

Bergmann's Rule

Named after Carl Bergmann, a German physiologist, Bergmann's rule is an empirical observation that morphological variation is related to physiogeographic features. Broadly stated, within a widely distributed genus, larger-bodied species inhabit colder environments and smaller-bodied species are found in warmer habitats (Rameaux and Sarrus 1838; Bergmann 1847). In Bergmann's words, *"If there are genera in which the species differ only in size, the smaller species would demand a warmer climate to the exact extent of the size difference . . . on the whole the larger species live farther north and the smaller ones farther south"* (James 1970, p. 390). This idea initially

described patterns observed for species within genera, but it turns out to also hold at other levels of the taxonomic hierarchy (Mayr 1956); even populations of animals within a species demonstrate a body size cline with temperature (Millien et al. 2006 and references therein; Fig. 5.5). Importantly, this rule *does not* suggest that large-bodied species cannot occupy warm environments. After all, many megaherbivores, such as elephants, giraffes, and rhinos, are found in sub-Saharan Africa. What it does suggest, however, is that *within* a particular clade, the smaller-bodied species are more likely to be present in hotter environments. Thus, within the elephant family Elephantidae, African elephants (*Loxodonta africana*, ~4 tons) are found, well, in Africa, but their much larger now extinct cousins, mammoths (*Mammuthus* sp., >10 tons), were largely restricted to the cooler northern latitudes. Similarly, the polar bear (*Ursus maritimus*, ~700 kg) is found exclusively in the Arctic, whereas other smaller bear species (e.g., *Ursus americanus*, the black bear, and *Ursus arctos*, the

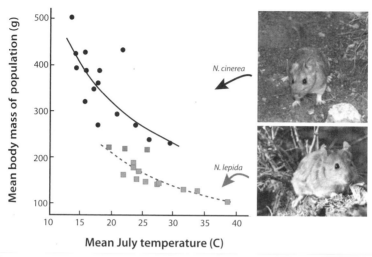

Fig. 5.5. Example of Bergmann's rule in woodrats (*Neotoma* sp.). Plotted is the relationship between mean July temperature and average population body mass of two species: *N. lepida* (desert woodrat, bottom picture, squares) and *N. cinerea* (bushy-tailed woodrat, top picture, closed circles). Both species demonstrate a strong response to temperature; body size of populations is smaller in hotter environments and larger in colder ones. Note also that the much larger bushy-tailed woodrat is found exclusively in much cooler habitats and does not occupy habitats if mean July temperature exceeds 30°C. July is the hottest month in the Western Hemisphere. Redrawn from Smith and Betancourt 2003; photos by author.

brown bear) range across much of the temperate latitudes.

Several comprehensive analyses have concluded that Bergmann's rule is valid for the majority of vertebrates (62%-83%; Ashton et al. 2000; Millien et al. 2006), including humans (Ruff 1994; Foster and Collard 2013). Interestingly, these ecogeographic clines are not just observed in endothermic birds and mammals but also in many ectothermic amphibians, reptiles, and invertebrates (Millien et al. 2006). The most notable exceptions among vertebrates are lizards and snakes (Millien et al. 2006). In mammals, larger-bodied species conform to the rule more often than smaller-bodied ones; this probably reflects their inability to avoid exposure to environmental extremes (Freckleton et al. 2003; Meiri and Dayan 2003). While small-bodied mammals can avoid stressful environments by denning in vegetation, a rock shelter, or burrow, or even by employing facultative torpor, these options are not generally available for larger mammals. There is some evidence that Bergmann's rule is stronger for Nearctic regions than tropical ones, perhaps because climate gradients are larger (Rodríguez et al. 2008).

The original descriptions attributed Bergmann's rule as a direct physiological response to environ-

mental temperature gradients (Rameaux and Sarrus 1838; Bergmann 1847; Rensch 1938; Mayr 1956). And this logic does make sense: assuming a similar geometric body shape, a larger animal has a reduced surface area relative to its volume, or mass (Thompson 1942; Gould 1966; Fig. 5.6). Because heat loss is proportional to surface area, this means that large-bodied animals lose less heat than do smaller-bodied animals per unit mass. Thus, having a larger body is a distinct advantage under cold environmental conditions. The reverse is also true: because smaller animals have a greater surface area relative to their mass (volume), they can more rapidly and efficiently dissipate heat under thermally stressful conditions.

This mechanistic explanation for Bergmann's rule has rarely been directly tested. One notable exception was an experiment where two lines of wild house mice were reared under different temperature conditions (-3°C and 21°C, respectively) in the lab. After eleven generations, the males reared under the colder conditions were about 30% larger (Barnett and Dickson 1984). Although this study was not replicated, other more comprehensive studies were carried out in the field and laboratory by Jim Brown and his colleagues in the 1960s (Brown 1968; Brown and Lee 1969). One of the first in-depth empirical

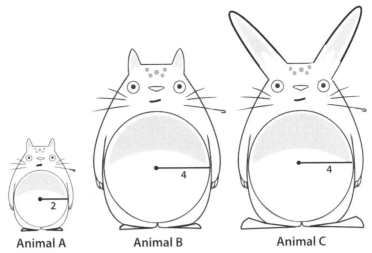

Fig. 5.6. Illustration of Bergmann's and Allen's rules. Here, we assume that a sphere is the geometric shape that best approximates the body mass of our three animal species: A, B, and C. We also assume that they are equivalent in all ways but size, and in the case of animal C, shape. The surface area (SA) of a sphere is given by: $SA = 4\pi r^2$, where r = radius of the animal. The volume (V) of a sphere is given by: $V = \dfrac{4\pi r^3}{3}$. Thus, the ratio of surface area to volume can be simplified to: $\dfrac{SA}{V} = \dfrac{4\pi r^2}{4\pi r^3} \text{ or} = \dfrac{3}{r}$. Note that animals A and B have radii of 2 and 4, respectively.

In our example, animal A has a ratio of 1.5 $\left(\text{e.g., } \dfrac{SA}{V} = \dfrac{3}{2}\right)$ and animal B has a ratio of 0.75 $\left(\text{e.g., } \dfrac{SA}{V} = \dfrac{3}{4}\right)$. This means that all else being equal, animal A will lose heat twice as rapidly than B (Thompson 1942; Gould 1966). This can be advantageous in a warm environment, but is a distinct disadvantage in cold conditions where heat conservation is important, particularly for endotherms.

Animal C demonstrates Allen's rule: another robust body size–related ecogeographic pattern, where animals in warmer environments have longer appendages (ears, limbs, and tails), presumably to improve their ability to dissipate heat (Allen 1877). Note that animal C has longer ears and feet than A or B. Conversely, animals in cooler environments tend to have small ears and thick, stocky limbs to conserve heat. This rule is fairly well supported, at least for mammals (see Millien et al. 2006 for further information).

explorations of Bergmann's rule, they demonstrated the body size of woodrats (*Neotoma* sp., a North American genus of medium-bodied herbivorous rodents) varied geographically with ambient temperature; this was true both intra- and interspecifically.

Importantly, they illustrated that both thermal conductance and lethal temperatures were negatively correlated with body size, leading to a mechanistic explanation (Brown 1968; Brown and Lee 1969). As a postdoc with Jim Brown, I was able to expand on these studies, using Brown's original data to demonstrate an inverse scaling of lethal temperature with individual body mass (Smith et al. 1995). Other studies have examined the basis for the rule in humans using experimental approaches (Roberts 1953; Ruff 1994). However, there has been a general lack of detailed

mechanistic studies, likely because of the difficulties in conducting thermal stress experiments.

As logical as a potential trade-off between body size, energetics, heat production, and heat loss seems as an explanation for Bergmann's rule (Fig. 5.6 and Fig. 5.7), as with many things in biology, it is likely to be more complicated (Millien et al. 2006). Some of the other causal mechanisms suggested have included gradients in primary productivity and humidity, success in mating or competitive interactions, nice breadth as well as selection acting on life history characteristics, development rates, or other factors that scale with body mass and are related to thermal characteristics of the environment (Scholander 1955; James 1970; McNab 1971; Calder 1984; Yom-Tov and Nix 1986, Rosenzweig 1995; Yom-Tov and Geffen 2006). Thus, while Bergmann's rule is

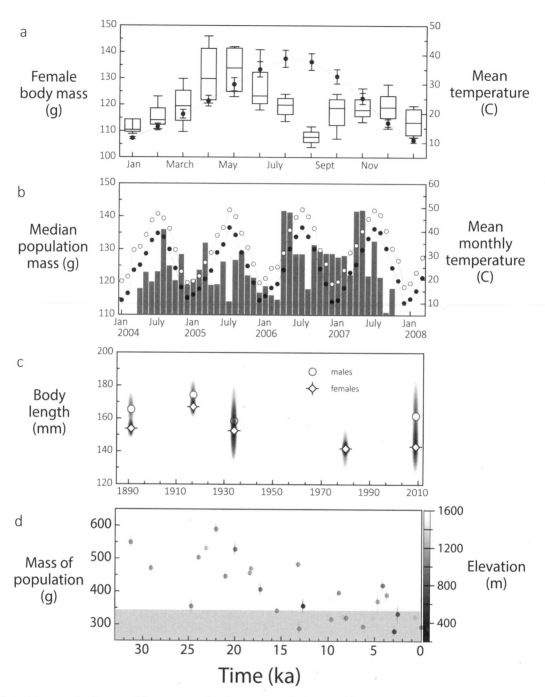

Fig. 5.7. Bergmann's rule across different temporal scales. **a**, Average body mass of female woodrats (*Neotoma lepida*) in Death Valley National Park in the 2000s as a function of the month of capture. Note the close correspondence with environmental temperature at the site (shown on the Y2 axis). **b**, Relationship between female woodrats and summer temperature at the same site in Death Valley as a function of year; again note the robust and significant relationship with temperature. This results from size-selective mortality; large-bodied woodrats experience higher mortality in the summer months and smaller ones in the winter. **c**, Body length of woodrats (mm) over the past 110 years at the same site in Death Valley as a function of winter temperatures at the site. Temperature anomalies are deviations from modern temperature. The body length of the population closely tracks environmental temperature. Body mass (or biovolume) scales as the cube of length, so these differences are even greater than shown. **d**, Average maximum body mass (g) of bushy-tailed woodrats (*N. cinerea*) over the past 30,000 years in Titus Canyon plotted against temperature; Titus Canyon is located on the east side of Death Valley. These latter data are derived from radiocarbon-dated woodrat paleomiddens. Note how closely woodrat body size tracks temperature at all spatial, temporal, and hierarchical scales, regardless of species. Woodrats are sometimes referred to as the poster children for Bergmann's rule. Redrawn from Smith et al. 2014.

considered valid, at least for mammals, no general consensus has been achieved about the mechanism(s) that actually underlie the pattern.

In addition to patterns across space, Bergmann's rule is valid across both historical and evolutionary time (e.g., Johnston and Selander 1964; Davis, 1977, 1981; Purdue, 1980; Klein and Scott, 1989; Smith et al. 1995; Smith and Betancourt 1998, 2003, 2006; Smith et al. 2009; Secord et al. 2012; Fig. 5.7; see Box 5.2). In a classic study, Johnston and Selander (1964) characterized adaptive differentiation in both color and size in house sparrows (*Passer domesticus*). The birds were introduced into North America in 1851 and again several times in the 1860s as a form of pest control. However, as is often the case in human-mediated introductions, this proved to be a disaster; the birds quickly established and soon became pests themselves. As they spread out across the continent, they adapted to local environmental conditions leading to a geographic cline in body size within a century (Johnston and Selander 1964). Such morphological change can occur even more rapidly: indeed, we were able to document morphological shifts in a rodent population over a decade. Using more than 50,000 historical records of animals trapped at the Sevilleta long-term ecological station in central New Mexico, my colleagues and I found that the average body size of white-throated woodrats (*Neotoma albigula*) had decreased by about 15% within eight years, tracking an increase in both winter and summer temperature (Smith et al. 1998). Both shifts favored smaller-bodied animals: warmer winters allowed animals to successfully overwinter when they might not normally do so, and warmer summers selected against larger animals because of problems with heat dissipation. The rapidity of this adaptive response reflected the strong selective pressure of temperature on woodrat populations: each one degree increase in temperature led to an approximate 10 g reduction of the average body mass of the woodrat population. This, and other work (e.g., Grant and Grant 2002; Yom-Tov et al. 2006), highlight the importance of considering animal body mass as a potential response to climate change (Dawson 1992; Murray and Smith 2012).

As concerns over anthropogenic climate change have intensified over the past few decades, scientists have turned to the fossil record for insights into how animals respond to the dynamic nature of climate. The climate shifts of the late Quaternary, in particular, are a good proxy for anticipated anthropogenic challenges (MacDonald et al. 2008). Given the robust influence temperature has on modern mammal body size, it is not much of a stretch to assume that morphological shifts occurred as a response to past environmental change. And indeed, this appears to have been the case. For example, woodrats adapted to late Quaternary climatic change through morphological shifts in body size: animals became smaller at the onset of the Holocene and during warming events in the middle and late Holocene and were considerably larger during the cool conditions of the full Glacial, the Little Ice Age, and the Younger Dryas (Smith et al. 1995; Smith and Betancourt 1998, 2003, 2006; Smith et al. 2009). Over the course of the late Pleistocene and Holocene, animals were able to adapt to virtually all climate shifts (Balk et al. 2019). Pocket gophers (*Thomomys talpoides*) in Wyoming displayed morphological and ecological sensitivity to middle and late Holocene climate shifts, decreasing in body size as climate warmed (Hadly 1997). Here, abundance shifts were also documented, which may have been related to alterations in vegetation composition. Morphological responses have also been documented in large mammals: gazelle, fox, wolves, and boars all dwarfed at the end of the Pleistocene in Israel (Davis 1981); the body size of deer, skunks, and archaic mammals tracked climate in North America (Koch 1986).

Bergmann's rule has even been demonstrated in deep time despite the coarser resolution of data. The early Cenozoic was a time of fluctuating climate. In particular, a well-known hyperthermal episode, the Paleocene-Eocene Thermal Maximum (PETM, ~55 Ma), led to abrupt shifts in Earth climate, including a more than 5°C ocean warming (Zachos et al. 2001). Studies of a North American early horse, *Sifrhippus*, demonstrate that PETM temperature shifts were significantly related to body size changes (Secord et al. 2012). The magnitude of the size changes was substantial; there was an approximate 30% reduction in size at the onset of the PETM, and body size increased by about 75% when climate cooled again at the end. And temperature has been implicated as an important factor influencing the body-size evolution of mammal clades (Saarinen et al. 2014). Thus, it is probably no exaggeration to say that Bergmann's rule is not only a robust spatial pattern but also holds over all temporal scales examined to date (Fig. 5.7). We discuss this further in later chapters, where we examine the utility of the fossil record to provide insights into anthropogenic climate change.

It should be clear from the previous discussion that one of the most important selective forces in evolution is energetics. After all, Bergmann's rule is essentially about energetics—the energy required to maintain a constant body temperature (homeostasis). For example, if a ready supply of high quality food were available, it might not be so critical for mammals to evolve larger body sizes under cold environmental conditions. They could potentially eat enough to offset the additional cost of homeostasis with a colder climate. In the real world, it is rarely the case that "free" high-quality food is sitting around, and thus there is strong selection for organisms to adapt to their local environment.

Foster's Rule, or the Island Rule

Imagine if you will, an adult elephant that stands 1.5-2 meters high, about as tall as a mop or broom. Such an animal would almost be small enough to have as a household pet (Fig. 5.8). Of course, that assumes that it could be domesticated. In the late Quaternary, a pygmy elephant (*Palaeoloxodon falconeri*) lived on Sicily. This rather adorable pachyderm was only about 1 m tall and probably weighed about 200-300 kg—more than an order of magnitude less than its closest mainland relative (Roth 1992; Palombo 2001, 2003). This diminutive proboscidean was not unusual: the modern and fossil record is replete with examples of extreme body size changes on islands. Indeed, dwarf elephants, hippos, and deer as well as giant rabbits, shrews, and mice evolved independently during the late Quaternary on many Mediterranean islands, including Sicily, Malta, Crete, Rhodes, and Cyprus (Palombo 2001, 2003). But, these body size oddities weren't confined to Europe; tiny mammoths were found on the Channel Islands off the coast of California and giant rodents

---

Box 5.2
Evolutionary Rates

How does one quantify the evolutionary rate of change in a trait such as body size? Often evolutionary rates are measured in darwins (d, Haldane 1949): the logarithmic change in a morphological trait (x) over two time periods (t_1 and t_2), standardized over one million years:

$$d = \frac{\ln \frac{x_2}{x_1}}{\Delta T}$$

However, darwins may not be appropriate if comparisons are being made among taxa that differ in proportions or that differ greatly in generation time (Gingerich 1993). Even when comparing rates within a lineage, it is important to remember that process rates are not independent of the measured time interval (Gingerich 1983, 1993; Gardner et al. 1987). Thus, when using darwins, standardized time bins are essential.

When making comparison among diverse taxa, it becomes necessary to use a biologically relevant time scale, such as generation time (Haldane 1949, Gingerich 1993, Evans et al. 2012). Evolutionary rates of morphological change often use the Haldane (h): the difference in the means of the natural logged measurements divided by the pooled standard deviation of the samples (S_p) divided by time, using a time scale in the number of generations, rather than chronological time:

$$h = \frac{\left(\ln \frac{x_2}{x_1} \right) / S_p}{\Delta T_g}$$

Newer metrics have recently been proposed to character major changes in traits within and across clades (Evans et al. 2012). Use of the clade maximum rate (CMR), for example, has revealed a basic asymmetry in macroevolution: large decreases in body size can evolve an order of magnitude faster than large increases (Evans et al. 2012).

Fig. 5.8. Comparison of mainland and island mammoth body size. During the late Quaternary, many islands around the world contained pygmy elephants or mammoths; some were even smaller than what is shown here. It is likely that human activities drove these insular populations to extinction (see text). A human is shown for scale. Mammoth drawing courtesy of public domain; human and mammoth silhouettes from PhyloPic (http://phylopic.org) under Creative Commons license.

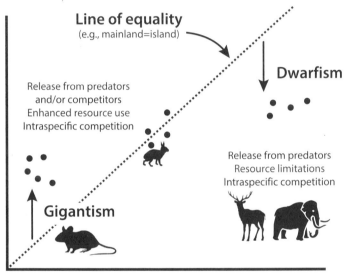

Fig. 5.9. Foster's rule, or the island rule. The x-axis represents the size of the ancestor on the mainland, the y-axis is the animal on the island. If they were the same body size, they would lie on the line of equality (shown by dotted line). Instead, smaller species tend to become larger, and larger ones tend to dwarf. A number of mechanisms have been proposed to explain this pattern; most invoke access to resources or energy constraints. See text for further details. Animal silhouettes from PhyloPic (http://phylopic.org) under Creative Commons license.

the size of an American black bear were found on islands in the West Indies. For example, the blunt-toothed giant hutia, *Amblyrhiza inundata*, which weighed as much as 200 kg, inhabited the islands of Anguilla and Saint Martin. And, dwarfed hominins lived on the island of Flores in Indonesia until about 50 ka (Morwood et al. 2004; Sutikna et al 2016). The "hobbit people" (*Homo floresiensis*), as they have been called, were only 1.1 m tall and probably evolved from a population of *Homo erectus* that became isolated during the Quaternary (Brown et al. 2004); on Flores they coexisted with pygmy elephants (*Stegodon*) and Komodo dragons. They, like many of the rest of these unusual mammals, appear to have been driven to extinction by the arrival of modern humans (Alcover et al. 1998; Sutikna et al. 2016).

The trend among mammals to evolve smaller or larger body sizes when isolated on an island is so prevalent over evolutionary history that it is known as Foster's rule, or sometimes simply, the island rule. The pattern is named after J. Bristol Foster, who originally described the pattern in a short paper published in the prestigious scientific journal *Nature* (Foster 1964). In it, Foster summarized the results of a survey of 116 insular species living on islands of the coast of western North America and Europe. He found an interesting pattern: certain mammal orders isolated on islands tend to become smaller, and others tended to be become larger (Foster 1964; Fig. 5.9). For example, he reported that rodents overwhelmingly tended toward gigantism, while carnivores and artiodactyls (e.g., deer, hippos, and other even-toed ungulates) were more likely to become dwarfed.

The interest in Foster's paper was heightened because it was followed shortly by the publication of *The Theory of Island Biogeography* by Robert MacArthur and Edward O. Wilson in 1967. This seminal book developed a theory explaining species richness on islands; they proposed this was a balance between colonization and extinction rates, which were influenced by the size of the island and the degree of isolation. The combination of these works inspired a generation of scientists to become fascinated with evolutionary trends on islands and the natural laboratory they provided. In the decades since Foster's paper, insular body size patterns were explored for many taxa (e.g., Van Valen 1973; Sondaar 1977; Case 1978; Heaney 1978; Lawlor 1982; Case and Cody 1983; Angerbjorn 1985; Lomolino 1985; 2005; Lister 1989, 1996; Roth 1990, 1992; Smith 1992; Whittaker 1998; Millien et al. 2006; Fig. 5.9), including hominins (Brown et al. 2004; Diniz-Filho and Raia 2017). Rather than being clade specific, the island rule became more generalized: larger mammals isolated on islands tend to become smaller, and smaller ones tend to be become larger (Lomolino 1985). Moreover, it has become clear that the island rule holds even deeper into the fossil record. For example, giant hedgehogs (*Deinogalerix*) over 60 cm in length and giant hamsters (*Hattomys*) were found on the island of Gargano during the Miocene when it was isolated from the boot heel of Italy (Villier and Carnevale 2013), and multiple lineages of dinosaurs during the late Cretaceous became dwarfed when rising sea levels isolated what became Hateg Island (Benton et al. 2010). Thus, the island rule in this more general sense is a robust and widespread pattern across both space and time.

So what causes this interesting pattern of body size change on islands? Most mechanistic explanations for the island rule have focused on limitations of resources or energy availability because of the well-established scaling of species richness with land area (Preston 1962; MacArthur and Wilson 1967). As habitats decline in size, there is a disproportionate loss of plant and animal diversity (Preston 1962; MacArthur and Wilson 1967; Rosenszweig 1995). Thus, the reduced area on islands means they can

typically support fewer resources. Moreover, they tend to also harbor fewer potential competitors or predators. Both of these factors are important: the removal of constraints normally imposed by the mainland suite of predators and competitors allows animals to evolve to a size that increases the net energy that they can obtain in the insular habitat (Foster 1964; Van Valen 1973; Sondaar 1977; Case 1978; Heaney 1978; Lawlor 1982; Lomolino 1985; 2005; Lister 1989, 1996; Smith 1992; Whittaker 1998; Millien and Damuth 2004; Millien et al. 2006). For large-bodied animals, where large size no longer reduces predation pressure or facilitates interspecific competition, this generally results in dwarfing; for small animals, gigantism may be favored because larger size increases ecological or physiological access to food resources (Fig. 5.9 and Fig. 5.10). Amelioration of predation pressure appears to be more important than the absence of competitors (Sondaar 1977; Smith 1992). Because fecundity scales negatively with body mass, this may suggest that dwarfism in low-mortality environments increases fitness (Raia and Meiri 2007).

Of course, the underlying causes are probably more complicated. The degree of divergence from mainland body sizes, and how fast evolution occurs, are inversely proportional to the size of the island (Heaney 1978; Millien 2011); that is, selective pressures may be higher on smaller islands. Not surprisingly, the strength of the island rule is also dependent on the degree of isolation (Foster 1964); islands located close to a mainland where episodic immigration is possible may display no pattern at all. The importance of size and isolation likely reflects the reduced resource base and species diversity concomitant with these factors (Preston 1962; MacArthur and Wilson 1967; Rosenszweig 1995), which can lead to intensified selection pressure. This can happen quickly if selection is intense. For example, the construction in 1996 of a hydroelectric plant in Brazil flooded an extensive area and formed three hundred mountain-top "islands" (de Amorim et al. 2017). Cut off from the mainland, the diversity of lizard species dropped as many were extirpated, leaving a single species of gecko (*Gymnodactylus amarali*) present on five islands. With reduced interspecific competition for food, the geckos rapidly evolved larger mouths

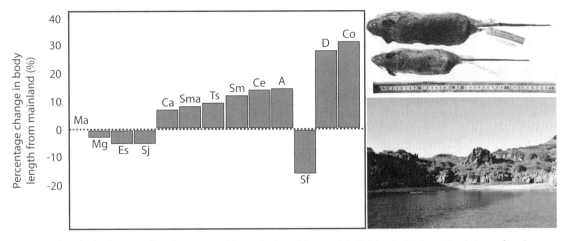

Fig. 5.10. The island rule in woodrats (*Neotoma* sp.) from the Sea of Cortez, Baja California, Mexico. Woodrats are found on thirteen landbridge islands surrounding Baja California: Santa Margarita (Ma), Magdalena (Mg), Espírtu Santo (Es), San José (Sj), Carmen (Ca), San Marcos (Sma), Todos Santos (Ts), San Martín (Sm), Cedros (Ce), Angel de la Guarda (A), San Francisco (Sf), Danzante (D), and Coronados (Co). The islands were isolated from the adjacent mainland by rising sea levels over the late Pleistocene and Holocene (around 11-6 ka). Both females and males have evolved giantism on most islands (Smith 1992); changes are shown in the bar graph as length relative to the mainland ancestor. Because mass scales as the cube of length, these represent *extremely* large body size shifts. For example, *Neotoma bunkeri* (top specimen in top photo), a now extinct woodrat from Isla Coronados (Co), weighed over 400 g, about four times the mass of the nearby mainland species (bottom specimen in top photo). This change occurred within 6,100 years of the island being isolated from the mainland. The *N. bunkeri* in the photo is the type specimen of the species, a male collected in 1932 by W.H. Burt. Only five of these woodrats have been preserved in museum collections; other than the type, they were all females. Pictured below the woodrats is Honeymoon Cove on Danzante Island (D), where the author conducted fieldwork on insular woodrats for several years. Woodrats on Danzante Island also evolved giant sizes, second only to those on Coronados (see bar graph). On most islands, woodrats also lost predator avoidance behaviors, becoming docile and active during the day (Smith 1992). In this system, the only ecological factor significantly associated with giantism was the absence of mammalian predators; the four instances where woodrats were smaller than on the mainland (i.e., Mg, Es, Sj, and Sf) were all islands where predators were still extant (Smith 1992). This observation, coupled with empirical evidence that larger body size allowed greater energy to be obtained from a given food source (Smith 1995), offered an explanation for the evolution of giantism in the genus. Photos by author.

and heads to utilize food resources that had not previously been available (de Amorim et al. 2017).

Moreover, whether the island rule holds for carnivores is unclear. While Foster (1964) reported dwarfing among carnivores, more recent work suggests competition and predation may have little influence on insular carnivore body size (Meiri et al. 2004; Raia and Meiri 2007, but see Lomolino 1985). Instead, variation in body size appears to be dependent on the abundance and body size of their prey. Finally, it should be noted that the unique selective pressures on islands lead to a number of other peculiar evolutionary transformations, including a marked tendency for the loss of costly traits such as flight ability in insects and birds, the complexity of bird song, and the dispersal ability of plants (Carlquist 1965).

Cope's Rule

A major question in paleontology is whether there are consistent directional changes throughout the history of life on Earth. But this begs the question: what exactly is a directional trend? This turns out to be more difficult to define. Clearly it must involve some pattern of large-scale change in a trait within a lineage in a given direction over time, but what "large-scale," "time," or "within a lineage" mean is unclear. For example, is the trend global in a taxonomical sense (i.e., does it apply to all major clades or all of life?) or in a temporal sense (i.e., is it exhibited over 1 million years? 10 million? 100 million? 1 billion?), or is it local in taxonomic and/or temporal scope? Moreover, is it the maximum, minimum, mean, or variance in a trait that is of relevance (Fig. 5.11)? And what drives or constrains large-scale trends? Are the same factors important?

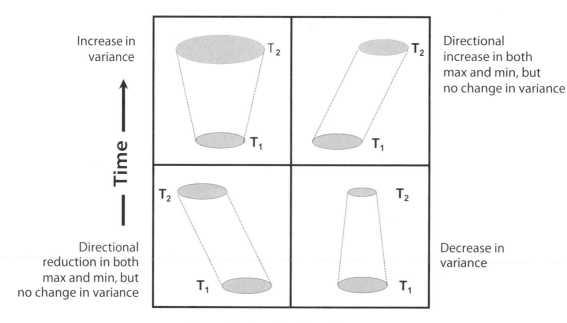

Morphological trait space

Fig. 5.11. The various ways a trait may change over time. Imagine that a trait starts at a point in morphological space. Over time (from T_1 to T_2) there might be an increase in the variance of the trait (gray oval) but no driven trend in the trait (e.g., no change in the mean; see top left quadrant), or there might be a reduction in the variance of the trait, due perhaps to strong selection, but again no change in the mean tendency (bottom right quadrant). Or the trait might be driven in a positive (top right quadrant) or negative (bottom left quadrant) direction. Of course, the trait might also exhibit absolute stasis, with no increase in variance and no change in the mean (not shown). Redrawn from Jablonski 1997.

Generally speaking, an active trend is when the changes occur consistently in one direction; this generally implies selection acting on the trait. In contrast, a passive trend is when the dynamics of the changes vary or are static. But, discovering that a trait is passive does not equate to it being random, nor does it suggest a lack of biologically important processes. Indeed, passive trends can actually result from complex and interesting causal mechanisms operating at different scales. In reality, most large-scale trends contain both driven and passive components; characterizing these depends on the taxonomic and/or temporal scale.

The canonical example of a large-scale trend is Cope's rule, a tendency for lineages to evolve larger body sizes over evolutionary time. Named after Edward Drinker Cope (see Box 2.1 in chapter 2), who was an important (and fascinating) figure in paleontology, it remains unclear whether it is a valid phenomenon or just a statistical inevitability of increasing hierarchical complexity and species

richness. A number of processes could lead to directional increases in body size over time. For example, if the founders of clades are small, if natural selection selects for larger-bodied organisms, or if there is among-lineage sorting (i.e., size-biased extinction and origination, and/or species selection) or some combination of all of these (Cope 1896; Stanley 1973; McShea 1994, 1998). Cope, himself, viewed the evolution of larger body size as part of the "doctrine of the unspecialized," the idea that clades originate from the average, or "golden mean," and then evolve over time to become increasingly more specialized and large-bodied (Cope 1896).

Recent work using a database of 17,208 different marine animal groups spanning the past 542 million years, found that Cope's rule was widely supported over many taxa and across broad time scales (Heim et al. 2015). That is, major clades over Earth history significantly increased in body size over the Geozoic. Indeed, the average animal in the oceans today is 150 times larger than the average one around 500 Ma.

Certainly there has been an increase in organismal complexity over time, and these major evolutionary innovations have allowed the exploration of more morphological space (Smith et al. 2016a; Heim et al. 2017). So, is this a driven trend?

If size evolves in manner similar to diffusion, both size increases and decreases are equally likely for any lineage over time. In this scenario, maximum size would be expected to increase with the square root of time elapsed, assuming constant diversity. Noel Heim and his colleagues found that the increase in body-size increase over the Phanerozoic was not explained by simple, random evolution from a small starting size. Rather, statistical tests indicated body-size evolution at the highest hierarchical level was an active evolutionary process that favored animals with larger sizes (Heim et al. 2015). So, in its broadest sense, the tendency of life to increase in body size over time fits the definition of Cope's rule.

However, when examined at the lineage level, there is no overwhelming trajectory of increasing body size for all clades. Numerous studies conducted at lower levels of the taxonomic hierarchy—on eurypterids (sea scorpions; an extinct group of arthropods that form the order Eurypterida), Cretaceous bivalve and gastropod genera, or fish, for example—show conflicting evidence for directional net increases in body size over time (Jablonski 1997; Lamsdell and Braddy 2009; Albert and Johnson 2011). Indeed, when examined at lower levels of the taxonomic hierarchy, Heim et al. (2015) found that the overall trend was one of stasis in many groups but driven changes in others; the trend toward greater size was driven primarily by increased diversification and survival of larger-bodied classes of animals (e.g., chordates; Heim et al. 2015). For example, anthropods showed stasis, and molluscs, brachipods, and echinoderms were best fit by an unbiased random walk. Others have demonstrated that directional selection is rarely observed within lineages in the fossil record (Hunt 2007). Only 13 of 251 studies examined, some 5%, displayed support for directional selection; the rest were about equally dividied between an unbiased random walk (i.e., where trait evolution is not directional but fluctuates around a central phenotype) and stasis. Hunt (2007) speculated that these results suggested that directional adaptative shifts occured over relatively short time frames, which were too brief to resolve in the fossil record. This makes sense in the context of the dynamic nature of most environments on Earth and the importance of local adaptation. It could prove interesting to model the environmental template and see if the fluctuations in climate matched the pattern of unbiased random walk or stasis seen in the studies.

The best examples of Cope's rule are for terrestrial vertebrate clades. Dinosaurs and mammals have long been the exemplar taxa (Alroy 1998; Hone et al. 2005). In part, this is because a reasonable phylogeny exists, and it has been possible to test directionality using ancestor-descendant pairs, the gold standard for assessing the directionality of a trend. For example, using a database of body-mass estimates for around 1,500 North American fossil mammals, John Alroy (1998) demonstrated that new species averaged 9.1% larger than their ancestors in the same genera. Moreover, he found that the pattern was stronger for larger mammals than small, leading to a gradual overall increase in average mass of mammals over the Cenozoic. Similarly, a set of sixty-five comparisons of ancestor-descendent non-avian dinosaur pairs found that later genera were about 25% larger (Hone et al. 2005). This pattern was found throughout their duration in the fossil record and across the clade. Interestingly, when examining the clade overall, we found evidence of a driven trend for mammals but not for non-avian dinosaurs (Smith et al. 2016a). These studies highlight the importance of examining patterns within a phylogenetic context.

So, why would larger body size be favored in the first place? As yet, there is no generally accepted underlying causal mechanism for why larger sizes should be favored over geologic time. If the environmental template were the main driver, there would be no clear pattern evident in non-avian dinosaurs or mammals since climate was fairly constant in the Mesozoic and highly variable in the Cenozoic (Zachos et al. 2001). Certainly, for mammals, large body size has a number of advantages: it may enhance the ability to avoid predators (or to capture prey); it may allow the exploitation of a wider variety of resources or more efficient use of resources; it may enhance mating and/or reproductive success, increase survival under low food or water conditions, or provide an edge in interspecific competition (Peters 1983; Calder 1984; Schmidt-Nielson 1984). In short, if larger body size leads to an increase in the resources and energy an mammal can acquire from the environment, it could be favored (Gould 1966; Stanley 1973; Brown and Maurer 1986). And in a meta-analysis of published studies on phenotypic evolution, Kingsolver and Pfennig (2004) demonstrated that larger-bodied animals did have greater fitness. However, they also found that direct positive selection on body size was greater than the estimate of total selection (i.e., the sum of direct selection plus corre-lated selection acting on other traits), implying positive selection was opposed by negative selection acting on traits that correlated positively with body size.

Indeed, larger size can be a distinct disadvantage at higher taxonomic levels. For example, longer genera-tion times and lower population densities might make larger animals more susceptible to extinction; this is a common assertion (Cardillo et al. 2005, 2008; Liow et al. 2008). Whether this is generally true, however, is unclear. Larger-bodied animals have larger geographic ranges (Brown 1995), and a large geographic range is a factor found to consistently buffer against extinction risk in the fossil record over the Phanerozoic (Jablon-ski and Raup 1995; Payne and Finnegan 2007).

Moreover, for mammals at least, there is no signature of size-selectivity of extinction over the Cenozoic fossil record (Smith et al. 2018, 2019). The size of mammals clearly depends critically on both the environment and on their evolutionary history; understanding the underlying mechanisms driving body size variation across time and space is a daunting task, requiring consideration of both ecological and evolutionary context (Fig. 5.12).

Testing for Evolutionary Trends

A major problem in evolution is determining the veracity of trends and, in particular, determining whether they are active or passive. An active or driven trend is one that deviates significantly from some null expectation—often some form of a random walk. This is taken as evidence of selection acting on the trait. In contrast, a passive trend suggests no increasing tendency or underlying driver. For example, if body size evolves in a manner analogous to diffusion, both increases and decreases should be equally likely. While in principle this seems an easy distinction, in practice demonstrating a trend exists and that it is not a statistical inevitability of random processes, or increasing species richness, can be difficult.

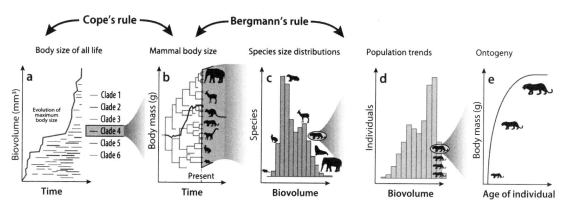

Fig. 5.12. Conceptual diagram of body size over various temporal and hierarchical scales. **a**, Body mass of all life over the Geozoic (the time period encompassing the entire history of life on Earth; see Fig. 2.4.1). The gray line represents the overall trend (Payne et al. 2009; Heim et al. 2015; Smith et al. 2016a). **b**, Macroevolutionary pattern of body size of mammals (clade 4) over their 210-million-year history. Trends in body size may be produced by Cope-style within-lineage trends, species sorting across subclades, or size-selective background and mass extinctions (Smith et al. 2010a, 2016a). **c**, Body size distribution of mammals at large spatial scale (continental or global). Patterns on the various continents are subsets of the global pattern (Smith et al. 2004; Smith and Lyons 2011, 2013). **d**, Body size distribution of individual organisms (local or global) for a population of a single species. A right-skewed distribution is common because of the contribution of juveniles; sometimes this pattern is more normally distributed. **e**, Ontogeny of an individual mammal. The rate of growth, maximum size, and age at death all reflect trade-offs between the genetic heritage and the environment. Redrawn from Smith et al. 2016a. Animal silhouettes from PhyloPic (http://phylopic.org) under Creative Commons license.

Another issue is deciding what metric is important—should we be concerned with changes in the mean, minimum, or maximum of the trait (Fig. 5.11)? Moreover, is a trend general? Or is it limited to specific taxonomic groups or time spans? The presence of boundaries, such as a lower size limit, can lead to a "trend," even if it is not driven. For example, the lower limit to mammal body size, probably set by physiological and ecological constraints (Smith et al. 2004; Smith and Lyons 2013), is about 1.8 g. Several different lineages have explored this threshold: today, the red-capped shrew (*Suncus etruscus*) and the bumblebee bat (*Craseonycteris thonglongyai*) occupy this morphospace. This is a hard-minimum boundary and we can reasonably conclude that it was true of earlier mammals in the Mesozoic (Smith et al. 2010a). Thus, even if speciation events over evolutionary time were random—meaning that descendants were equally likely to be larger or smaller than their ancestors—the mean body size of mammals would necessarily increase. This is because the lower bound prevented variance from increasing in the direction of smaller body size. Consequently, observing an increasing trend in mean body size does not *necessarily* suggest it is a driven trend. It could be passive process arising solely because diversity increased in the presence of a lower limit on body size (Stanley 1973). Trends can also be produced by lineage-level biases, such as differences in speciation and extinction rates. Imagine, for example, that originations were higher for large-bodied species than for small-bodied species. Thus, interpretation of even passive trends is complicated because they can arise from biologically meaningful processes.

So, how do we test for trends? There are a number of methods that depend on the type of data available and whether a good phylogeny exists (McShea 1994, 1998; Wang 2001; Hunt 2006, 2007; Marcot and McShea 2007). Generally, when examining patterns at a clade level, the maximum, mean, and minimum of a trait are compared to expectations generated by simulations under various types of evolutionary branching models (e.g., Hunt 2006, 2007; Heim et al. 2015). Models employed generally invoke some sort of driven trend (i.e., selection), a random walk (i.e., Brownian motion, with or without boundary), and stasis. Departure from, or adherence to, one of these models provides strong evidence of underlying processes. Of course,

large-scale trends may consist of both driven and passive components, which vary depending on the taxonomic or temporal scale examined.

Other commonly employed statistical techniques include (1) the minimum test, which as its name suggests examines whether the minimum value of a trait increases along with the mean and maximum values; (2) the subclade test and analysis of skewness tests, which subsample data within the distribution and characterize similarities in the statistical moments; and (3) the ancestor-descendant test (McShea 1994, 1998, 2001; Wang 2001). Of these, the ancestor-descendant test is considered the most robust because it directly compares changes in a trait along a lineage (Alroy 1998). However, the need for a detailed phylogeny limits its use in most studies (but see Alroy 2000). As discussed earlier, the evolution of body size in mammals is a good example of a directional and driven trend (Smith et al. 2010a, 2016a).

FURTHER READING

Brown, J.H. 1995. *Macroecology*. Chicago: University of Chicago Press.

Calder, W.A. 1984. *Size, Function, and Life History*. Cambridge, MA: Harvard University Press.

Damuth, J.D., and B.J. MacFadden. 1990. *Body Size in Mammalian Paleobiology: Estimation and Biological Implications*. Cambridge: Cambridge University Press.

Foster, J.B. 1964. Evolution of mammals on islands. *Nature* 202:234–235.

Gould, S.J. 1966. Allometry and size in ontogeny and phylogeny. *Biological Reviews of the Cambridge Philosophical Society* 41:587–640.

Kleiber, M. 1932. Body size and metabolism. *Hilgardia* 6:315–351.

Mayr, E. 1963. *Animal Species and Evolution*. Cambridge, MA: Harvard University Press.

Niklas, K.J. 1984. *Plant Allometry: The Scaling of Form and Process*. Chicago: University of Chicago Press.

Peters, R.H. 1983. *The Ecological Implications of Body Size*. Cambridge: Cambridge University Press.

Smith, F.A., and S.K. Lyons. 2013. *Animal Body Size: Linking across Space, Time, and Taxonomy*. Chicago: University of Chicago Press.

Smith, F.A., et al. 2016. Body size evolution across the Geozoic. *Annual Review of Earth and Planetary Sciences* 44:523–553.

Van Valen, L. 1973. Pattern and the balance of nature. *Evolutionary Theory* 1:31–49.

6 Show Me Your Teeth, and I Will Tell You What You Are

Teeth are the most enduring legacy of mammals. Their importance in mammalian paleoecology stems from several characteristics. First, teeth are one of the most common and resilient fossil types in the geologic record (Ungar 2010). This is because they pretty much start out as rocks; mammal tooth enamel is about 98% mineral, mostly in the form of hydroxyapatite ($Ca_5(PO_4)_3OH$). Thus, more than bones, teeth are highly resistant to chemical and physical degradation and other forms of diagenesis. Second, mammal teeth are highly diagnostic, allowing fairly easy identification to order, family, genus, and sometimes even species. And finally, teeth record wear and diet and are a proxy for body mass. Thus, analyses of the physical structure, wear pattern, chemical composition, and size of teeth can yield insights into the diet, body size, and general ecology of the animal, which is especially important for extinct species. In short, teeth are arguably the most useful part of the fossil record for mammals.

The study of animal teeth has a long history. For example, in his epic work *Historia animalium*, the generally acknowledged father of biology Aristotle wrote, *"Again, in respect to the teeth, animals differ greatly both from one another and from man. . . . Further, some animals are saw-toothed, such as the lion, the pard, and the dog; and some have teeth that do not interlock but have flat opposing crowns, such as the horse and the ox; and by 'saw-toothed' we mean such animals as interlock the* sharp-pointed teeth in one jaw between the sharp-pointed ones in the other" (Thompson 1910, p. 68). And, while many later scientists and philosophers contributed to the study of the structure and function of animal teeth, it was the French scientist Georges Cuvier who really pioneered the development of comparative anatomy and vertebrate paleontology with his meticulous and extensive studies of animals, including their teeth. Indeed, his first paleontological paper (*Mémoires sur les espèces d'éléphants vivants et fossiles*) compared living elephant teeth and bones with materials obtained by collectors from Siberia and Ohio (Cuvier 1796). In it, he conclusively demonstrated not only that African and Indian elephants belonged to distinct species but also that the Siberian and Ohio forms he examined represented unique and extinct species (Fig. 6.1)—what he called *des espèces d'éléphants perdues* (species of lost elephants). He called these animals *Elephas mammonteus* and *Elephas americanus*. Of course, we now recognize these as mammoth (*Mammuthus* sp.) and mastodon (*Mammut americanum*), respectively. This and subsequent papers by Cuvier were seminal in demonstrating the veracity of extinction, which was hotly contested at the time.

Cuvier wrote extensively about animal form and function (e.g., 1829-1844; Fig. 6.2). Given Cuvier's unparalleled ability to reconstruct the form of an animal from its fossil remains (Rudwick 1997), it is not surprising that the quote *"show me your teeth, and I will*

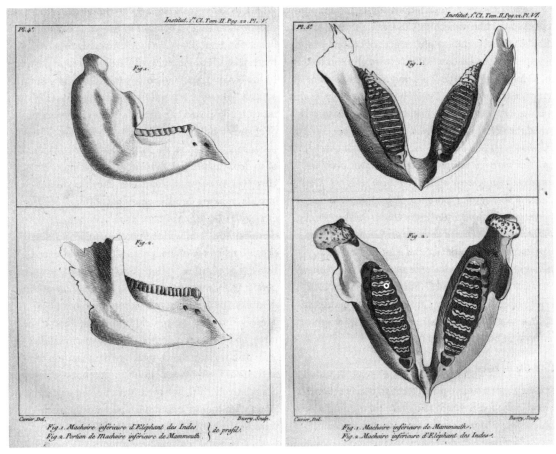

Fig. 6.1. Plates from Cuvier's first paleontological publication, *Mémoires sur les espèces d'éléphants vivants et fossiles*, illustrating the mandibles of several species of Proboscidea. His paper was read in April 1796. In this work, he clearly distinguished living (left top panel, right bottom panel) from extinct (left bottom panel, right top panel) forms. Cuvier 1796.

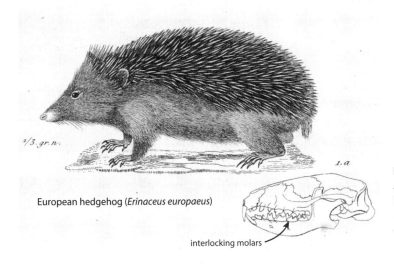

European hedgehog (*Erinaceus europaeus*)

interlocking molars

Fig. 6.2. Portion of a plate from Cuvier's *Iconographie du Régne Animal*, where he illustrated the dentition and form of many animals. Note how his illustration captures the interlocking dentition of the lower and upper jaw. Cuvier 1829, p. 37.

tell you who you are" is generally attributed to him. And it does sound like something Cuvier might or, at least, should have said. But the phrase actually appears to have entered the lexicon from a lecture given in 1860 at the Odontological Society of London meeting, some twenty-eight years *after* Cuvier had passed away. In this speech, T.R. Mummery, a British dentist and naturalist, stated rhetorically, *"Cuvier said, 'Give me a tooth, and I will show you the animal;' but men like Professor Owen can say, 'Give me a microscopic fragment of a tooth, and I will tell you to what genus of animals the tooth belonged.' The science of histology has now attained, in the hands of such indefatigable students as Professors Owen, Quekett, Huxley, and other eminent men, a degree of perfection which was altogether unimagined at first; and it is found, that the minute structure of the teeth presents peculiar and characteristic features in every animal"* (Mummery 1860, p. 290).

Regardless of whether Cuvier actually made this claim or not, it *is* true that teeth are highly diagnostic in terms of both form and function.

The Structure of Teeth

The very aspects of mammal teeth that make them so useful in the fossil record also make them complicated to describe. Unsurprisingly, then, an entire scientific vocabulary has been developed to characterize tooth morphology (Table 6.1). A typical mammal tooth has three main parts: the crown, neck, and root (Fig. 6.3). The crown is the part of the tooth normally exposed above the gum line and used for chewing, crushing, or tearing. There is considerable variation in the shape and function of crowns among mammals (Ungar 2010). The neck extends from the crown to the root. As the name implies, the root is the portion of the tooth that fits into the alveolus, or socket, of the jaw. While fish, amphibians, and reptiles have a single root, the number of roots in cheek teeth varies among mammals (Seo et al. 2017). For example, within rodents, the first upper molar of the spiny rat (*Maxomys*) has three roots, while the black rat (*Rattus*) has five (Bienvenu et al. 2008). What drives differences in root morphogenesis is poorly understood, but it may be related to the number and orientation of molar cusps (Seo et al. 2017).

A tooth is made up of four main materials: enamel, dentin, cementum, and pulp. The center of the tooth, or the pulp cavity, is composed of pulp, a nonmineralized and soft gelatinous connective tissue. It is where blood vessels and nerves are located; they exit the tooth through a conduit in the roots called the apical foramen (Fig. 6.3). Pulp and other organic materials, of course, typically are not well preserved in fossil materials. The pulp is surrounded by dentin, a hard, light yellow bone-like substance. Most of the tooth is composed of dentin, which is regenerated over time by the pulp. Dentin is fairly strong. This is because it, as well as enamel, is partially constructed of hydroxyapatite, a calcified inorganic material. However, in contrast to enamel, the crystals are smaller, making dentin somewhat softer (Goldberg et al. 2011). By far the major protein contained within dentin is collagen, which makes up to 90% of the organic matrix (Goldberg et al. 2011); under exceptionally good preservation conditions, collagen can persist in fossils as long as 20,000 years. As we discuss in chapter 7, stable isotope analysis of preserved collagen can yield information about the nitrogen and carbon composition of mammal diet.

Dentin is generally capped by enamel in the crown of the tooth and by cementum in the root. Cementum is a relatively thin, light yellow bone-like tissue that, along with the periodontal ligament, helps anchor teeth to the bony socket of the jaw (Table 6.1). It is worn down and reformed continuously (Ungar 2010).

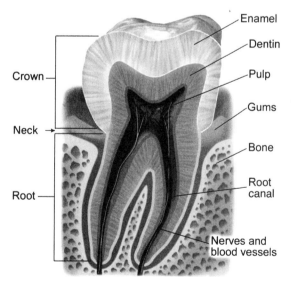

Fig. 6.3. Basic structure of a mammal tooth. While the shape and function of mammal teeth vary across the jaw, the internal structure is much the same. See text for details. OpenStax College, CC BY 4.0.

Table 6.1. Terminology frequently used in dental morphology

Term	Description
Parts of the jaw	
Alveolar process of the mandible	Upper border of mandibular body; anchors lower teeth
Alveolar process of the maxilla	Curved, inferior margin of the maxilla; anchors the upper teeth
Alveolus/alveoli	Bone of tooth socket
Angular process	Lowest projection on posterior portion of the mandible
Coronoid process	Flattened anterior (forward-facing) bony projection of mandibular ramus; attachment on lower jaw for biting muscles
Mandible/mandibular	Lower jaw bone; the only moveable part of the skull
Mandibular condyle	Point of articulation (joining) with upper jaw; top of upward projection from posterior margin of mandibular ramus
Mandibular foramen/foramina	Opening(s) on surface of the mandibular ramus; houses sensory nerves and blood vessels
Mandibular fossa	Deep, oval-shaped depression at base of skull; mandible joins skull at this site
Mandibular notch	Broad U-shaped curve between coronoid and condylar processes
Mandibular ramus	A curved process projecting upward and backward from the posterior of the body of the mandible
Maxilla/maxillary	Upper jaw bone
Mental foramen	Opening on each side of the anterior lateral mandible; exit site for sensory nerve
Mental protuberance	Forward projection from inferior margin of anterior mandible
Mylohyoid line	Bony ridge along the medial surface of the mandible
Palate	Bony plate that forms roof of mouth and floor of the nasal cavity; divided into soft and hard palate
Palatine process	Paired maxillary bones; join at midline to form the posterior quarter of the hard palate
Temporomandibular joint	Point of opening and closing of mouth; position dependent on trophic guild
Anatomy of a tooth	
Cementum	Bone-like tissue covering the root of the tooth
Carnassial teeth	Cheek teeth that are modified with sharp cutting edges for shearing flesh; sometimes called sectorial teeth
Cingulum/cingulid	Shelf-like ridge around the outside of an upper or lower molar, respectively
Crista/cristid	Crest or ridge on tooth; often used with prefix to indicate location
Crown	Portion of tooth above gum line
Cusp/cone	Raised portions on the crown (occlusal surface) of a molar; also called a cone; prefixes (proto-, para-, meta-, hypo-, and ento-) often added to indicate relative position along the molar; minor cones often have "-ule" added to the name; "-id" indicates tooth on lower jaw
Dentin	Bone-like tissue of tooth; found between the pulp and enamel and the cementum
Enamel	Tissue covering the crown of the tooth; hard mineral substance
Gums/gingiva	Tissue surrounding teeth
Occlusion	Joining of upper and lower teeth
Periodontal ligament	Anchors tooth to bone
Pulp	Innermost part of tooth; soft gelatinous tissue containing blood vessels and nerves
Root	Portion of tooth that fits into the alveolus; opening at end serves as conduit for blood vessels and nerves
Loph/lophid	Enamel ridge formed by elongation or fusion of cusps; often modified by prefixes to indicate location
Function or identification of teeth	
Apex/apices	Tip of tooth root
Apical	Toward the root of the tooth
Brachyodont	Low-crowned teeth
Buccal	Adjacent or facing the cheek
Bunodont	Molar with low, rounded cusps
Deciduous teeth	First set of teeth; replaced by permanent teeth

(continued)

Table 6.1. (continued)

Term	Description
Diastema	Space between teeth types
Diphyodont	Animal with two successive sets of dentition
Heterodont	Different types of teeth present in jaw
Homodont	Teeth similar throughout jaw
Hypselodont	Ever-growing with open roots
Hypsodont	High-crowned teeth with rough, flat occlusal surface adapted for grinding; tall relative to area of surface; enamel typically extends past the gum line
Incisal	Cutting surface of teeth
Labial/facial	Adjacent or facing the lips/face
Lingual	Adjacent or facing the tongue
Lophodont	Tooth with flat, fused cusps and elongated ridges or lophs; typical of herbivorous mammals that grind food
Mesial; distal	Toward the midline of skull; away from midline of skull
Occlusal	Chewing surface of cheek teeth
Polyphyodont	Animal with more than two successive sets of dentition
Proximal	Toward the adjacent tooth
Quadrate or quadritubercular	Molars with four cusps
Selenodont	Crescent-shape ridges on teeth formed by elongation of single cusps; typical of many members of the order Artiodactyla
Tribosphenic molar	Three-cusped molar, with trigonid (shearing end) and talonid (crushing end/heel); typical of early therian mammals
General	
Anterior/posterior	Forward; backward
Superior/inferior	Higher; lower
Dorsal/ventral	Toward the top surface; toward the bottom surface
Lateral/medial	Toward the side of the body/skull/jaw; toward the interior of the body/skull/jaw
Fissure	Groove in tooth/bone
Foramen/foramina	Bone opening(s)
Fossa	Shallow bone depression
Ramus	"Branch"; e.g., ramus of the mandible
Process	A projection from the bone. Often a site of muscle attachment

The point of transition between enamel and cementum is known as the cementoenamel junction, where they may overlap or meet in a discrete line. The crown of the tooth is covered with a thick layer of enamel, which provides exceptional durability; enamel is the strongest material in the body. It is largely (95%-98%) inorganic and is made up of hydroxyapatite as well; the remaining 2%-4% contains organic substances and water. Because enamel is largely inert, with no blood or nerve supply, it does not regenerate. Given that the occlusal surface of teeth is where most mastication of food occurs, it makes evolutionary sense that delicate blood vessels or nerves would not be potentially exposed.

Across the diversity of mammals, the crown of the tooth is folded in a wide array of complex shapes, which reflects adaptation to habitat and diet (Davit-Béal et al. 2009; Ungar 2010). Projections that arise from the occlusal surface are called cusps, which can be modified or elongated to form various types of crests or ridges. There is a complex terminology for labeling parts of the cusps that relies on position, location in upper jaw (maxilla) or lower jaw (mandible), and shape (see Table 6.1). In herbivores, the enamel on the top of ridges or cusps wears away as the animal ages, which exposes the underlying dentin. The differential hardness of the two tissues (enamel and dentin) form an effective grinding surface for mastication of plant vegetation. These specialized adaptations for diet are covered in the following sections.

Edentulous Mammals

While all mammals had teeth, some have lost them over evolutionary history (Davit-Béal et al. 2009; Fig. 6.4). This has occurred independently in several lineages, including anteaters, pangolins, and baleen whales (Davit-Béal et al. 2009). But this is clearly a secondarily derived characteristic: molecular and phylogenetic analyses suggest that edentulous species descend from ancestral forms with teeth (Davit-Béal et al. 2009; Meredith et al. 2009). Indeed, baleen whales, pangolins, and anteaters all apparently still form tooth buds during ontogeny that are resorbed prior to birth (Peyer 1968; Deméré et al. 2008; Davit-Béal et al. 2009). Some mammals, such as the duck-billed platypus, have teeth as embryos and as juveniles but not as adults (Davit-Béal et al. 2009). For example, tribosphenic molars are present in embryo and juvenile platypuses but are lost before young leave the burrow. In adults, they are replaced by keratinized pads, similar in composition to the duck-like beak (Davit-Béal et al. 2009). The echidna has a single egg tooth, which allows it to exit the eggshell and is lost soon after (Davit-Béal et al. 2009).

It is likely that highly specialized dietary adaptations, which increased the efficiency of foraging and digestion of prey, predated the loss of teeth (Davit-Béal et al. 2009; Meredith et al. 2009). Platypuses collect food in the water using their beak, which have sharp ridges to help capture prey (Davit-Béal et al. 2009).

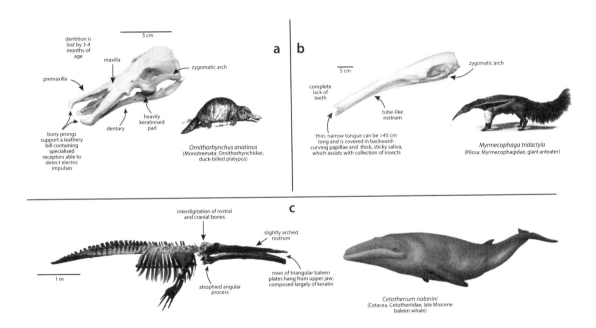

Edentulous mammals

Fig. 6.4. Examples of Edentulous mammal teeth. **a**, Duck-billed platypus (*Ornithorhynchus anatinus*). While juvenile duck-billed platypuses develop three cheek teeth in the upper and lower jaws, these are lost by three to four months of age and adults have no teeth. Instead, adults rely on heavily keratinized pads to crush or grind food. **b**, Giant anteater (*Myrmecophaga tridactyla*). Not only does the giant anteater have no teeth, but also its jaw is capable of very limited movement. Instead, it opens and closes its mouth using the rotation of the two halves of its lower jaw, which are held together by a ligament at the tip. The jaw houses a slender tongue that can be up to 60 cm long, coated in thick, sticky saliva and covered in backward-curving papillae. As its name implies, the giant anteater specializes on collecting ants. **c**, An example of a baleen whale (Mysticeti)—here, the late Miocene *Cetotherium riabinini*, which was related to the pygmy right whale. Instead of teeth, baleen whales have a series of keratinous plates attached to the upper jaw; plates decrease in size as they go further back into the jaw. These are used to sieve or filter small prey items such as krill, copepods, and zooplankton from seawater. Whale skeleton by Pavel Gol'din, Dmitry Startsev, and Tatiana Krakhmalnaya, CC BY 2.0, courtesy of Wikimedia Commons. Whale image by Nix Illustration (https://alphynix.tumblr.com), CC BY-NC 4.0. All other images from Beddard 1902.

Both edentulous anteaters and echidnas use a long slender tongue to extract insects from holes. The tongue is sticky and contains keratinized spines to help capture and hold prey (Davit-Béal et al. 2009; Fig. 6.4b). The morphogenesis of an elongated snout to accommodate this specialized tongue probably led to a reduction and eventual loss of teeth in both genera (Davit-Béal et al. 2009). In addition to the elongated, sticky tongue, pangolins house keratinized spines in their stomach (along with stones and sand), which effectively grind their highly chitinous ant and termite prey (Davit-Béal et al. 2009). And early fossils suggest that the evolution of filter feeding in whales was a progressive transition, with intermediate forms having both teeth and baleen plates on the jaw (Davit-Béal et al. 2009). Note that baleen is not homologous to teeth but is instead made of a tough keratinous material (Deméré et al. 2008).

Moreover, several clades of mammals possess teeth without enamel, including the sloths, aardvarks, and pygmy and dwarf sperm whales. In sloths, the enamel is replaced by two kinds of dentin, called orthodentin and vasodentin, and capped with an outer layer of cementum (Hautier et al. 2016). Why these species "lost" enamel is unclear; it may have been related to the their highly abrasive diet and both hypsodont and hypselodont dentition (Jernvall and Fortelius 2002). In rodents, only the front surface of the incisors is covered with enamel, while the back is composed of softer dentin. This facilitates the maintenance of a sharp edge on the continuously growing incisors as they are worn during chewing (Ungar 2010).

The Dental Formula

Mammal teeth are heterodont, which means that they differ in form and function across the jaw. This is distinct from other toothed vertebrates, such as fish, sharks, lizards, snakes, or dinosaurs, all of which tend(ed) to be homodont, with teeth varying in size but not in shape (Ungar 2010). Heterodont morphology allows identification of taxonomy and diet fairly readily from the number, shape, and relative size of the different teeth. It also allows better articulation between teeth on the upper and lower jaws, increasing the efficiency of mastication (Ungar 2010).

Mammals have four basic types of teeth: incisors, canines, and two types of cheek teeth (premolars and

molars). Not all mammals have all types of teeth, nor do they have the same number of each (Table 6.2; Fig. 6.5). For example, while anteaters have no teeth, humans have thirty-two, giant armadillos can have up to one hundred, and dolphins more than two hundred (Ungar 2010; Carter et al. 2016). Fish can have even more; catfish (Siluriformes) reportedly have more than 9,200 teeth. The numbers of the four types of teeth (incisors, canines, premolars, and molars) provide something referred to as the dental formula, which is conserved among taxa and, thus, can be diagnostic. The dental formula is obtained by dividing the mouth into quadrants: upper versus lower, and right versus left. Because vertebrates are bilaterally symmetrical, teeth on the left side of the jaw are the same as on the right. The general formula for placental mammals is written as I3/3, C1/1, P4/4, M3/3, or alternatively as $\frac{3143}{3143}$, with the numerator representing the numbers of each tooth type in the maxilla, and the denominator, the teeth in the mandible. By convention, the upper dentition is capitalized when referring to individual teeth (e.g., M1, I2) and the lower is not (e.g., m1, i2). The dental formula for an adult human is written as I2/2, C1/1, P2/2, 3/3 or $\frac{2123}{2123}$, which reflects that most of us have two incisors, one canine, two premolars, and three molars in both the maxilla and mandible—at least before our wisdom teeth (i.e., m3/M3) are removed! In both these instances, the upper and lower jaws have the same number of each type of teeth, but this isn't always the case. Sometimes there is a space between teeth types; this is called a diastema. Rodents, for example, have a large diastema between their incisors and cheek teeth. Telling premolars from molars can be tricky if a mammal does not have the maximal number of cheek teeth; size is not always a good indicator. But, an ontological sequence can help since molars are present only in the permanent dentition. Finally, reduction of the dental formula is considered a secondary adaptation, with premolars lost from the front to rear and molars the opposite (Ungar 2010). Thus, we number teeth as P3-P4, if a mammal has only two upper premolars, and m1, if it has only a single lower molar.

Most mammals have two sets of teeth over their lifetime. As juveniles, mammals have deciduous milk teeth, which are replaced by the permanent teeth during ontogeny; deciduous teeth are generally indicated by using the prefix "d" before the tooth (e.g., dp2, dI1). This type of tooth replacement is called diphyodont, referring to the two sets of teeth. The dental formula for deciduous and permanent teeth are different: molars, for example, only form in the permanent teeth after the jaws have grown long enough to accommodate them (Ungar 2010). Other animals, such as sharks, have polyphyodonty, or many sets of teeth over their lifetime. Indeed, shark teeth are grown in multiple rows inside their mouth and are only anchored to the tissue covering the jaw, rather than set in alveoli within the jaw. This leads to a sort of perpetual conveyor belt that allows the

near-continuous replacement of teeth over time. Thus, at any given time, the great white shark, *Carcharodon carcharias*, has about 300 teeth in its mouth, and it may go through more than 20,000 teeth over its lifetime.

Most mammal teeth stop growing in the adults; in some species, however, the roots are replaced by active stem cells and the teeth grow continuously over the lifetime of the animal. This is hypselodont dentition, and it may be an adaptation for a coarse or abrasive diet (Ungar 2010; Tapaltsyan et al. 2015). The incisors and/or cheek teeth of many rodents are a good example. Because they are ever growing, malocclusion can occur if rodents are restrained from gnawing on hard surfaces.

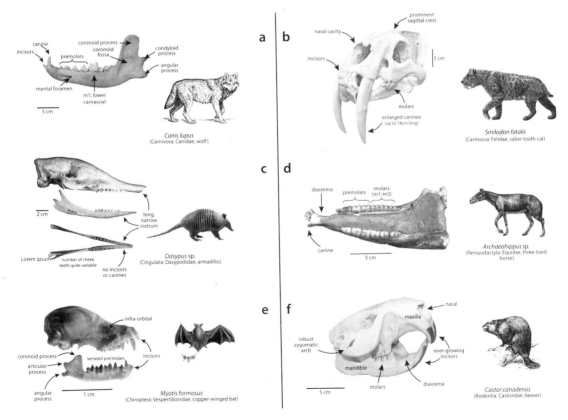

Diversity of mammal teeth

Fig. 6.5. Examples of dentition seen among mammals. See Table 6.2 for dental formulas. **a**, Wolf (*Canis lupus*). **b**, Extinct saber-tooth cat (*Smilodon fatalis*). **c**, Armadillo (*Dasypus* sp.). **d**, Extinct early horse (*Archaeohippus* sp.). **e**, Copper-winged bat (*Myotis formosus*). **f**, Beaver (*Castor canadensis*). Note how the size and shape of teeth vary among lineages and with diet. While carnivores generally have quite enlarged canines (a and b), they can be present even in herbivores (d). Skull and mandible of the holotype of *Dasypus beniensis* courtesy of Feijó et al. 2018 (https://doi.org/10.1371/journal.pone.0195084.g010), CC × 4.0. Image of bat skull courtesy of Ruedi et al. 2015 (http://doi.org/10.5281/zenodo.287927), CC × 4.0. All other images from Beddard 1902.

Table 6.2. Dental formulas for various extinct and extant mammals

Description: Family (Subfamily)	Genera	Dental Formula
Humans: Hominidae	*Homo*	I 2/2, C 1/1, P 2/2, M 3/3 = 32
Carnivores: Canidae	*Canis, Vulpes, Urocyon*	I 3/3, C 1/1, P 4/4, M 2/3 = 42
Carnivores: Didelphidae	*Didelphis*	I 5/4, C 1/1, P 3/3, M 4/4 = 50
Carnivores: Felidae	*Lynx*	I 3/3, C 1/1, P 2/2, M 1/1 = 28
Carnivores: Felidae	*Panthera*	I 3/3, C 1/1, P 3/2, M 1/1 = 30
Carnivores: Felidae	*Smilodon[†], Homotherium[†]*	I 3/3, C 1/1, P 2/1, M 1/1 = 26
Carnivores: Mephitidae	*Mephitis, Spilogale*	I 3/3, C 1/1, P 3/3, M 1/2 = 34
Carnivores: Mustelidae (Guloninae)	*Martes, Gulo*	I 3/3, C 1/1, P 4/4, M 1/2 = 38
Carnivores: Mustelidae (Lutrinae)	*Lutra*	I 3/3, C 1/1, P 4/3, M 1/2 = 36
Carnivores: Mustelidae (Mustelinae)	*Mustela, Mephitis, Spilogale*	I 3/3, C 1/1, P 3/3, M 1/2 = 34
Carnivores: Mustelidae (Taxidiinae)	*Taxidea*	I 3/3, C 1/1, P 3/3, M 1/2 = 34
Carnivores: Procyonidae	*Procyon*	I 3/3, C 1/1, P 4/4, M 2/2 = 40
Carnivores: Ursidae	*Ursus*	I 3/3, C 1/1, P 2-4/2-4, M 2/3 = 34-42
Insectivores: Soricidae	*Blarina*	I 4/2, C 1/0, P 2/1, M 3/3 = 32
Insectivores: Soricidae	*Sorex*	I 3/1, C 1/1, P 3/1, M 3/3 = 32
Large herbivores: Antilocapridae	*Antilocapra, Capromeryx[†], Stockoceros[†], Tetrameryx[†]*	I 0/3, C 0/1, P 3/3, M 3/3 = 32
Large herbivores: Bovidae	*Bison, Ovis, Bos*	I 0/3, C 0/1, P 3/3, M 3/3 = 32
Large herbivores: Cervidae	*Cervus, Rangifer, Megaloceros[†]*	I 0/3, C 1/1, P 3/3, M 3/3 = 34
Large herbivores: Cervidae	*Odocoileus, Alces*	I 0/3, C 0/1, P 3/3, M 3/3 = 32
Medium herbivores: Leporidae	*Lepus, Sylvilagus*	I 2/1, C 0/0, P 3/2, M 3/3 = 28
Large herbivorous rodents: Castoridae, Erethizontidae	*Castor, Erethizon*	I 1/1, C 0/0, P 1/1, M 3/3 = 20
Rodent granivores: Heteromyidae	*Perognathus, Dipodomys*	I 1/1, C 0/0, P 1/1, M 3/3 = 20
Rodent herbivores: Sciuridae	*Marmota, Spermophilus, Tamias, Tamiasciurus, Sciurus*	I 1/1, C 0/0, P 2/1, M 3/3 = 22
Rodent carnivores: Cricetidae	*Onychomys*	I 1/1, C 0/0, P 0/0, M 3/3 = 16
Rodent carnivores: Talpidae	*Scalopus*	I 3/2, C 1/0, P 3/3, M 3/3 = 36
Rodent herbivores: Cricetidae	*Phenacomys, Reithrodontomys, Clethrionomys, Microtus*	I 1/1, C 0/0, P 0/0, M 3/3 = 16
Small rodent mixed feeders: Muridae	*Peromyscus, Rattus, Mus*	I 1/1, C 0/0, P 0/0, M 3/3 = 16

Note: There is a large variety in the dental formula for carnivores, even among the same family. In contrast, rodents generally show much more similarity in their dental formula despite their disparate diets.

[†] Extinct species

Teeth and Diet

Its molar teeth have flat crowns, which show compartments of bony substance and of enamel. This is the structure of molars found in all animals that feed on plants, because they have to have a kind of millstone for grinding, rather than the sort of scissors for cutting that carnivores have.

—G. Cuvier, from *"A memoir on an animal of which the bones are found in the plaster stone (pierre à platre) around Paris, and which appears no longer to exist alive today,"* 1798 (translation by Rudwick 1997)

As Cuvier and other early naturalists noted, the number and shape of teeth reflect the diet of the animal (Fig. 6.6). After all, the point of teeth is to process food and masticate it into small enough pieces to swallow and digest. Thus, large canines and razor-sharp molars are characteristic of mammals in the order Carnivora (e.g., cats, mountain lions, bears, dogs, wolves, raccoons, foxes; Figs. 6.5a and 6.5b; Figs. 6.6a and 6.6c). This morphology helps capture and process prey. In contrast, insectivores tend to have extremely slender, sharp teeth, which are well designed for piercing chitin, the tough polysaccharide that helps form rigid insect exoskeletons (Fig. 6.5e). And herbivores such as deer, elk, bison, horses, and many rodents typically have flat and well-developed premolars and molars for grinding fibrous plant

Mammal teeth and diet

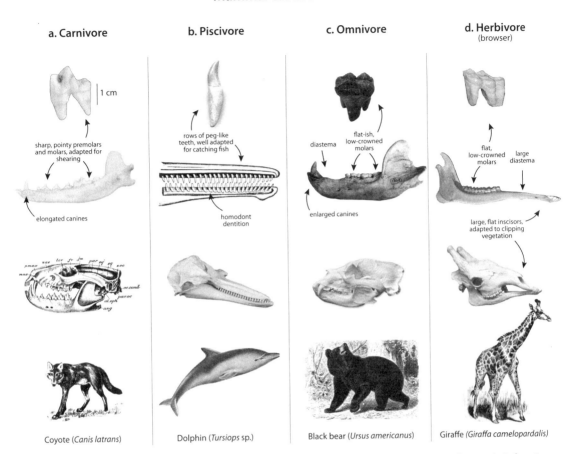

Fig. 6.6. Mammal teeth and diet. Most mammals have heterodont dentition, with the form of teeth reflecting their function. **a**, Carnivores, such as the coyote (*Canis latrans*) depicted here, have sharp, pointy incisors, canines, and cheek teeth that are used to shear flesh (see also Fig. 7.7). **b**, Dolphins, such as the bottlenose (*Tursiops* sp.) shown here, are unusual in their homodont dentition. The rows of similarly shaped peg-like teeth are ideal for catching fish. **c**, Omnivores tend to have a combination of flat molars and jagged premolars, canines, and incisors. Despite being grouped in the Carnivora, black bears (*Ursus americanus*) consume a variety of foods, including vegetation, berries, and roots. **d**, Herbivores, such as giraffe (*Giraffa camelopardalis*), have flat molars for grinding vegetation. Browsing and grazing herbivores vary in the shape and height of their cheek teeth (see also Figs. 6.7, 6.8). All images from Beddard 1902.

materials (Figs. 6.5d and 6.5f; Fig. 6.6d). Omnivores, such as pigs and humans, have a combination of sharp canines and flat molars. The close occlusion of the upper and lower teeth in mammals has made it possible to efficiently process food items.

Anterior Teeth

The "showy" part of the dental battery, at least in carnivores, are often the anterior teeth—the incisors and canines. These tend to have specialized functions and may be used for display, defense, or food con-

sumption (Van Valkenburgh 1989; Ungar 2010). The shape and orientation of the anterior dental arcade reflects use; while burrowing mammals have large robust front teeth, the incisors of grazers tend to be broad and flat and those of browsers may be angled or round (Ungar 2010). In rodents, the incisors are often used for ingestive behaviors, including gnawing, nipping, or scraping of materials; indeed, the order is named from the Latin words *rodere* (to gnaw) and *dentis* (tooth) (Ungar 2010). And, their incisors are well suited for gnawing; ever growing, they lack enamel on the lingual surface. The different wear pattern between the dentin and enamel surfaces results in the characteristic chisel-shaped tooth.

Sometimes the front teeth are co-opted for other purposes. The lower incisors of lemurs and tree shrews have become toothcombs, which are used for grooming (Ungar 2010). The most extreme example is with proboscideans, where over evolutionary time the second incisors have transformed into ever-growing enormous tusks, which can weight more than 200 kg (Lister and Bahn 2007). Tusks are used for a variety of activities, including defense and foraging (Ungar 2010). Interestingly, the intense poaching of elephants over the past half century has not only led to selection for smaller tusks but has also favored a higher frequency of tuskless elephants in modern populations; a rare genetic trait leading to enhanced survival in our anthropogenic world (Jachmann et al. 1995; Whitehouse 2002; Chiyo et al. 2015). Among mammals, elephants and mammoths are not the only species with tusks; walruses, warthogs, and even some species of deer have tusks, although usually these develop from the elongation of canines and not incisors (Ungar 2010).

In many hypercarnivores, the canines are greatly enlarged and recurved. This is particularly true of the feliforms, many of whom possess extreme morphological adaptations for hypercarnivory (Cope 1880; Schaller 1972; Ewer 1973; Van Valkenburgh 1989, 1999). While the lion is an obvious modern example, the iconic hypercarnivore is the late Quaternary felid, *Smilodon fatalis*, or the saber-tooth cat (Fig. 6.5b). The common name of this animal provides a graphic—and quite accurate—description of its dental battery. The elongated recurved canines of *Smilodon* could exceed 20 cm in length (Van Valkenburgh 1989, 1999; Meachen-Samuels and Van Valkenburgh 2010). Moreover, the canine teeth were also laterally compressed, or bladelike, and serrated along the entire length, thus providing an effective cutting edge (Van Valkenburgh and Ruff 1987; Van Valkenburgh and Sacco 2002; Slater and Van Valkenburgh 2008; Meachen-Samuels and Van Valkenburgh 2010). The extreme elongation of the upper canines in the sabertooth linage was accompanied by changes in the skull, jaw, and neck bones and musculature, which allowed wide gaping while

maintaining bite force (Emerson and Radinsky 1980; Werdelin 1983; McHenry et al. 2007; Slater and VanValkenburgh 2008; Meachen-Samuels and Van Valkenburgh 2010). Interestingly, this sabertooth morphology evolved independently at least four times over mammal evolutionary history, suggesting it was effective (Simpson 1941; Emerson and Radinsky 1980; Van Valkenburgh 1995, 1999).

Considerable debate has raged over the function of these large and formidable teeth (Emerson and Radinsky 1980). Some paleontologists have argued that they were an adaptation for stabbing through the tough and armored hides of large-bodied prehistoric herbivores, while others speculate that *Smilodon* suffocated prey, as do modern lions (e.g., Cope 1880; Simpson 1941; Emerson and Radinsky 1980; Van Valkenburgh and Ruff 1987; Van Valkenburgh and Sacco 2002; Wroe et al. 2005; Freeman and Lemen 2007; McHenry et al. 2007; Slater and Van Valkenburgh 2008). However, suffocating prey through the throat or muzzle can be a lengthy and noisy process (Schaller 1972), which could attract unwelcome competitors or scavengers (Slater and Van Valkenburgh 2008); competition was intense among the late Quaternary mammalian carnivore guild (Van Valkenburg 1989, 1999, 2001). Moreover, the elongated recurved canines were prone to breakage, which was more likely to occur if dealing with struggling prey (Van Valkenburgh and Ruff 1987; Freeman and Lemen 2007). These observations, coupled with *Smilodon's* exceptionally strong muscular forelimbs and neck, suggest saber-tooth cats most likely killed through the use of a piercing bite, unlike modern felids (Freeman and Lemen 2007; Slater and Van Valkenburgh 2008; Meachen-Samuels and Van Valkenburgh 2010).

Cheek Teeth

Cheek teeth show a great deal of diversity in size and shape across mammal groups. For example, in addition to the enlarged and recurved canines, terrestrial carnivores usually have scissorlike carnassials; a term that refers to the dental complex of the fourth (last) upper premolar and the lower first molar (P4/m1), collectively (e.g., Cope 1880; Osborn 1907; Ewer 1973; Van Valkenburgh 1989; Ungar 2010; Fig. 6.7). These cutting teeth are paired such that the inside of the upper molar (e.g., the crest connecting

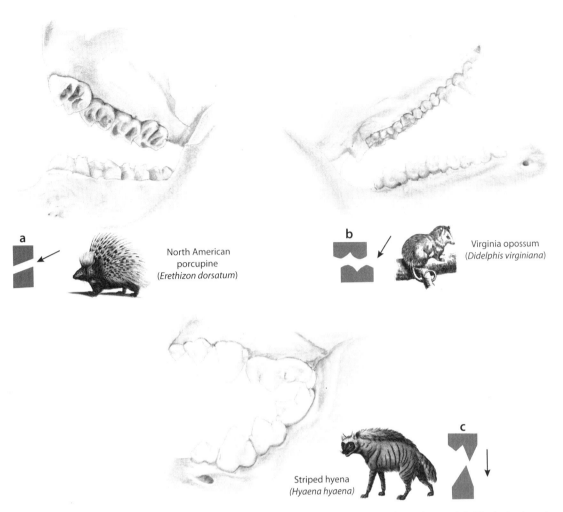

Fig. 6.7. Tooth function. **a,** Mostly lateral grinding (e.g., North American porcupine, *Erethizon dorsatum*). **b,** Vertical to lateral crushing (e.g., Virginia opossum, *Didelphis virginiana*). **c,** Vertical shearing (e.g., striped hyena, *Hyaena hyaena*). Many carnivores have modified molars and/or premolars specially adapted to allow for shearing flesh and enhancing the consumption of meat. These are most developed in the case of true carnivores, such as the hyena. Here, the scissor-like carnassials (P4/m1) form a formidable dental battery. Drawings of animal jaws courtesy of Kat Schroeder, MSc; porcupine and hyena images from *Zoological Lectures Delivered at the Royal Institution in the Years 1806 and 1807* by George Shaw 1809, courtesy of PICRYL; opossum from Beddard 1902.

the paracone and metacone) passes by the outer surface of the lower molar (e.g., the crest connecting the paraconid and protoconid). The crests are reciprocally concave, thus effectively shearing meat and increasing consumption efficiency (Ungar 2010). Found in most terrestrial carnivores, carnassials are absent in pinnipeds and in some early Cenozoic meat-eating lineages, such as the mesonychids. The latter is interesting considering this clade included one of the largest and most impressive terrestrial

mammalian carnivores of all time, *Andrewsarchus*. This hoofed predator, which was present in the middle Eocene, had a skull that was 83 cm long by 56 cm wide and weighed more than 1,900 kg (Smith et al. 2010a). Other archaic orders, such as Creodonts, did evolve carnassial teeth (Gunnell 1998), although the considerable variation in the molars and/or premolars involved in the complex underscores the claim that the clade is likely polyphyletic. The size and shape of carnassials varies among carnivore lineages, mirroring the wide diversity of prey consumed (Van Valkenburgh 1989): from seeds, fruits, insects, fish, and birds to those who specialize on crushing bone. For example, virtually all felids have buccolingually compressed blade-like carnassials, but some ursids have much more generalized dentition with poorly developed carnassials (Ungar 2010). Of course, felids tend to be hypercarnivorous, while most bears are highly omnivorous.

Herbivores face a different problem. Plants are largely composed of cellulose—the most abundant organic compound on Earth (Van Soest 1982). Together with hemicellulose and lignin, cellulose forms the structural component of plant cell walls. Because cellulose is refractory to enzymatic digestion, mammalian herbivores rely on symbiotic microflora in a fermentation chamber in the stomach or cecum to break it down; this has required the evolution of an entire suite of morphological and behavioral specializations (Janis 1976, 1988; Van Soest 1982). Among these are adaptations of the cheek teeth—the premolars and molars.

The name "molar" comes from the Latin *mola* (millstone) and *dens* (tooth); thus, molar means "millstone tooth." This is a pretty accurate description for herbivores where the primary function is to reduce particle size of plant materials to enhance digestion by microbes. With some fibrous materials, this can take a lot of chewing, which can wear teeth down. Hence, there is a gradient in the relationship between the heights of the crown and the root, which is characteristic of adaptation to different diets (Fig. 6.8). For example, if the ratio of crown to root height is from 0.3 to 0.9, the tooth is termed brachyodont, or low-crowned (Tapaltsyan et al. 2015). Molars of this sort are often found in omnivores and some browsers, including humans (Ungar 2010). Mesodont (middle or intermediate) molars have a crown to root height ratio of 0.9 to 1.5 and tend to be characteristic of browsers. Hypsodont, or high-crowned teeth, have a ratio more than 1.5 (Tapaltsyan et al. 2015), although some authors refer to anything above 1.0 as hypsodont (Williams and Kay 2001).

The selection for teeth to be high-crowned among herbivores is strong. Most mammals only have one set of permanent teeth and once they are worn away, chewing is difficult, if not impossible. Thus, dental

senescence leads directly to reduced fitness (Kojola et al. 1998). Consequently, it is not surprising that hypsodonty is particularly prevalent in mammals that eat fibrous or abrasive forage, as well as those found in arid environments (Jernvall and Fortelius 2002; Ungar 2010). Indeed, the expansion of grasslands in the Miocene may have been a strong selective force on tooth morphology; grass contains silica along with fiber. But others have pointed out that the endogenous and exogenous grit on plants also may be an important factor (Janis 1988; Sanson et al. 2007; Mendoza and Palmqvist 2008; Madden 2014). Considerable speculation remains whether hypsodonty is a response to the abrasive qualities of food or to grit in the environment, both of which are tough on teeth (e.g., Osborn 1907; Janis 1988; Fortelius and Solounias 2000; Jernvall and Fortelius 2002; Sanson et al. 2007; Mendoza and Palmqvist 2008; Strömberg et al. 2013). While both factors are likely important, diet appears to play a more critical role (Merceron et al. 2016).

In addition to the height of the tooth crown, other adaptations for the mastication of plant materials include changes in the orientation and shape of cusps on the molars (Fig. 6.7 and Fig. 6.8). For example, in several clades of mammals, the cusps are fused to form elongated ridges and broaden the grinding surface. Such lophodont dentition is typical of horses (Perissodactyla) as well as elephants and mammoths within the order Proboscidea. Along with extreme hypsodonty, it may be associated with intensive grass feeding to cope with the silica within grass blades (Osborn 1907; Ungar 2010). When many lophs are formed in parallel rows, this is often called loxodont; this molar characteristic gave rise to the generic name for African elephants, *Loxodonta*. Similarly, single cusps may be elongated to form crescent-like ridges (selenodont teeth); these are seen in many species of ruminant mammals, including bison, elk, and deer (Figs. 6.8k, 6.8m, and 6.8n). The crescent shape is made up of layers of enamel, dentin, and cementum, which wear at different rates, leading to a heterogeneous grinding surface well adapted for the mastication of plant materials. In contrast, omnivores such as bears, pigs, and humans, who process a variety of food types, tend to have bunodont or rounded dentition (Fig. 6.6c).

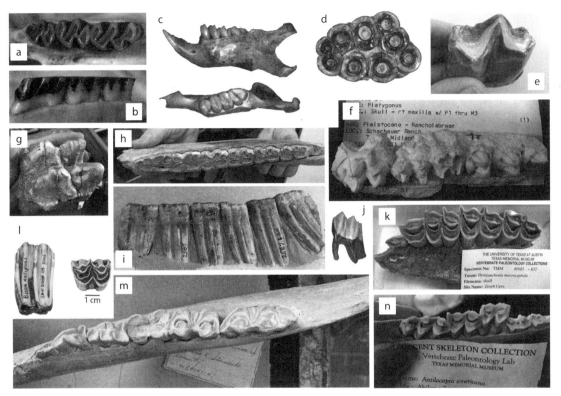

Fig. 6.8. Specialized molars for different types of mammalian herbivores. Note the high diversity in the arrangement and shape of cusps among clades. **a,** Bushy-tailed woodrat (*Neotoma cinerea*) right maxilla, occlusal view. The three distinctive high-crowned molars are reflective of the high degree of herbivory in this extant rodent clade. **b,** Lateral view of the molars; note the wear patterns and flat occlusal surface. **c,** Lateral and occlusal views of the left mandible of a prairie dog, *Cynomys gunnisoni*. Prairie dogs are colonial burrowing rodents native to the grasslands of North America. **d,** The very oddly shaped molar of *Desmostylus hesperus*, an extinct desmostylid that lived from the late Oligocene to Miocene. This large-bodied animal weighed up to 1.2 tons and was aquatic; its molar resembles that of modern sea otters. **e,** An isolated right first molar of the extinct *Brontops*, a species of Perissodactyla present in the late Eocene of North America. *Brontops* were one of the first truly large-bodied mammals of the Cenozoic at around 2-3 tons. **f,** Right maxilla of extinct *Platygonus compressus*, the flat-headed peccary. This approximately 30-kg mammal was widespread throughout North America during the Pleistocene. **g,** Premolar of the extinct American mastodon, *Mammut americanum*. Note the pointed cones, or cusps, called zygolophodont molars, which are characteristic of these large-bodied browsers. **h,** Occlusal view and **i,** lateral view of dentary of an extinct horse, *Equus* sp. Extinct and modern horses all possess high-crowned lophodont teeth, which assist in the mastication of abrasive materials like grasses. **j,** Isolated molar and **k,** maxilla of extinct llama, *Hemiauchenia macrocephala*. Note the selenodont, or crescent-shaped, molars. **l,** Lateral and occlusal view of molar and **m,** dentary of the extinct Pleistocene *Bison antiquus*. Bison molars have a stylus between the cusps of the molar, which assists in identification from other Artiodactyla. **n,** Dentary of the extant American pronghorn, *Antilocapra americana*, which also displays selenodont dentition. All photos by author.

Several quantitative indices have been developed to describe the relationship between dental morphology and diet. For example, the Occlusal Enamel Index (OEI) represents the length of occlusal enamel bands relative to the tooth area (Famoso et al. 2013; Famoso and Davis 2016). Mammals who forage on more complex and fibrous plant materials tend to have higher OEI values. Similarly, the relief index measures the variation in height over the occlusal tooth surface and may well be related to the type of food consumed (Ungar and Williamson 2000). Use of more sophisticated 2-D and 3-D imaging technologies is leading to new ways to characterize the degree and type of herbivory in mammals (Evans et al. 2001, Evans and Sanson 2003; Evans et al. 2007; Pampush et al. 2016). Moreover, increasingly, studies are focusing on the actual biomechanics of teeth and their relation to different dietary resources; such approaches may yield powerful functional understanding of dental performance in mammals.

Elephants, Mammoths, and Mastodons

Then the Elephant's Child sat back on his little haunches, and pulled, and pulled, and pulled, and his nose began to stretch. And the Crocodile floundered into the water, making it all creamy with great sweeps of his tail, and he pulled, and pulled, and pulled.

And the Elephant's Child's nose kept on stretching; and the Elephant's Child spread all his little four legs and pulled, and pulled, and pulled, and his nose kept on stretching; and the Crocodile threshed his tail like an oar, and he pulled, and pulled, and pulled, and at each pull the Elephant's Child's nose grew longer and longer.
—Rudyard Kipling, "The Elephant's Child," *Just So Stories*, 1902

The elephant lineage is striking, not just because of the extremely large body sizes they have attained over evolutionary history but also because of their unique and conspicuous trunks and tusks. Only three species of Proboscidea remain today, but the fossil record records over 160 species during the Cenozoic (Shoshani and Tassy 1996; Shoshani 1998). These ranged in body size from just under 10 kg to in excess of 16 tons (Smith et al. 2010a; Larramendi 2016). Some early proboscideans evolved fantastical forms. For example, the aptly named shovel-tusker, *Platybelodon*, had a wide, shortened trunk tipped with extended and flattened lower incisors, which formed a sort of shovel used to scoop up vegetation and/or strip bark from trees (Lambert 1992). Some *Gomphotherium* had four tusks, and the gigantic *Deinotherium* had tusks curved downward instead of upward (Larramendi 2016).

The namesake feature of the group, the proboscis, or trunk, evolved from the fusion of the nose and upper lip (Shoshani 1998). The trunk is very dexterous and has numerous functions, including breathing, touching, grasping, foraging, drinking, smelling, and sound production. The selective forces driving the evolution of the trunk remain unclear, despite what Kipling might have speculated, but ever-larger body sizes may have required adaptations leading to greater foraging efficiency. Enlarged mouthparts, such as a trunk or large tongue, increase bite volume and the quantity of forage that can be obtained (Pretorius et al.

2016). Consequently, there were changes in the skull shape and tooth morphology to accommodate the evolution of the trunk as well as the transformation of the incisors into tusks. In particular, the number of teeth was reduced over time and the molars became large and complex to assist in the grinding of fibrous plant materials (Fig. 6.9). Ultimately, molars became anteroposteriorly elongated and loxodont—that is, large and flat with dozens of parallel rows of compressed folds of enamel filled with cementum.

Modern elephants, and their extinct cousins the mammoths, generally have only one or two cheek teeth in the jaw at any time (Fig. 6.9a). This is because they are polyphyodonts with multiple dental sets, unlike most other mammals whose baby teeth are replaced over ontogeny with a single set of permanent teeth. While the elongated second incisors, or tusks, are permanent and grow continuously throughout life (Ungar 2010), in elephantids the molars are replaced six times. Moreover, molars do not emerge vertically from the jaw as they do in other mammals, but rather they have a horizontal progression somewhat like a conveyor belt (Fig. 6.10). As new teeth grow in at the back of the jaw, they push the older molar toward the front where it is eventually lost. Generally, an elephant loses their first set of molars between 2 and 4 years old, the second set between 4 and 6 years old, the third between 9 and 15, the fourth between 18 and 28, and the fifth in their early 40s (Shoshani 1998). When the final molars wear out, the animal generally starves to death because it can no longer chew its food (Shoshani 1998; Ungar 2010).

The evolutionary history of mastodons (Family Mammutidae) took a different path around 25 Ma, when they split from the elephantids (Osborn 1936; Shoshani and Tassy 1996; Shoshani 1998). As proboscideans, mammutids also featured a large proboscis and large tusks, although the latter were less curved than in elephants and mammoths. To accommodate these features, the dentition was also much reduced, but whereas elephantids became polyphyodonts, the mastodon lineage did not. Instead, mammutids had a single set of permanent teeth, with distinctive cone-shaped cusps adapted for crushing leaves and twigs (Ungar 2010); these high-pitched transverse ridges are referred to as

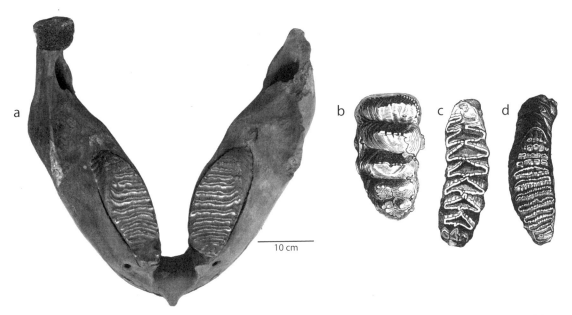

Fig. 6.9. Molars of different species of Proboscidea. **a,** Location of molars in the jaw of a mammoth. **b-d,** Occlusal views of the lower molars of an Asian elephant, *Elephas asiaticus*; an African elephant, *Loxodonta* sp.; and a mastodon, *Mammut americanum*. Both mammoth and elephants have ridged, hypsodont molar teeth as opposed to the zygolophodont molars of mastodon. Images from Leunis and Ludwig 1891.

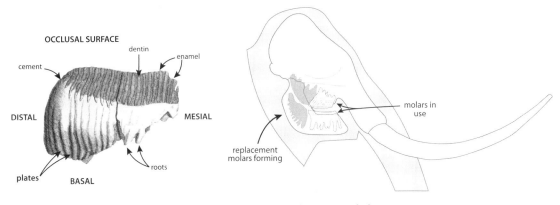

a. Structure of a mammoth tooth   b. Conveyor belt

Fig. 6.10. Details of mammoth dentition and tooth replacement. **a,** Structure of a typical mammoth tooth; note the series of plates that make up the tooth. **b,** A "conveyor belt" of new teeth produced in the upper and lower jaw and pushing forward in a horizontal progression, similar to a factory conveyor belt. The molars could (and can in modern-day elephants) be replaced up to six times over a lifetime.

zygolophodon. The common name "mastodon" used today refers to these molars; μαστός (breast) and ὀδούς (tooth) means "nipple tooth" or "breast tooth" in Greek. Cuvier rather aptly named the genus in 1817 to describe the characteristic nipple-like projections on the crowns of the molars. However, as a genus epithet, *Mastodon* was subordinate to *Mammut*, which had already been used to describe what turned out to be the same mammal (e.g., *Mammut ohioticum*; Blumenbach 1799). The etymology gets even more confusing because *Mammut* comes from the French and Russian words for "mammoth" (French *mammouth*, Russian мámонт, *mámont, or mammoth*), which a mastodon is only distantly related to. However, the rules of nomenclature are rigid, and thus the more descriptive term "mastodon" remains a junior synonym and the common name for the genus.

Using Teeth to Reconstruct Past Environments

The robust relationship among the teeth of modern mammals and their diet, and by association, their environment, can be employed with fossil animals to reconstruct ancient ecosystems (Fig. 6.11). Such analogies have been long employed to speculate about the diet of extinct species (e.g., Gregory 1920), but more recently, studies of entire communities have yielded a better understanding of paleocommunity structure and function. Increased hypsodonty, in particular, is particularly informative because it appears to be common among mammals occupying open and dry environments (Janis et al. 2002). Thus,

the proportion of mammals with hypsodont dentition has been used to reconstruct early Cenozoic environments (Damuth et al. 2002; Jernvall and Fortelius 2002; Damuth and Janis 2011). For example, using a hypsodonty index developed from the Neogene mammal fossil record, Kaya et al. (2018) explored the development of environmental biomes in the Old World. These authors were able to demonstrate the widespread extent of savannah across much of the Eurasian and African continents during the middle and late Miocene. In other work, the abundance of mammals in the Neogene has been connected to their degree of hypsodonty: those mammals who were more common tended toward increased hypsodonty, which may have allowed them to more effectively exploit food resources under changing environmental conditions (Jernvall and Fortelius 2002).

The mean hypsodonty within a community can even serve as a paleoclimate proxy (e.g., Damuth

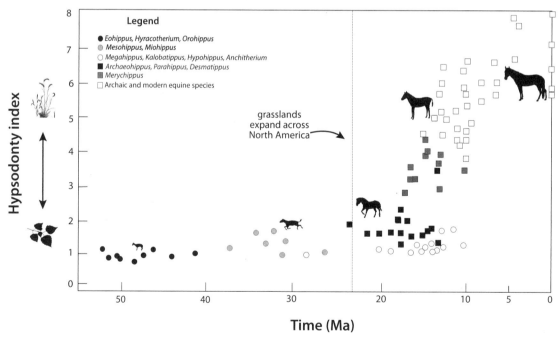

Fig. 6.11. Changes in the hypsodonty (high-crownedness) index over time in the horse lineage. Dentition of mammals can be used to infer climate conditions. The shifts seen correspond with changes on the landscape and, especially during the early Miocene, the expansion of grasslands across temperate North America. Values over 5 suggest grazing, or a diet predominately made up of grasses; those below 2 suggest browsing, or a diet made up largely of leaves. Many herbivores are mixed feeders and fall somewhere in between. True grazing only developed in the late Miocene with the advent of grasslands. The relationship between hypsodonty and diet was ground-proofed using modern mammals; when extrapolated to fossils, it provides valuable insights into past environments. Modified from Damuth and Janis 2011. Organism silhouettes from PhyloPic (http://phylopic.org) under Creative Commons license.

et al. 2002; Fortelius et al. 2002, Eronen et al. 2010a, 2010b). Indeed, a validation study on modern mammal communities demonstrated that more than 60% of the variance in mean annual precipitation could be explained simply by the average hypsodonty of the species present (Damuth et al. 2002). Further studies (e.g., Eronen et al. 2010b) substantiated this relationship; comparison of the ecomorphology of fossil mammals with independent paleobotanical proxies suggests tooth morphology is well correlated with both paleotemperature and precipitation at a site (Eronen et al. 2010b). Thus, it has been possible to reconstruct spatial variation in aridity patterns through the Cenozoic using fossil mammal teeth (e.g., Fortelius et al. 2002; Eronen 2006; Eronen et al. 2010b).

Isotopes

While the structure and composition of cheek teeth may be a powerful way to reconstruct paleoclimates and paleoenvironments, a more direct method is through the stable isotopic analysis of enamel or collagen preserved within the tooth. The food and water mammals consume over their lifetime leaves a geochemical fingerprint in their tissues and bones. By measuring the ratios of different isotopes in bones or teeth, we can gain dietary and environmental information. As we discuss in chapter 7, the isotope geochemistry of fossil remains is a powerful tool, which can provide not only diet or environmental reconstruction but also allow the provenancing of migration patterns in terrestrial mammals. Strontium, carbon, oxygen, nitrogen, and hydrogen isotopes have all been employed with the bone and/or dental enamel of fossil mammals to great effect. For example, the proportion of the different isotopes of carbon is often used to characterize the major dietary distinction between browsers, who eat mostly nongrassy vegetation, and grazers who specialize on grass. Thus, we can retrace the dietary shifts of ancient mammals as grasslands evolved and spread (Harris and Cerling 1996; Cerling et al. 1997, 1998). For example, about 8 Ma, isotope analysis suggests, the elephant lineage switched to eating mainly grass (Harris and Cerling 1996). In time, this dietary change led to changes in the morphology of their cheek teeth, which became much taller, with a proliferation of enamel ridges. But,

this adaptation arose after the switch from browsing to grazing. Isotope geochemistry is reviewed extensively in chapter 7.

The Last Supper

While stable isotope analysis of collagen or enamel from bones and teeth can provide an integrated overview of an animals diet, the shape and/or wear marks on the occlusal surface of teeth can also be informative (Walker et al. 1978; Fortelius and Solounias 2000; Ungar and Williamson 2000). For example, the shape of molar cusps (sharp, round, or blunt) and occlusal relief (the relative difference in height between the tips of the cusps and the valleys between) yield information about the general diet of an animal. Termed "mesowear," such analyses can be a quick and reliable method to accurately assign general dietary categories (Fortelius and Solounias 2000; Kaiser and Solounias 2003). Herbivores can generally be well stratified into grazer, browser, or mixed feeders—as long as at least twenty individuals/taxa are available for analysis.

Similarly, the pits and scratches left by various types of food resources can be quantified to get a picture of an animal's last few meals, or as it is sometimes referred to, its "last supper" (Grine 1986; Fig. 6.12). The study of the microscopic abrasion patterns on the occlusal surface of teeth is called dental microwear analysis (DMA). DMA represents a snapshot into dietary behavior, rather than a long-term record, as is recovered with isotopes or other techniques (Walker et al. 1978; Ungar and Williamson 2000; Scott et al. 2005, 2006; Ungar 2010; Rivals and Semprebon 2011). While both environmental dust and food mastication can influence tooth wear, the intrinsic properties of the food is the primary signal recovered (Merceron et al. 2016). Thus, daily—or at most, seasonal—behavior is recorded in the pits and scratches left on the enamel of the tooth, with the temporal span dependent on the rate of microwear turnover of a particular mammal.

Dental microwear analysis is based on characterizing the abundance, size, distribution, orientation, and general morphology of abrasion patterns left on teeth during chewing (Walker et al. 1978; Ungar and Williamson 2000; Scott et al. 2005, 2006; Ungar 2006, 2010;). These tend to be distinctive between different

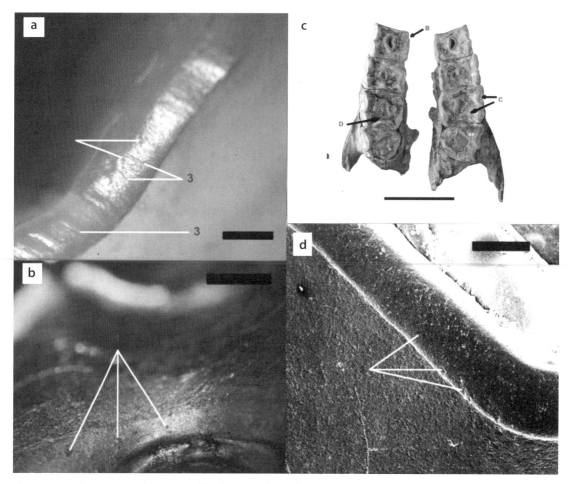

Fig. 6.12. Dental microwear analysis can yield information about the most recent diet of an animal. Panels **a**, **b**, and **d** show tooth surfaces with microwear scars; 1 = pits, 3 = scratches. Panel **c** illustrates the occlusal view of cheek teeth of the extinct South American lipotern, *Neolicaphrium recens*. The arrows point to microwear scars. The scale bar represents 3 cm. Modified after Corona 2019 under CC BY 3.0 license.

dietary resources. For example, a higher number of scratches relative to pits might indicate the consumption of phytolith-containing resources such as grasses (Ungar 2010). In contrast, a high abundance of pits may indicate consumption of brittle foods such as seeds and nuts (Solounias and Semprebon 2002; Ungar 2010). Folivores tend to have few pits and narrow scratches (King et al. 1999; Ungar 2010). These assumptions are typically ground-proofed with studies on modern animals with known diets (DeSantis 2016).

The traditional method of DMA is straightforward: tooth surfaces are cleaned and molded, and a cast is made. The occlusal surface is examined under some type of microscopy: originally this was a scanning electron microscope, but confocal or stereomicroscopic techniques are now more commonly employed (Gordon 1988; Ungar 2006, 2010; Semprebon et al. 2011). Then the scar topography and abrasion patterns on the grinding facet are quantified: the number of pits (i.e., rounded features) and number of scratches (i.e., elongated features) are counted within a standardized area. Generally, several areas are examined per tooth. Because this is a fast and relatively low-cost method, this approach is still commonly used despite concern about the often high variation in feature assignment between observers (e.g., Ungar and Williamson 2000; Scott et al. 2005, 2006; Ungar 2006; DeSantis 2016).

More recently, a more quantitative method has been developed that automates the characterization of the occlusal surface in 3-D. Dental microwear texture analysis (DMTA), pioneered by P. Ungar and his colleagues, uses confocal profilometry coupled with automated scale-sensitive fractal analysis by computer, which is less subject to intra- or interobserver variation (Scott et al. 2005, 2006; Ungar 2006, 2010; DeSantis 2016). A number of standardized variables are used to describe the tooth topography in this technique, commonly including (1) complexity, which measures the change in the roughness of the tooth surface with scale; (2) anisotropy, which quantifies the extent that features are oriented in a similar direction; and (3) textural fill volume, which measures the difference in volume filled by large and small cuboids. The values of these variables reflect diet; for example, higher values of complexity are interpreted to mean consumption of hard, brittle foods, while high levels of anisotropy might indicate a high degree of meat consumption (Scott et al. 2005, 2006; Ungar 2006, 2010). A problem with this method, however, is its cost. Not only does it require 3-D images, but it also uses commercial software packages (e.g., ToothFrax and SFrax) that are expensive. Recent innovations in free R-based software (Pampush 2016; Strani et al. 2018) may help with the latter issue.

Dental microwear has been used effectively in a number of studies. For example, it has been possible to distinguish grazers, folivores, and flesh-consumers from woody browsers, frugivores, and bone-consumers across a number of disparate phylogenetic clades (Ungar 2006; DeSantis 2016). DMTA analyses suggest that increased anisotropy and decreased complexity is associated with increased grass consumption in bovids (Ungar 2006). In other work, dental microwear was used to resolve the uncertain use of the lower tusks in *Platybelodon*, a shovel-tusked gomphothere from the middle Miocene. Using molar and tusk microwear analysis, Semprebon et al. (2016) found that microscopic abrasion patterns on the occlusal surfaces were consistent with browsing on leaves and/or twigs and microwear on the lower tusks was inconsistent with using them to shovel aquatic or terrestrial materials. Studies of fossil hominins have extensively employed dental microwear methods to investigate diet and foraging patterns (e.g., Walker et al. 1978; Grine 1986; Grine and Kay 1988; Scott et al. 2005, 2006); this work has yielded the insight that

Australopithecus had a tougher diet than did *Paranthropus*, despite the massive chewing apparatus found in the latter (Scott et al. 2005).

Teeth and Body Size

One of the most common ways of estimating the body size of extinct mammals is by measuring the length of the cheek teeth (Damuth and MacFadden 1990). While the length or width of a number of postcranial elements may be related to body mass, these are not always encountered in excavations. Teeth are. And, as discussed earlier, there is a pretty good predictive relationship between the length (or area) of the first molar (M1) and mammal body mass (see Box 5.1 for an example). These allometries are based on relationships constructed on a range of extant taxa; as one might expect, the slopes and intercepts vary between different mammalian orders. Use of fossil teeth has allowed the development of databases of mammal body size over the Cenozoic (Smith et al. 2010a, 2018) and studies examining morphological variation in relation to large-scale climatic fluctuations (Millien et al. 2006; Secord et al. 2012; Saarinen et al. 2014). Thus, it is possible to examine patterns of body-size evolution and diet across time based solely on recovered teeth (Liow et al. 2008; Evans et al. 2012; Smith et al. 2010a, 2018).

Postcranial Elements

While our discussion has focused on mammal teeth, the study of postcranial elements has yielded considerable insights into the ecomorphology of ancient mammals and the link with environment (e.g., Van Valkenburgh 1987, 1999; Lagaria and Youlatos 2006; Samuels and Van Valkenburgh 2008; Polly 2010; Meachen et al. 2015). Because there is a relation between form and function, the biomechanical analysis of the size and shape of limbs can reveal information about locomotion, movement, and other important aspects of mammal ecology (Alexander 1971, 2003). For example, the epicondyle index, which measures the width of the distal humerus divided by the length of the entire humerus, yields information about how well an animal can dig. Because this metric measures the relative area of the origin of wrist and digital flexors or extensors, higher values indicate a greater muscle volume, which is important for fossorial activities (Lagaria and Youlatos 2006).

The mechanical advantage of limbs measures trade-offs between the strength and speed of movement. For example, the brachial index measures the length of the ulna divided by the length of the humerus. If the relative length of the distal forelimb is higher, this suggests an animal can attain higher velocities while running, and it is likely more maneuverable. A similar pattern is also indicated with the crural index: the length of the tibia divided by the length of the femur. The joints are also informative. Not only does the mean calcaneal gear ratio provide information about carnivore foraging patterns, but it can even tell us about the local environmental conditions (Polly 2010); similarly the brachial index is correlated with both temperature and precipitation in the habitat (Meachen et al. 2015). Thus, postcranial scans provide much more information than even how cursorial, or ambulatory, a mammal was. There is a huge body of literature on postcranial elements and their analyses.

Limb shaft measurements are also often used for the estimation of body size, especially for large-bodied mammals (Gingerich 1990; Ruff 1990; Damuth and MacFadden 1990). For example, the width of the midshaft of the femur or humerus is directly related to load-bearing pressures and, thus, must be related to the body mass of a quadruped. Other elements, such as the width or cross-sectional area of the distal or proximal end of limb bones can also be used to estimate body mass (Damuth and MacFadden 1990).

The focus on mammal teeth within this chapter is because of their durability and ubiquity in the fossil record; when postcranial material are present in sufficient abundance for analysis, they can yield invaluable information.

FURTHER READING

Damuth, J., and C.M. Janis. 2011. On the relationship between hypsodonty and feeding ecology in ungulate animals, and its utility in palaeoecology. *Biological Reviews* 86:733-758.

DeSantis, L.R.G. 2016. Dental microwear textures: reconstructing diets of fossil mammals. *Surface Topography: Metrology and Properties* 4:023002.

Evans, A.R., and G.D. Sanson. 2003. The tooth of perfection: functional and spatial constraints on mammalian tooth shape. *Biological Journal of the Linnean* Society 78:173-191.

Janis, C.M. 1988. An estimation of tooth volume and hypsodonty indices in ungulate mammals, and the correlation of these factors with dietary preferences. *Mémoirs de Musée d'Histoire Naturelle Paris* 53:367-387.

Ungar, P.S. 2010. *Mammal Teeth: Origin, Evolution, and Diversity*. Baltimore: Johns Hopkins University Press.

Van Valkenburgh, B. 1989. Carnivore dental adaptations and diet: a study of trophic diversity within guilds. In *Carnivore Behavior, Ecology, and Evolution*, vol. 1, edited by J.L. Gittleman, 410-436. Ithaca, NY: Cornell University Press.

Van Valkenburgh, B., and C.B. Ruff. 1987. Canine tooth strength and killing behaviour in large carnivores. *Journal of Zoology* 212:379-397.

7 Stable Isotopes and the Reconstruction of Mammalian Movement, Diet, and Trophic Relationships

Stable isotopes are a valuable way of obtaining insights into the diet of mammals, both living and dead. Moreover, because there are geographic signatures present in isotopes, one can potentially gain insights into the source of food and/or water (e.g., pelagic organisms appear to have different isotopic signature than benthic ones). Further, different body tissues and organs (e.g., teeth, hair, bones) turn over at different rates, which makes it possible to examine variation in the isotopic signature over time within a single individual. Here, we review how isotopes work and provide examples of how they have been employed in the paleoecological and modern literature.

What Is a Stable Isotope?

As you probably remember from freshman chemistry, matter is made up elements, or atoms. These are defined by their atomic number, Z, which represents the number of protons contained within the nucleus (Fig. 7.1). Carbon, for example, has six protons, so its atomic number is 6. And, oxygen contains eight protons, leading to an atomic number of 8. But the nucleus of an atom also contains neutrons; collectively, the number of neutrons and protons (the nucleons) is equivalent to the atomic mass number. The normal atomic mass of carbon is 12, and that of oxygen is 16. But, the number of neutrons can vary for a particular element, leading to what are known as isotopes. In the

case of carbon, for example, there are fifteen known isotopes, ranging from ^8C to ^{22}C, although the only naturally occurring ones are ^{12}C, ^{13}C, and ^{14}C. While the vast majority of carbon (98.9%) exists in the form of ^{12}C, about 1.07% is in the form of ^{13}C, and the tiny remainder is in the form of unstable ^{14}C. Oxygen has three stable isotopes: ^{16}O, ^{17}O, and ^{18}O; of these, ^{16}O is by far the most abundant (Table 7.1).

Isotopes generally have somewhat similar chemical properties, although they can vary in their stability. Carbon-14, for example, is radioactive and, consequently, spontaneously decays into nitrogen with a half-life of about 5,700 years. This turns out to be particularly useful for paleoecologists since it provides a means of aging organic materials from the late Quaternary (see chapter 4 on dating). Other isotopes, such as ^{12}C and ^{13}C, are considered stable and do not undergo detectable radioactive decay. Of the first eighty-two naturally occurring elements in the periodic table (Fig. 7.1), all but two have at least one stable isotope; the exceptions are technetium (43) and promethium (61). The number of stable isotopes may be related to the fill and stability of the quantum shells. For example, tin (Sn) has a "magic number" of fifty protons, which refers to their arrangement into complete shells within the nucleus (Scerri 2007). Tin also has the largest number of stable isotopes with ten; it also has at least twenty-nine unstable isotopes. Other elements with large numbers of stable isotopes such as

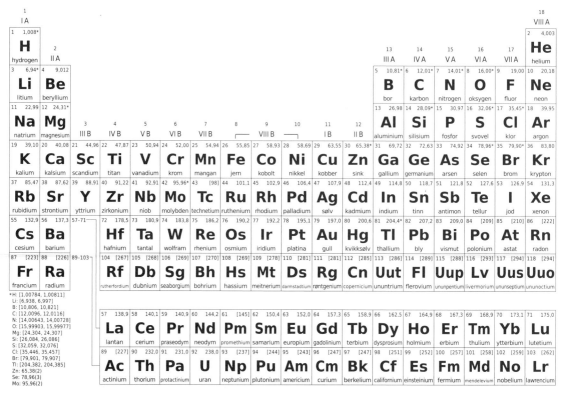

Fig. 7.1 Periodic table of the elements. The atomic number is found in the upper left corner of each element box. Of the ninety-four naturally occurring elements on Earth, most of the first eighty-two have at least one stable isotope; this is less common for the heavier elements. Usually, the heavier isotopes of an element are rarer. Courtesy of Wikimedia Commons.

xeon (Xe; eight stable isotopes plus over forty radioactive forms) also have a magic number of nucleons. And it has been noted that odd-numbered elements tend to have fewer stable isotopes than do even-numbered elements. For a variety of reasons, including their importance as the main components of life, the most commonly employed stable isotopes are those of hydrogen, carbon, nitrogen, and oxygen (Table 7.1).

The main difference between the stable isotopes of an element is in atomic mass and bond strength. This turns out to be important. While heavier isotopes undergo the same chemical and physical reactions as do their lighter counterparts, they do so at slightly slower rates in part because more energy is required to break chemical bonds. The slower reaction rate of the heavier isotope leads to isotopic separation or discrimination in substances, including living organisms (Sharp 2017; Ben-David and Flaherty 2012). Thus, the products of reactions can have different isotopic ratios than the reactants, with the fraction-

ation proportional to the mass difference between the isotopes. For example, ^2H and ^1H differ by 100% in mass (2 g mole^{-1}/1 g mole^{-1} = 1.0) but ^{13}C and ^{12}C differ by about 8% (13 g mole^{-1}/12 a g mole^{-1} = 1.08) and ^{87}Sr and ^{86}Sr differ by only 1.1% (87 g mole^{-1}/86 g mole^{-1} = 1.01). This makes the latter two isotopes ratios more difficult to detect analytically.

Isotopes are present everywhere on Earth. Moreover, fairly predictable isotopic signatures, which can vary geographically, can often characterize different terrestrial and aquatic environments; there are also predictable differences between living organisms and inorganic materials (Fig. 7.2; Hobson 1999; West et al. 2006; Hobson and Wassenaar 2008). Thus, the natural fractionation of isotopes produces a signal that can be detected by analytical means and is exceptionally useful as a research tool. Over the past fifty years, application of stable-isotope analyses has led to important insights in many fields, including planetary science, hydrology, climatology,

Table 7.1. Common stable isotopes used in ecology and paleontology

Element	Symbol	Natural abundance (%)	Stable isotope(s)	Natural abundance (%)	Example of scientific use
Carbon	^{12}C	98.89	^{13}C	1.11	Characterizing diet; identifying use of C3 vs. C4 plants, marine versus terrestrial sources, habitat use, migration
Nitrogen	^{14}N	99.63	^{15}N	0.37	Characterizing trophic position and diet; habitat use, physiological status (i.e., starvation, reproduction, etc.)
Hydrogen	^{1}H	99.985	D or ^{2}H	0.015	Migration, habitat use, diet, trophic level, hydrological tracers
Oxygen	^{16}O	99.759	^{17}O	0.037	Paleoclimatology; migration patterns; hydrological tracer
			^{18}O	0.204	(usually $^{18}O/^{16}O$), diet, thermoregulation
Sulfur	^{32}S	95.02	^{33}S	0.75	Diet, particularly marine from terrestrial sources; habitat
			^{34}S	4.21	use and movement (usually $^{34}S/^{32}S$)
			^{36}S	0.02	
Chlorine	^{35}Cl	75.53	^{37}Cl	24.47	Hydrological tracer
Strontium	^{88}Sr	82.58	^{84}Sr	0.56	Population movement or migration; hydrological tracer
			^{86}Sr	9.86	(usually $^{87}Sr/^{86}Sr$)
			^{87}Sr	7.00	

Less commonly: Lithium (Li), Boron (B), Magnesium (Mg), Calcium (Ca), Titanium (Ti), Vanadium (V), Chromium (Cr), Iron (Fe), Copper (Cu), Zinc (Zn), Selenium (Se), and Molybdenum (Mo)

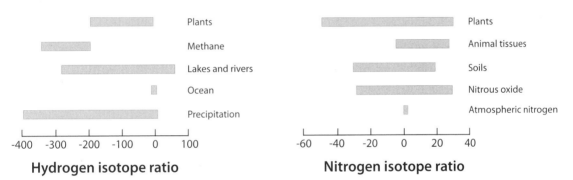

Fig. 7.2. The natural range of isotopic values of hydrogen and nitrogen across living organisms and Earth features.

petrology, oceanography, archaeology, and more recently ecology and paleoecology (see references in Hobson 1999; Koch 2007; Ben-David and Flaherty 2012). Indeed, the application of this technique has allowed for the characterization of food webs; tracing of pollutants; determination of field metabolic rate of fish; study of migration patterns of animals, whales, fish, and even shrimp; and recreation of ancient ecosystems (Fry 1981; Koch et al. 1994, 1995, 1998; Best and Schell 1996; Cerling et al. 1997; Kennedy et al. 1997; Cerling and Harris 1999; Hobson 1999; Hoppe et al. 1999; Feranec 2003; Chamberlain et al. 2005; Feranec and MacFadden

2006; Hoppe and Koch 2007; Hobson and Wassenaar 2008; Clementz 2012).

Stable-Isotope Analysis

While the idea behind stable-isotope analysis is straightforward—the isotopic composition of a compound is measured relative to that of a standard—in practice, it is more difficult. Most analyses are conducted using mass spectrometers, which measure the mass-to-charge ratio (m/q) of different molecules within a sample (Fig. 7.3). Although mass spectrometers can vary substantially in their capabilities and cost, they all have three main

Detection

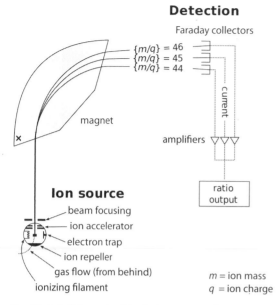

Fig. 7.3. *Left*: Schematic of the basic structure of a mass spectrometer typically used to measure the isotopic ratios of compounds. A small aliquot of sample is first pretreated through a series of chemical procedures to remove organics and/or contaminants, and then the isotopic composition is measured. In biological applications, the compound of interest is often collagen extracted from dentin or bone, or apatite from tooth enamel. *Right*: Dr. Emma Elliott Smith cleans the ion source of one of the many mass spectrometers used at the Center for Stable Isotopes at the University of New Mexico. *From left to right*: Schematic courtesy of the US Geological Survey; photo by author.

components in common: an ion source, a mass analyzer, and an ion detector. First, samples are combusted and/or reduced into a gas by a peripheral instrument. Then, gaseous samples are ionized within the source and accelerated into a finely focused beam with the same kinetic energy. The individual ions are then deflected by a magnetic field according to their m/q; the lighter and/or the more charged the particle, the more they are deflected. To visualize this, consider two cars going around a tight turn: a race car (i.e., the lighter isotope) can make the turn more sharply than can a minivan (i.e., the heavier isotope). While the first mass spectrometers used photographic film to detect the different ions, most modern machines use an electronic detector, such as a Faraday cup (Fig. 7.3). As the focused beam of ions is intercepted by the detector, they transfer their charge, and the current created can be precisely measured. This entire process is performed under vacuum (10^{-6} to 10^{-8} torr) to remove contaminating nonsample ions.

Delta notation (δ) is typically used to represent the difference in the isotopic ratio between a sample and a universally agreed upon international standard (Table 7.2). The delta value is reported in parts per thousand (per mil, ‰) and is computed with the equation:

$$\delta = \left| \frac{R_{sample} - R_{standard}}{R_{standard}} \right| \cdot 1000.$$

Here, R_{sample} is the empirically derived ratio of the heavy to light isotope in the sample, and $R_{standard}$ represents the ratio of the heavy isotope to light isotope in the reference gas. If the sample yields a positive δ value, it means that the material contains a *higher* ratio of the heavy to light stable isotope than does the reference standard. There may be a number of reasons for a higher isotopic ratio. For example, foraminifera (single-celled marine protists) incorporate oxygen into their calcium carbonate shells; this process stops when they die. Because the ratio of ^{18}O to ^{16}O is known to vary with environmental temperature, characterizing the variation in oxygen isotopic ratio through a sediment column can yield estimates of ocean

Table 7.2. International standards for the most common elements used in stable-isotope analyses. In practice, laboratories calibrate working standards against these internationally accepted universal benchmarks as a reference for analyses and to facilitate accurate comparisons between laboratories.

Element	Standard	Abbreviation	$R_{standard}$ [a]
^2H	Vienna Standard Mean Ocean Water	VSMOW	0.0001557
^{13}C	Vienna PeeDee Belemnite	VPDB	0.011056
^{18}O	Vienna Standard Mean Ocean Water	VSMOW	0.0020004
^{15}N	Atmospheric nitrogen	N AIR	0.003663

[a] Ratio of heavy to light isotope

temperature over time. Let's suppose you analyzed a sample of fossil forams and they yielded a $\delta^{18}O$ value of +12.5‰. What this means is that these forams recorded a $^{18}O/^{16}O$ ratio that is 12.5 per mil, or 1.25%, *higher* than that of the standard; conversely, if you had obtained a negative value, that would mean your forams had a $^{18}O/^{16}O$ ratio *lower* than the standard. Since warmer water tends to evaporate more of the lighter isotopes, foram shells that were formed when oceans were warmer than today will be enriched in the heavier isotope. It is probably not surprising that both forams and diatoms are commonly used paleoclimate proxies (Hays et al. 1976; Zachos et al. 2001).

While isotopes can be used to infer a number of geochemical processes, including past climate and ocean chemistry, of more interest here is the insights they provide in the ecology of modern and extinct plants and animals. As animals grow, the food and water they consume leaves a geochemical fingerprint in their tissues and bones. By measuring the ratios of different isotopes in bones or teeth we can gain dietary and environmental information. Different tissues integrate consumer diets at different time scales: from a few hours in the case of blood or breath, to a lifetime in the case of skeletal materials (Tieszen et al. 1983; Koch 2007). In paleoecology, diet is typically reconstructed using the isotopic composition of purified collagen extracted from dentine or bone, or purified apatite from tooth enamel; these both provide integrated long-term records. Metabolically active tissues are commonly used in modern studies since they are typically easier to sample from living organisms. Moreover, sampling multiple different tissues can provide insights into dietary changes over time. For example, blood plasma may turn over in just

a few hours or days, whereas muscle may take weeks to months to fully turn over (Hobson 1999). It should be noted that nitrogen isotopes can only be measured in protein-rich tissues such as collagen or keratin; the latter is a fibrous structural protein found in fingernails/claws, feathers, hair, and horns.

Unfortunately, these soft tissue biogenic materials degrade rapidly over time (Behrensmeyer 1978; Koch 2007). Consequently, it is rare for collagen to be preserved for more than about 10,000 years, although this can depend on environmental conditions. In our own work, we have been able to obtain collagen from subfossils between 16,000 and 20,000 years old (Tomé et al. 2020). In contrast, the structurally bound carbonates in enamel undergo minimal diagenetic alteration and can be preserved for many millions of years, making them the material of choice for most paleontologists (Lee-Thorp and Sponheimer 2003; Koch 2007).

Applications of Stable Isotopes

Living things are largely composed of just four basic elements: carbon, hydrogen, oxygen, and nitrogen. Indeed, these elements alone make up about 96% of your body. Consequently, it should come as no surprise that the stable isotopes of greatest utility to ecologists and paleoecologists are forms of these, specifically, $\delta^{13}C$, $\delta^{15}N$, $\delta^{18}O$, and δ^2H (Table 7.1). An advantage of stable-isotope analysis is that it provides information about the diet of long-dead animals that is independent of other proxies or morphology. Over the past few decades, such analyses have been increasingly important in reconstructing the paleoecology of mammals, including extinct species for which we have limited other options for understanding their biology (e.g., Koch et al. 1994, 1998;

MacFadden et al. 1994, 1996, 1999; Cerling et al. 1997, 1998; MacFadden 1998; Hoppe et al. 1999; Franz-Odendaal et al. 2002; Clementz et al. 2003; Fox and Koch 2004; MacFadden and Higgins 2004; Koch 2007; Clementz 2012). The loss of organics from deep-time fossil materials means that paleontologists generally employ bioapatite from tooth enamel, which undergoes limited diagenesis of biogenic carbonates (Koch 2007; Clementz 2012). Standardizing for fractionation effects can facilitate comparisons between the two sources. In mammalian herbivores, for example, carbonate $\delta^{13}C$ values are typically about 6.8‰ higher than those derived from collagen (Lee-Thorp et al. 1989; Tykot 2004; Passey et al. 2005); this difference is less for carnivores (~4.3‰). There also are a few differences in the information that can be gleaned from the use of bone versus tooth enamel. The first is that nitrogen isotopes cannot be measured from tooth enamel, which hinders the interpretation of trophic level. Another important difference is that while collagen and apatite in bone are constantly being resorbed, and thus reflect the average diet of the last few years of an individual's life, tooth enamel is a more static estimate that reflects the diet during crown formation.

Following is a more in-depth review of the main isotopes commonly employed in ecology and paleoecology (e.g., Table 7.1).

Carbon ($^{13}C/^{12}C$)

Carbon is a ubiquitous element within plant and animal tissues: it forms the building blocks for all biological macromolecules. Within animals, carbon can be found in virtually every tissue type. Not surprisingly then, the analysis of carbon isotopes can yield important information about diet and habitat, especially for primary consumers. This stems from some very basic differences in how primary producers capture energy from the sun and convert it into carbohydrates (Fig. 7.4).

There are two main photosynthetic pathways among terrestrial plants: C3 and C4. These vary in the enzymes they use to capture and fix atmospheric carbon dioxide. A third type, CAM, or crassulacean acid metabolism, which is found in cacti and succulents, is similar to C4. All terrestrial plants preferentially take up CO_2 containing the lighter carbon isotope (^{12}C) during photosynthesis, but the

degree of preference depends on both water availability and the particular photosynthetic pathway employed (O'Leary 1988; Ehleringer and Monson 1993). C3 photosynthesis is the most energy efficient and hence the most common among terrestrial plants (~82%); it is also the ancestral state and evolved during a time when CO_2 in the Earth's atmosphere was higher and O_2 lower (Kohn 2010). All trees, most shrubs, and grasses in cool-growing seasons are C3 (Ehleringer and Monson 1993; Koch 2007). The C4 pathway evolved in multiple lineages in response to rising oxygen levels and aridity in the late Miocene; plants in hot and xeric environments tend to use the C4 (or CAM) photosynthetic pathway (Ehleringer and Monson 1993; Kohn 2010). About 3% of plants—mostly grasses and sedges—are C4, and the remaining 15% employ CAM (Koch 2007). While the overall abundance of C4 plants is low, they include some important agricultural crops, such as maize, sugarcane, and sorghum.

The enzyme responsible for CO_2 uptake in C3 photosynthesis discriminates strongly against $^{13}CO_2$, whereas the biochemical pathways in C4 plants are less "picky" about CO_2 isotopic composition. Consequently, both C4 and CAM pathways have reduced carbon fractionation than does C3, and their carbon isotopic signatures differ significantly (O'Leary 1988; Ehleringer and Monson 1993). Typically, C3 plants generally have mean isotopic values of around −27‰ (ranging from −21‰ to −37‰), while C4 plants average −13‰ (ranging from −6‰ to −19‰), closer to the value of atmospheric CO2 (~−7‰) (Fig. 7.4; O'Leary 1988; Tykot 2004; Koch 2007; Kohn 2010; Ben-David and Flaherty 2012). These values vary spatially and temporally and are influenced by a number of factors, including light availability, soil moisture, stoichiometry (C:N:P ratios), and temperature (Kohn 2010; Ben-David and Flaherty 2012).

As carbon moves up the food web from producers to consumers, there is limited alteration in the isotopic values (DeNiro and Epstein 1978; Peterson and Fry 1987), although recent work suggests discrimination factors may be more variable than previously thought (Caut et al. 2009; Ben-David and Flaherty 2012). Regardless, the carbon isotopic values of consumers largely reflect their diets, leading to the oft-used phrase *"You are what you eat"* (sort of) (Tykot

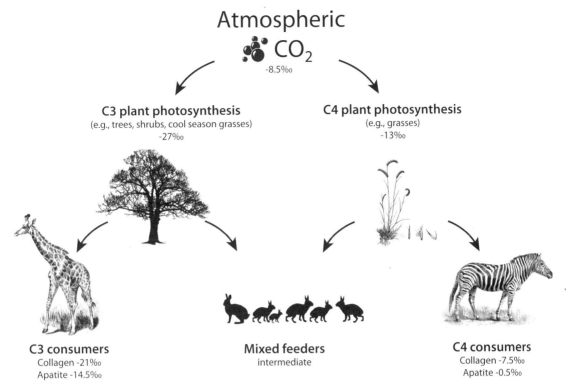

Fig. 7.4. $\delta^{13}C$ values from atmosphere to primary consumers. Note the clear separation between C3 and C4 plants; plants utilizing the CAM pathway have isotopic signatures that are intermediate between C3 and C4. Among C3 plants, tropical rain forests show low $\delta^{13}C$ values, while those in dry ecosystems are much higher. C3 plants in the Atacama Desert, for example, typically have isotopic values greater than $-23‰$, approaching (although still distinct from) the range of values of C4 plants in wetter environments (Kohn 2010). Consumers are enriched in nitrogen from the autotroph baseline by about 3‰ per trophic level. Animal silhouettes from PhyloPic (http://phylopic.org) under Creative Commons license. Giraffe and zebra from Beddard 1902. Grass from USDA-NRCS PLANTS Database 1950. Tree courtesy of the public domain.

2004). As it turns out, most grasses are C4, thus it becomes possible to distinguish grazers from browsing herbivores, even in the fossil record (Fig. 7.4).

Some of the first studies employing carbon stable isotopes came from archaeologists studying early human diets in the New World (Vogel and Van der Merwe 1977; Van der Merwe and Vogel 1978). In attempting to reconcile the discrepancy between radiocarbon dates obtained on corn (maize) and other materials within the same excavation, investigators realized that maize used a C4 photosynthetic pathway, which was different than the other materials dated. Thus, maize had different ratios of ^{13}C and ^{14}C, leading to the different radiocarbon dates. More relevant to our discussion, however, is scientists soon realized that the variation in stable carbon isotopes provided a way to characterize the

importance of maize in early human diets (Vogel and Van der Merwe 1977; Van der Merwe and Vogel 1978; Tykot 2004); this was quickly extended to include nitrogen isotopes, which provided information about the trophic position of consumers.

Over the past few decades, many studies have employed carbon isotopes to examine dietary shifts and resource partitioning in fossil mammals. In a comparison of Miocene ungulates between Florida and California, for example, Feranec and MacFadden (2006) found strong evidence of resource partitioning among species in Florida but not in California. They speculated that this might be because the California site had more abundant resources, which facilitated coexistence. Feranec and MacFadden (2006) also found that equids had more positive $\delta^{13}C$ values than other co-occurring species of antilocaprids or camelids, suggesting that they occupied more open grassy habitats. Others have employed isotopic analysis to characterize the diet of extinct megafauna. Carbon-isotope analysis revealed that the Pleistocene giant ape, *Gigantopithecus blacki*, fed on a heavily C3 diet and lived in a subtropical forest (Zhao and Zhang 2013). Koch (2007) demonstrated that mammoth (*Mammuthus* sp.) were predominately grazers who consumed largely (~83%) C4 diets, with an average enamel $\delta^{13}C$ of −2.4‰ in Texas and −1.6‰ in Florida. In contrast, *Paleolama* were primarily browsers, utilizing a limited range (23%) of C4 resources. Consequently, their $\delta^{13}C$ values were −11.1‰ and −14.8‰ in Texas and Florida, respectively. Some authors have demonstrated that morphology and stable-isotope analysis yield conflicting results about the paleoecology of extinct mammals: contrary to expectations from modern work, during the Cenozoic, horses in some habitats were browsing, not grazing, herbivores despite their high-crowned teeth (MacFadden et al. 1999; Koch et al. 2004). Similarly, Feranec (2003) found that hypsodonty is not always suggestive of a strict grazing diet; the extinct llama, *Hemiauchenia*, was a mixed feeder. And deep-time studies have demonstrated that Desmostylians, an extinct clade of enigmatic mammals, may have been largely dependent on marine resources such as sea grass and kelp (Clementz et al. 2003). At a community level, examining the diets of coexisting species can tell us about the habitat. For example, assuming an assemblage is not time averaged, if animals with diets sourced from both C3 and C4 food webs coexist, we can assume they occupied a habitat mosaic containing both woodlands or forests and grasslands.

Isotopes can also shed light on large-scale trends. For example, carbon isotopes were used to document the late Miocene expansion of C4 plants in low and intermediate latitudes. Herbivorous mammals from Asia, Africa, and the Americas exhibit no significant C4 component to their diet at 8 Ma (Cerling et al. 1997, 1998). However, this changed rapidly coincident with CO_2 "starvation" of the terrestrial biosphere between 5 Ma and 8 Ma (Cerling et al. 1998). This atmospheric shift, which favored C4 plants, may have been caused by a combination of seafloor spreading, silicate weathering, and sediment burial,

compounded by increased weathering because of the uplift of the Himalayas. Mammalian turnover followed these faunal shifts. For example, in Pakistan, the C3 ecosystem was almost completely replaced by a C4 grassland. In Africa, by 6 Ma, both browsers and grazers were present with only a few animals sourcing from both C3 and C4. Because cool-season C3 grasses were also present in North America, the expansion of C4 grasses led to many mixed feeders (Cerling et al. 1997, 1998). Moreover, in Africa at least, the dietary shift from C3 to C4 appeared to be sequential, with perissodactyls the first to become grazers, followed by proboscideans, suids, and bovids (Harris and Cerling 1996).

In addition to characterizing the consumption of plants C3 from C4 photosynthetic pathways, carbon stable isotopes can help distinguish between the use of terrestrial and marine resources by consumers (Box 7.1). Although most marine producers utilize the C3 photosynthetic pathway, they have distinct signatures from terrestrial plants because of differences in growth rates, water temperature, levels of dissolved gases, and ocean chemistry (Clementz and Koch 2001; Ben-David and Flaherty 2012). Moreover, even within marine systems, pelagic and intertidal systems generally have stable isotope ratios that are quite different. A number of studies have exploited these differences to characterize migratory movements of animals in marine systems. For example, because inshore sea-grass meadows are about 10‰ more enriched in ^{13}C than phytoplankton-dominated open bays, shrimp feeding nearshore have different carbon isotope signatures than those offshore (Fry 1981). These shifts in the carbon isotopic signature allowed for the reconstruction of the migration pattern of Texas brown shrimp (*Penaeus aztecus*), a commercially valuable species (Fry 1981). Marine isotope gradients have been used to trace the migration and diet of many animals, including whales, seabirds, and seals (Schell et al. 1989; Mizutani et al. 1990, Hobson et al. 1994; Best and Schell 1996; Smith et al. 1996; Hobson and Wassenaar 2008).

Nitrogen ($^{15}N/^{14}N$)

Within animal tissues, only two macromolecules (proteins and DNA) contain nitrogen. While nitrogen stable-isotope analysis has proven extremely useful for

Box 7.1
Conservation Implications of Stable Isotope Analysis

Sometimes stable isotope analysis yields unexpected results. California condors (*Gymnogyps californianus*) were the largest extant land bird in North America when they went extinct in the wild in 1987. A remnant of a Pleistocene genus that was once widespread across the continent, they were the victims of a deadly combination of poaching, habitat destruction, and lead poisoning that led to a dramatic decline over the last century (Walters et al. 2010; IUCN 2018). In a controversial effort to save the species, the remaining twenty-seven birds were captured and bred at several facilities in southern California. Surprisingly, this turned out to be effective. From 1992 to 2003, some 150 birds were reintroduced to northern Arizona, southern Utah, and the coastal mountains of central and southern California (IUCN 2018). But some of the same threats that led to their decline still remained, which continued to hamper their recovery (Walters et al. 2010).

Condors are scavengers and thought to have been dependent on mammalian megafauna, which went extinct at the terminal Pleistocene. Comparing carbon and nitrogen stable isotopes from modern, historical, and Pleistocene condors with potential marine and terrestrial food sources, Chamberlain et al. (2005) found large-scale changes in the diet of condors over time, presumably reflecting shifts in the availability of scavenged prey. In the Pleistocene, California condors foraged heavily on marine resources (see figure) although they also consumed animals dependent on C3 plants. Surprisingly, there was no signal of the use of megafauna carcasses, which were largely C4 (grazing) herbivores. Whether this pattern was generally true of Pleistocene California condors, unique to the population found at the Rancho La Brea Tar Pits, or because the samples postdated the megafauna extinction is unclear (the time range of fossils employed in the study was 11-36 ka). However, marine resources were clearly important. There was a switch to terrestrial dietary sources in the historical samples, which may have reflected human activities: specifically, the hunting of marine mammals that led to drastic declines from the 1770s to 1990s as well as the soaring number of domesticated livestock in terrestrial ecosystems (Chamberlain et al. 2005). The stark increase in the use of C4 resources in modern birds reflects the active management of the reintroduced populations (Walters et al. 2010); because carcasses of large-bodied mammals are scarce on the landscape, condors are provisioned with stillborn dairy calves.

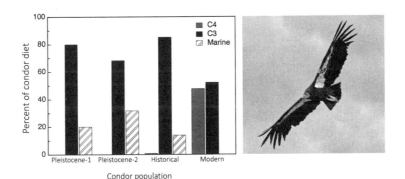

BOX FIG 7.1. The proportion of condor diet made up of marine C3- and C4-derived sources over time. The Pleistocene samples yielded somewhat different results when analyzed using $\delta^{15}N$ ("Pleistocene-1") or $\delta^{13}C$ ("Pleistocene-2"). Nonetheless, the overall pattern is clear: condors in the Pleistocene mostly relied on marine sources. While they consumed some C3 browsers, they did not appear to scavenge on mammalian grazers. This changed over time as human activities led to a drastic decrease in marine mammals and an increase in terrestrial livestock. Modern condor diet is supplemented with stillborn dairy cows, hence the strong C4 signature Left: Plotted from data in table 1 in Chamberlain et al. 2005; Right: Photograph of California condor in flight, August 2006, by Don Graham (CC BY-SA 4.0).

An important insight that Chamberlain et al. (2005) provides for the conservation of modern condors is that the carcasses of marine mammals, particularly beached whales and pinnipeds, may provide an important and unrecognized food resource. But most coastal environments are heavily urbanized and dead marine mammals are often butchered, buried, or towed out to sea because of the intense and unpleasant smell associated with their decomposing bodies. Effective management of reintroduced condors may require agreements with the National Oceanic and Atmospheric Administration, who since 1992 has been legally responsible for stranded or entangled marine mammals, and local governments and landowners.

modern and/or historic samples (e.g., Chamberlain et al. 2005; Newsome et al. 2010), it is unfortunately of limited use for most paleoecological studies. $\delta^{15}N$ is typically extracted from protein-rich tissues such as collagen or keratin, which degrade rapidly over time (Behrensmeyer 1978; Koch 2007). Thus, as mentioned earlier, it is rare for collagen to be preserved for more than about 10,000 years (Koch 2007; Clementz 2012).

As mentioned above, as carbon moves up the food web from producers to consumers, there is limited alteration in isotopic values, which means that the $\delta^{13}C$ of consumers largely reflects that of their diet. In contrast, the $\delta^{15}N$ of consumers are typically enriched by 3‰–4‰ relative to their diet as one moves up each link of the trophic chain (DeNiro and Epstein 1981; Peterson and Fry 1987; Caut et al. 2009). Of course, this assumes that consumers are only feeding from the trophic level directly below them, and this is not always the case. The difference between the $\delta^{15}N$ values of an animal and its diet is known as trophic enrichment, or trophic discrimination; it is influenced by factors such as diet quality, environmental conditions, marine or terrestrial source, nitrogen excretion pathways, and factors influencing the nitrogen balance of the animal (e.g., lactation, pregnancy, or starvation) (Ben-David and Schell 2001; Koch 2007; Caut et al. 2009).

Environmental $\delta^{15}N$ values vary spatially and are related to soil composition and age. Thus, different geographic features, such as ridge tops and valley bottoms, can have very distinct nitrogen isotope signatures (Koch 2007). And foliar nitrogen is negatively related to rainfall, rooting depth, and especially symbioses with nitrogen-fixing bacteria (Handley et al. 1999; Koch 2007). These differences have been used by plant physiological ecologists to estimate the importance of nitrogen fixation in ecosystems (Vallano and Sparks 2013). Because $\delta^{15}N$ is influenced by protein catabolism, it can be used to evaluate whether animals are feeding while they migrate (Ben-David and Schell 2001; Koch 2007; Caut et al 2009). Bowhead whales, for example, feed during their southward migration, but fast when they migrate to the north (Best and Schell 1996).

A particularly important application of nitrogen stable-isotope analysis is the ability to characterize the trophic level and food sources of species. This can have important conservation uses, such as in reconstructing the historic diets of a species of concern (Box 7.1). Here, mixing models are often employed, which use the isotopic ratios of consumers and their food to estimate dietary sourcing of different resources (Ben-David and Flaherty 2012; Newsome et al. 2012; Bowen et al. 2013; Philips et al. 2014). A fascinating example of the use of mixing models was the study by Yeakel et al. (2009) who quantified the diet of a coalition of two reputed man-eating lions in the late 1890s. These lions, known as the "man-eaters of Tsavo" had reputedly killed upwards of 135 construction workers in Kenya during the building of the Uganda Railway in March to December of 1898 (Patterson 1907; Patterson 2004). Workers were dragged from their tents at night by the lions, a grisly story that has been the subject of numerous books and half a dozen Hollywood movies. The lions were ultimately killed by Lieutenant Colonel J.H. Patterson, a British officer and engineer, who wrote a book about his exploits (Patterson 1907). The lion skins spent over twenty years as floor rugs in Patterson's house before being sold to the Field Museum in Chicago in 1924. The lions were ultimately reconstructed and put on display, where they remain today (Fig. 7.5).

Yeakel et al. (2009) employed a Bayesian isotope-mixing model using $\delta^{13}C$ and $\delta^{15}N$ that was extracted from bone, teeth, and hair keratin of the lions. Their results revealed that one of the lions had indeed progressed from a diet largely sourced from grazers (multiyear average: $\delta^{13}C = -12.1$‰; $\delta^{15}N = 14.3$‰; 55% grazers) to one sourced from browsers, mixed feeders, and ultimately humans toward the end of 1898 (final-months average: $\delta^{13}C = -13.4$‰; $\delta^{15}N = 13.0$‰; 30% humans). The other lion was not as heavily dependent on humans (~13% sourced from humans). These values were quite distinct from modern lions in Tsavo, which specialize on grazers ($\delta^{13}C = -8.9$‰; Yeakel et al. 2009). The dietary shift in the primary man-eating lion may have been precipitated by severe craniodental deformities, which included a fractured lower right canine, root-tip abscess, missing lower incisors, and rotation and malocclusion of the upper right canine (Yeakel et al. 2009). These injuries, coupled with a decline in natural herbivore abundance over this time due to rinderpest and drought as

Fig. 7.5. The infamous man-eaters of Tsavo. *Left*: The first lion killed by Colonel Patterson (FMNH 23970) on December 9, 1897. This lion had considerable craniodental abnormalities, which must have made hunting difficult. Based on bulk stable-isotope analysis, by the end of his life, the lion's diet was sourced from approximately 30% human prey (Yeakel et al. 2009). *Right*: Photograph of the man-eaters in their diorama inside the Hall of Mammals at the Field Museum of Natural History in Chicago; FMNH 23970 is the crouching one on the right. Taxidermic mounts were made after the skins were acquired from Colonel Patterson in 1924; the skins were in poor condition when purchased because they had been used as floor rugs for more than twenty-five years. It is estimated that the lions were about seven or eight years old at time of death; notice the peculiar lack of manes. *Left:* Photo from January 1898, courtesy of Wikimedia Commons; *Right:* Image by Jeffrey Jung; CC BY-SA 3.0, courtesy of Wikimedia Commons.

well as local burial practices, may have made it difficult to capture more traditional prey. The estimated number of humans killed by these lions was thirty-five (range 4-72; Yeakel et al. 2009), somewhat less than that reported by popular accounts (e.g., Patterson 1907).

In our own work in the Edward's Plateau of Texas, we are fortunate to work at a site that has unusually good collagen preservation (Smith et al. 2016d). Thus, we have been able to extract $\delta^{15}N$ for a wide variety of mammal species spanning the past 20,000 years, allowing the reconstruction of ancient food webs (Fig. 7.6).

Hydrogen ($^2H/^1H$)

Deuterium (^2H, or D) is the heavy isotope of hydrogen (Table 7.1). In the past, it was largely used to study the hydrological cycle (Clark and Fritz 1997; Gibson et al. 2010) as well as water use and physiology in plants (Vander Zanden et al. 2016). More recently, hydrogen ratios have been employed to investigate the geographic origin, migration, and habitat use of animals (Estep and Dabrowski 1980; Hobson 1999; Koch 2007; Ehleringer et al. 2008; Hobson and Wassenaar 2008; Voigt et al. 2015; Vander Zanden et al. 2016). Because δ^2H is strongly influenced by temperature and

precipitation, it varies predictably by latitude and altitude (Craig 1961; Hobson 1999; Koch 2007; Bowen 2010; Vander Zanden et al. 2016). Thus, it is possible to construct continental isotopic maps and use this characteristic signature to trace large-scale movement patterns of animals. This work was pioneered by Hobson (1999), who was able to track the migration of terrestrial animals using δD from keratin. They found that the deuterium signature in feathers and claws of birds reflected their breeding latitude (Hobson and Wassenaar 2008). However, the isotopic fractionation of hydrogen and oxygen is complex: the source of these elements within animal tissues can arise from a variety of sources, including drinking water, water from the diet, and molecular exchange with atmospheric gases (Koch 2007; Ben-David and Flaherty 2012). This leads to heterogeneities in the isotopic landscape, which can complicate the interpretation of movement; moreover, isotopic maps can have low accuracy if there is a limited number of sampling stations (Ben-David and Flaherty 2012).

To date, the use of δ^2H in food-web analysis and in evaluating the physiological condition of animals is fairly limited, in part because of analytical challenges and in part because of the difficulty of

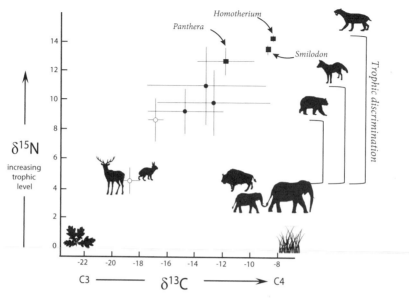

Fig. 7.6. The δ^{13}C and δ^{15}N "niche" (Newsome et al. 2007) for a late Pleistocene mammal community in the Great Plains of Texas. This area has been a focus of study in my lab since 2016. To assess potential trophic relations and reconstruct animal diets, we first had to characterize the prey space by measuring the isotopic ratios of all potential food sources. Here, the approximate δ^{13}C and δ^{15}N position of typical browsing (deer, lagomorphs) and grazing (bison, mammoth) herbivores in this habitat are depicted by animal silhouettes. Not all herbivores in the ecosystem are presented here. Three predator guilds are shown—canids with filled circles, felids with filled squares, and ursids with open circles; the maximum trophic offset is shown with brackets. Of the predators, the bears were the least carnivorous; indeed, one species (black bear, *Ursus americanus*) had an isotopic niche that was actually quite similar to that of the herbivores. This suggested it was feeding on similar foods. The other bear species, the now extinct short-faced bear (*Arctodus simus*), was enriched by one trophic level (~3.5‰) above the herbivores and had a mean δ^{13}C of about −16.5‰. From this, we inferred it was a predator that specialized on browsers in this ecosystem. The canids averaged about 5‰ higher in δ^{15}N than that of the prey, and the felids, about 8‰ higher. We suspect that the much higher trophic position of felids suggests that they were feeding on baby mammals, which tend to be higher in δ^{15}N because they are essentially "consuming" their mothers during lactation (Fogel et al. 1989; Tykot 2004). Moreover, the enriched δ^{13}C values for the saber-tooth and scimitar cats (*Smilodon fatalis* and *Homotherium serum*, respectively) suggested they were specializing on grazers, perhaps baby mammoth. Although the smaller American cave lion (*Panthera leo atrox*) was also feeding on grazers, it had a more varied diet. In contrast to the felids, the canids had a much more variable diet, which appeared to be sourced from both C3 and C4 food webs. Redrawn from Smith et al. (in prep). Organism silhouettes from PhyloPic (http://phylopic.org) under Creative Commons license.

characterizing the metabolic routing of hydrogen (Podlesak et al. 2008; Vander Zanden et al. 2016). Unlike δ^{13}C and δ^{15}N, which are solely derived from the diet in animals, δ^2H (and δ^{18}O) values reflect a blend of dietary and body water composition. Moreover, numerous processes influence the integration of these isotopes from environmental and dietary sources; many are not well character- ized (Koch 2007; Podlesak et al. 2008; Vander Zanden et al. 2016). Even factors such age, body mass, and growth rate may influence δ^2H and δ^{18}O values (Soto et al. 2011, 2013; Wolf et al. 2012; Graham et al. 2013). A better understanding of the

factors influencing the complex flux of hydrogen and oxygen within animals may allow insights into physiological water use, nutrient transport, better dietary sourcing, and connectivity of marine and terrestrial food webs (Podlesak et al. 2008; Vander Zanden et al. 2016). For example, because the δ^2H of body water is positively correlated with evaporative water loss in rock doves (McKechnie et al. 2004), it provides a way to characterize the physiological condition. Other work has demonstrated that aquatic and terrestrial primary producers vary significantly in their δ^2H values (Doucett et al. 2007); such differences can be used to trace nutrient flow and quantify land to water or (vice versa) energy flow (Doucett et al. 2007; Myers et al. 2012; Vander Zanden et al. 2016).

Oxygen ($^{18}O/^{16}O$)

As mentioned earlier, stable-isotope analysis of oxygen has been extremely important for paleoclimate reconstruction. These analyses take advantage of the selective fractionation of the lighter isotope ($\delta^{16}O$) during evaporation. Because shelly marine animals incorporate oxygen into their calcium carbonate shells, this means they incorporate more $\delta^{18}O$ during cold glacial intervals than they do during warmer intervals. Thus, foraminifera, diatoms, and other shelly marine organisms can effectively record ocean temperatures over time (e.g., Zachos et al. 2001).

In animals, characterizing the routing of oxygen isotopes is difficult because of the many pathways by which oxygen enters or leaves the body. For example, intake isotopic ratios are strongly influenced by both diet and drinking water, which are not fractionated during uptake, and by the respiration of atmospheric oxygen and water vapor, which do undergo fractionation (Kohn 1996; Vander Zanden et al. 2016). Oxygen leaves animals through the respiration of carbon dioxide, excretion pathways, and evaporation from the skin; all of these processes are influenced by the physiological state of the animal and ambient environmental conditions (Kohn 1996).

The sourcing matters in quantifying the flux of oxygen through an animal. For example, drinking water is usually supplied by meteoric water—that is, water derived from snow and/or rain. Because meteoric water varies in $\delta^{18}O$ both spatially and temporally (e.g., higher values in warm conditions, and lower values in colder conditions), use of this isotope can allow study of migratory patterns in both marine and terrestrial systems. While the isotopic ratio of water in plant stems resembles that of meteoric water, the leaves of plants may be enriched because of evapotranspiration; factors such as temperature, humidity, and the photosynthetic pathway influence the degree of $\delta^{18}O$ enrichment (Helliker and Ehleringer 2000). Thus, food and water can have quite different $\delta^{18}O$ even within the same environment. Because of these issues, oxygen isotopes are probably most useful when the scale of

the study is large such that variation in meteoric water is much greater than that of other sources. For example, in a large-scale study spanning Canada to Argentina, MacFadden et al. (1999) found a strong correlation between the $\delta^{18}O$ of Pleistocene horse tooth enamel and latitude, which they used as a proxy for environmental temperature.

Despite the difficulty in characterizing the metabolic routing of oxygen isotopes, they have often been employed in paleoecology studies because they can be measured from apatite in enamel, even for deep-time fossils (Koch 2007). For example, Secord et al. (2012) used oxygen and carbon isotopes to examine the response of a local mammal community in Wyoming to an abrupt warming and cooling event in the early Cenozoic known as the Paleocene-Eocene Thermal Maximum (PETM, ~55 Ma). While the underlying cause of this event is still not clear, the PETM is recognized globally by a negative $\delta^{13}C$ carbon isotope excursion; sea surfaces may have warmed by more than 5°C (Zachos et al. 2001). Because temperature is an important selective force on mammal body size (Millien et al. 2006; Smith et al. 2010a; Secord et al. 2012) hypothesized that this temperature shift led to changes in the body size of *Sifrhippus*, an early horse. Using $\delta^{13}C$ and $\delta^{18}O$, they developed a local paleoclimatic record, which mirrored the negative carbon excursion associated with the PETM in other locations. They used oxygen isotopes from several mammal species that varied in their affinity for aquatic environments as a way to distinguish between temperature and habitat productivity. They found $\delta^{18}O$ values were significantly related to morphology of *Sifrhippus*: body size decreased by approximately 30% at the beginning of the PETM, and increased abruptly by greater than 75% at its end, as climate cooled.

Other Isotopes

There are a number of other stable isotopes less commonly used that have nonetheless proved useful in paleoecological studies. These include isotopes of sulfur (S), calcium (Ca), and strontium (Sr). The latter, in particular, has been employed to investigate large-scale movement and migration patterns of mammals, particularly ancient peoples (Bataille and Bowen 2012). As it turns out, because of differences

Box 7.2

Dr. Marilyn Fogel, Isotope Pioneer

Advances in the field of biogeochemistry have revolutionized modern ecology and paleoecology, allowing us to study ancient animal behavior, ecology, and the environment. Dr. Marilyn Fogel, who has been a pioneer in the use of carbon, oxygen, hydrogen, and nitrogen isotopes to investigate modern and ancient ecosystems, drove many of these advances.

Born in Moorestown, New Jersey, Marilyn received a bachelor of science in biology from Penn State University and later a PhD in botany and marine science from the University of Texas at Austin's Marine Science Institute in Port Aransas. She started graduate school just as the Vietnam War was winding down, a turbulent time in American history. During graduate school, she spent her evenings and weekends creating jewelry out of old eyeglass lenses to sell at tourist shops, running an ice cream truck, and working at a hotel desk (isotopequeen.blogspot.com/2019/08/the-other -side-of-grad-school.html). Nonetheless, Marilyn finished her degree in less than three years; in 1977, she went on to postdoc research at the Carnegie Institution of Washington's Geophysical Laboratory.

Much of Dr. Fogel's almost fifty-year career was spent at Carnegie. She began as a postdoc but was hired as a staff member after the president of the institute heard her give a talk at a scientific conference. When she joined the lab in 1977, biogeochemistry was just taking off, thus Marilyn was at the forefront of developing high-precision measurements of natural isotopes to understand biological, environmental, and ecological processes in new and innovative ways. Using these techniques, she was able to lay the groundwork for understanding how nitrogen and carbon enter the food web and how they are transformed and/or fractionated by living organisms. Along the way, she pioneered how to characterize the isotopic signatures of the synthesis of biomarkers such as amino acids, fatty acids, and lignin.

A quick search of her articles on Google Scholar reveals an inspiring breadth of interests and contributions that ranges from astrobiology to plant physiology. For example, using collagen extracted from ancient human remains, Dr. Fogel analyzed the carbon isotopes in individual amino acids as well as the overall bulk values (Fogel and Tuross 2003). She found that the offset between compound-specific and bulk isotopic analyses could be used to reveal the extent of omnivory in these early hominins. Similarly, compound-specific analysis of lipids allowed for the characterization of climate change in Ethiopia over the past five thousand years (Terwilliger et al. 2008).

Of relevance here is her paleoecological work on mammals. For example, along with her former postdoc Paul Koch and others, Marilyn used carbon, nitrogen, and strontium isotope compositions to examine historic changes in abundance, diet, and habitat use among elephants in Amboseli National Park in Africa (Koch et al. 1995). While normally such investigations would necessitate long-term observational studies, isotopic markers allowed the characterization of important shifts in ecology with conservation implications. They observed a change from browsing to grazing that was coincident with a shift in movement. As migration patterns were disrupted, elephants began foraging inside the park instead of moving seasonally. This led to a heavy reliance on grasses rather than browse; indeed, the proportion of C3 plants in the diet dropped from 75% to 40% over several decades (Koch et al. 1995).

In addition to her incredible achievements as a scholar, one of Dr. Fogel's greatest legacies are the countless postdoc, graduate, and undergraduate students, many of whom went on to be famous in their own right. Her aim as a mentor was in demonstrating that creative, innovative science was fun and that *"the people, friendships, and relationships that are built along the way are equally as important as the work that is being pursued"* (Fogel 2019). Her blog, *Isotope Queen* (isotopequeen.blogspot.com), which she started several years ago, provides lively anecdotes about science, discovery, and importantly, being a female in a world dominated by male scientists. For example, she was only the second

BOX FIG 7.2 Marilyn Fogel in the field. Photo by Kevans27, Courtesy of Wikipedia, CC BY-SA 4.0.

woman staff member hired at the Carnegie Geophysical Laboratory; it took thirty years before another was hired. Her years there were filled with exciting discoveries and innovation but also discrimination and rampant underappreciation (isotopequeen.blogspot.com/2019/09/before-metoo-era.html).

In 2013, Marilyn moved to the other side of the United States, first becoming chair of the Department of Life and Environmental Sciences at University of California, Merced, and then in the fall of 2016 accepting the inaugural Wilbur W. Mayhew Endowed Chair of Geoecology at University of California, Riverside. Unfortunately, these moves were complicated by a life-altering diagnosis of amyotrophic lateral sclerosis (ALS) in May 2016. This upended her world, curtailed many of her activities, and ultimately led to her retirement in June 2020. She has written extensively about her scientific and personal experience coping with a terminal disabling diagnosis (e.g., isotopequeen.blogspot .com/2020/05/four-years-into-als-and-still-running.html).

As is fitting given her long and distinguished career, Marilyn is a fellow of the Geochemical Society, the American Association for the Advancement of Science, and the American Geophysical Union. In 2014, she was awarded the Alfred Treibs Medal in Organic Geochemistry for lifetime achievement in the field—the first women to have received this honor. Marilyn also has been a Fulbright Scholar and a National Science Foundation program director in geobiology and low temperature geochemistry, as well as served on the National Research Council's Space Studies Board and as president (2013-2018) of the biogeosciences section of the American Geophysical Union. She was elected to the United States National Academy of Sciences in 2019.

in age and elemental composition, continental rocks vary a great deal in their δ^{87}Sr ratios; for example, granites have enriched δ^{87}Sr ratios relative to limestones and basalts. Because soil types are strongly influenced by the bedrock, and to some extent by atmospheric deposition of dust, this means that the δ^{87}Sr ratio in plants (and consumers) reflects that of the underlying bedrock. Unlike carbon and nitrogen, there is no fractionation along the food chain; the δ^{87}Sr of a consumer directly reflects that of bedrock (Koch 2007). Of course, it is slightly more complicated and other factors, such as rooting depth and photosynthetic pathway, may influence the strontium isotope ratios within plants. Thus, this tracer works best for mammals that migrate long distances, because differences in basal rock composition swamp local variation in δ^{87}Sr values. Some particularly interesting studies have reconstructed the ancient migration of mastodons and mammoth using δ^{87}Sr (Hoppe et al. 1999; Hoppe and Koch 2007). These workers developed an isoscape of Sr ratios across Florida and Georgia using modern rodents, plants, and surface water. Comparing samples from mastodons and mammoths with this template revealed that while mastodons migrated relatively long distances, mammoths did not range more than a few hundred kilometers (Hoppe et al. 1999).

FURTHER READING

Ben-David, M., and E.A. Flaherty. 2012. Stable isotopes in mammalian research: a beginner's guide. *Journal of Mammalogy* 93:312-328.

Clementz, M.T. 2012. New insight from old bones: stable isotope analysis of fossil mammals. *Journal of Mammalogy* 93:368-380.

Hobson, K.A., and L.I. Wassenaar. 2008. *Tracking Animal Migration with Stable Isotopes*. London: Academic Press.

Koch, P.L., et al. 1994. Tracing the diets of fossil animals using stable isotopes. In *Stable Isotopes in Ecology and Environmental Science*, edited by R. Michener and K. Lajtha, 63-92. Boston: Blackwell Scientific.

Newsome, S.D. et al. 2012. Tools for quantifying isotopic niche space and dietary variation at the individual and population level. *Journal of Mammalogy* 93:329-341.

Sharp, Z. 2017. *Principles of Stable Isotope Geochemistry*. 2nd ed. University of New Mexico Digital Repository. Open Textbooks. Accessed April 1, 2019. doi:10.5072/FK2GB24S9F.

8   Nontraditional "Fossils"

Not all fossil data come in the form of a long bone or tooth. There are a number of other preserved materials that can yield valuable information about past ecology of mammals and their environments (Fig. 8.1). Often these are the traces or remains of the activities of animals, such as their walkways, burrows, or even excrement. More precisely, trace fossils are produced when an organism interacts with a medium in an environment; this process generates a three-dimensional physical structure (Hasiotis et al. 2007). Some of the most useful insights into the past (particularly the late Quaternary) have come from these nontraditional nonbody "fossils."

Trace Fossils

How or where an animal lives can leave a trace in the geological record, even if the original animal is not preserved (Fig. 8.2). The field of ichnology refers to these records of long-ago behavior. They can include footprints or impressions made by walking (or running from a predator), burrows or borings made for shelter, foraging traces, and fossilized waste material known as coprolites (Buckland 1829). Because these traces all reflect the behavior of animals and, with the exception of footprints, are normally not attributable to any specific animal, a whole vocabulary has developed to describe them. Thus, as noted in chapter 2, we often get a peculiar situation where a walkway of a particular mammal has a different

scientific name than its bones. Interpretation of traces in the geologic record is facilitated by a good understanding of those made by modern animals, particularly mammals (i.e., Elbroch 2003).

Classification of Ichnofossils

Classification of trace fossils is based on the shape, form, and implied behavior each one represents (Seilacher 1967; Bromley 1996; Hasiotis 2003). The highest taxonomic level represents the broad behavioral mode. Not all of these are relevant to mammals, but the five main classifications are as follows:

1. Domichnia, dwelling traces that can include many morphologies, such as vertical or U-shaped burrows
2. Fodinichnia, three-dimensional structures left by animals foraging through sediments
3. Pascichnia, grazing traces made while feeding
4. Cubichnia, resting traces, typically formed when an organism pauses during feeding or hides
5. Repichnia, movement traces produced as organisms crawl, walk, or run

Trace fossils are further divided into *form genera*, and if they are distinct enough, they are identified as an ichnospecies; this is a separate nomenclature than that used for body fossils. Sometimes, but not always, ichnospecies can be referred back to the genus or species of animal that created them. For example, *Lamaichnum* represents a somewhat

Fig. 8.1. Examples of nonbody trace fossils. **a,** Dinosaur footprint belonging to the ichnotaxa *Eubrontes*. From the Lower Jurassic and found at the St. George Dinosaur Discovery Site, southwestern Utah. **b,** Woodrat (or packrat) paleomidden recovered in Death Valley National Park; inset picture of fossilized fecal pellets found in abundance within these debris piles. **c,** *Sporormiella*, a genus of fungi found in abundance and related to megafauna presence in an area. **d,** A legendary Miocene-aged coprolite (fossilized feces) nicknamed Precious. It was found in South Carolina and measures some 195.9 mm by 125.4 mm and weighs 1.92 kg. **e,** Indurated Pleistocene packrat midden from Burro Canyon, Arizona. Note the piñon leaves exposed, which are completely intact. **f,** Pollen grains from a variety of common plant, magnified by 500; actual length of pollen grains is about 50 μm. Panel a: Photo by Mark A. Wilson, courtesy of Wikimedia Commons. b: Main photo and inset by the author. c: Courtesy of public domain. d: "Precious the Coprolite" by Poozeum, CC BY-SA 4.0. e: Courtesy of public domain. f: Courtesy of Dartmouth College Electron Microscope Facility.

Fig. 8.2. A 280-million-year-old fossil walkway from Grand Canyon National Park, Arizona. These tracks were produced by a primitive tetrapod during the Permian period and assigned to the ichnotaxa *Ichniotherium*. These were primitive tetrapods that possessed characteristics of both amphibians and reptiles. *Left*: General view of the track-bearing sandstone boulder and the tracks. *Right*: False color depth map showing details of footprints (depth shown in mm). Scale bar is 50 cm. Courtesy of the National Park Service, www.nps.gov/articles/grca-fossil-footprints.htm.

rounded rectangular artiodactyl track with two hooves in both the manus and pes. At the highest level, it is classified as repichnia (a movement trace), with a geologic range from the middle Miocene to the Recent and a geological setting of terrestrial and continental. However, here we know that *Lamaichnum* are probably made by the extinct camel of North America, *Camelops* (Lucas et al. 2007). A good source for exploring ichnofossils is the IchnoBioGeoScience research group's website KU Ichnology at the

University of Kansas (http://ichnology.ku.edu). This site provides a visual guide to the various types of trace fossils as well as detailed descriptions.

It should be noted that a single species can be responsible for multiple types of ichnofossils. For example, a ground squirrel could potentially leave a burrow, traces of feeding, a walkway, and/or coprolites, which represents four different behaviors. Likely when found, these would be interpreted as representing the behavior of four different animals.

Ichnofossil Localities

There are a number of sites that house fabulous collections of trackways and other ichnofossils in North America, particularly those of Cenozoic mammals (Lucas et al. 2007). While some localities do date to the Paleogene, most are Miocene or younger in age (Lucas et al. 2007). An exception is Clayton Lake State Park (www.emnrd.state.nm.us /SPD/claytonlakestatepark.html) in the northeastern corner of New Mexico (Fig. 8.3), which has some of the most extensive Mesozoic dinosaur trackways in North America. These occur along the shore of the lake and are easily accessible to visitors. While other localities also contain the trackways of dinosaur (Fig. 8.4), lizards, or insects, our main interest here is in the ichnofossils of mammals.

In addition to dinosaurs, mammal trackways are found throughout New Mexico. These include White Sands National Monument, where Pleistocene mammoth and camels walked repeatedly along the shore of a shallow saline lake around 30,000 years ago; Santa Fe, where fifty-two tracks of a large camel (ichnospecies *Lamaichnum macropodom*) were found just west of the airport; and Rio Rancho, where five long trackways record the movements of a large felid and multiple species of camels in the Benavidez Ranch local fauna (Lucas et al. 2007).

The Riverbluff Cave Site near Springfield, Missouri, contains thousands of trackways, claw marks, coprolites (and bones) of extinct Pleistocene mammals, including mammoth, felids, short-faced bears, horses, and bears (Forir et al. 2007). The most impressive are the abundant claw marks some 20 cm wide, which are found about 4 m above the surface of the cave, along with what appear to be bear beds (Forir et al. 2007). The claw marks are thought to

have been made by the extinct short-faced bear, *Arctodus simus*, a voracious predator whose mass exceeded 1 ton (Smith et al. 2003). There are also trackways of large cats—thought to be either the American lion, *Panthera leo atrox*, or the saber-tooth cat, *Smilodon fatalis*—on high terraced areas within the cave, although no bones of these species have yet been recovered (Forir et al. 2007).

One of the most abundant and diverse fossil track localities in North America is in Death Valley National Park. A 3,000-meter lacustrine deposit within Copper Canyon preserves the trackways of hundreds of ancient mammals and birds within large slabs of fine-grained mudstones (Curry 1939; Scrivner and Bottjer 1986; Nyborg 2009). These include llamas and camels (one with a newborn), mastodon, horses, tapir, and multiple felids, all of which roamed along the shores of a spring-fed lake that existed at the time. The slabs are now almost vertically tilted and thus highly suspectable to erosional processes and landslides. For example, only 20% of identified tracks from 1999 could be relocated in 2003 (Nyborg 2009), suggesting extremely high turnover. However, there appear to be multiple track-bearing layers, which are exposed as the surface is eroded (Nyborg 2009). To date, thirty-six ichnospecies of vertebrates have been described from the Pliocene lakeshore sediments (Curry 1939; Scrivner and Bottjer 1986; Nyborg 2009). Sadly, despite legal and pragmatic protections that include not advertising the exact location of the site, limiting access, and requiring a permit to visit the site, some of these fossil trackways were recently stolen from Copper Canyon (www.nps.gov/deva /learn/news/fossils-stolen-from-death-valley-national -park.htm), underscoring the precarious state of these unusual and valuable ichnofossils.

Pleistocene mammal coprolites are locally common in the Southwest United States (Lucas et al. 2007). Many are found within the hundreds of caves in Grand Canyon National Park, where preservation potential is high. Indeed, their abundance led Hunt et al. (2005) to propose this area as a speleological Lagerstätte. Mammal remains include those of woodrats and other rodents, ringtail cats, bighorn sheep, and the extinct Shasta ground sloth and Harrington's mountain goat (Lucas et al. 2007). Other cave sites in the Southwest contain fossilized dung of

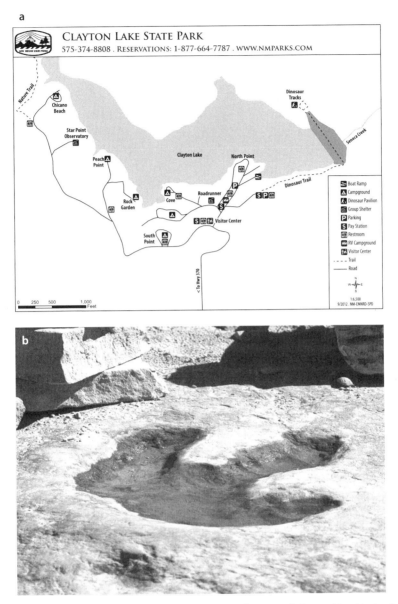

Fig. 8.3. a, Location of fossil trackway in Clayton Lake State Park, near the town of Clayton, New Mexico. **b**, The early Cretaceous sandstones in this area preserve the remains of more than five hundred fossil dinosaur footprints, all dating to around 100 million years ago. These are some of the best-preserved and most extensive dinosaur track sites in the United States. Map courtesy of Clayton Lake State Park; photo of large dinosaur track by Greg Willis, CC BY-SA 2.0.

extinct large-bodied megafauna, such as the mammoth, bison, shrub-ox, and horse (Lucas et al. 2007). Assigning coprolites to species is usually done based on the morphological similarity with living species; with the exception of some of the largest bovids and cervids, mammals have fairly characteristic excrement (Davis et al. 1985; Elbroch 2003). For example, lagomorphs (bunnies, hares, and pikas) have small

round fecal pellets, while many rodents have elongated Tylenol-shaped pellets; the sizes of these scale isometrically with animal body mass. At present, little ichnotaxonomy has been conducted on coprolite despite their potential utility for paleoecological reconstructions.

The first paleoburrows attributed to mammals appear shortly after the clade evolved—about 190

Fig. 8.4. Dinosaur tracks at Dinosaur Ridge, Morrison Fossil Area National Natural Landmark, Colorado. Photo by Chris Light, CC BY-SA 4.0.

Ma (Cardonatto and Melchor 2018; Fig. 8.5). However, other tetrapods, particularly dinosaurs, also constructed burrows, so identifying the animal can be difficult. The burrow trace-fossil record is most abundant for the Miocene, perhaps because of the development of grasslands and other open habitats suitable for fossorial animals (Lucas et al. 2007). There is also a diversification of architectural structures by the Miocene, which likely reflects both greater diversity of fossorial animals and a greater array of behavioral strategies (Cardonatto and Melchor 2018). The oldest food cache within a burrow also dates to the early Miocene and was found in the Rhine Embayment near Cologne, Germany (Gee et al. 2003). Containing preserved nuts (*Castanopsis pyramidata*) from chinquapin trees (a type of evergreen belonging to the beech family), the food was probably hoarded by a large hamster or ground squirrel (Gee et al. 2003).

Inferring from Geological Traces

So what information do ichnofossils provide? Well, actually, a surprising amount. For example, the microscopic analysis of materials preserved within the coprolites of extinct Shasta ground sloth (*Nothrotheriops shastensis*) identified more than seventy species of plants, providing information on the environmental context as well as the diet of the sloth (Martin et al. 1961; McDonald 2003). Because multiple stratigraphic levels were present at Rampart Cave in Arizona, potential changes in the ecological niche of this browser could be examined over the Last Glacial Maximum (Martin et al. 1961). Similar analyses

identified grasses and sedges as the most abundant macrofossils in mammoth dung (Davis et al. 1985).

Analysis of coprolites often reveals information about the environment the animals lived in. By comparing the abiotic temperature and precipitation tolerances of the plants identified, it is possible to construct an abiotic "envelope" where all could co-occur. Thus, we know that sloths in South America occupied open, somewhat mesic environments (Hofreiter et al. 2003), and mammoths foraged extensively on riparian plants (Davis et al. 1985; Mead et al. 1986).

Moreover, it has proved possible to extract ancient DNA (aDNA) from coprolites to obtain higher-resolution data (i.e., Poinar et al. 1998; Hofreiter et al. 2003). For example, using 20,000-year-old dung recovered from Gypsum Cave, near Las Vegas, Nevada, researchers were able to not only verify it was deposited by the extinct Shasta ground sloth but also analyze its diet in considerable detail. Shasta ground sloths apparently were particularly fond of capers and mustards (family Brassicaceae) and yucca and agave (family Liliaceae, including Agavaceae), which made up about 24% and 19% of the plant DNA, respectively (Poinar et al. 1998). Other plants identified in the sloth coprolite included grasses (family Poaceae), borages and mints (order Lamiales, formerly Scrophulariales), wild grape (family Vitaceae), mallows (family Malvaceae), and saltbush (family Chenopodiaceae) (Poinar et al. 1998). Two particularly interesting conclusions of this analysis were that (1) some species identified by molecular techniques were not identified in earlier studies that relied on macrofossils, and (2) some of these plants (grape, yucca, and agave) do not occur anywhere nearby today; their modern elevational range has shifted upward by more than 800 m (Poinar et al. 1998). Similarly, molecular analysis of other sloth coprolites has revealed surprising findings, such as new species not represented by fossil bones within the cave (Hofreiter et al. 2003).

Coprolite analysis has been most extensively employed in archaeology. The oldest hominin coprolites recovered are from the Olduvai Gorge of Kenya and are more than a million years old (Leakey 1971). While rare, coprolites attributed to *Australopithecus* and Neanderthals have been found, but most are produced by archaic and anatomically modern humans (Reinhard and Bryant 1992). In some more recent sites, such as Salmon Ruin, a late Holocene site in New Mexico, the abundance of coprolites in a single latrine made formal excavations using horizonal and vertical sampling possible (Reinhard et al. 1987). Thus, coprolite analysis provides an important window to investigate the diet, environment, and population dynamics of hominins over time (Callen and Cameron 1960; Clary 1984, Reinhard et al. 1987; Reinhard and Bryant 1992; Shillito et al. 2020). Studies of coprolites have recovered bacteria and fungi, insects, phytoliths pollen, parasites, and macrofossils as well as characterized the mineral and chemical compositions of the droppings (Reinhard and Bryant 1992). They reveal the influence of many important evolutionary transitions in hominin history—such as plant domestication and the adoption of a meat-based diet—as well as provide a way to investigate population movement (as indicated by seasonal changes in the diet), nutritional changes, and incidences of disease (Reinhard and Bryant 1992; Shillito et al. 2020).

Trackways provide unique insights into behavior and population dynamics. Wally's Beach, located on the eastern shore of the St. Mary Reservoir in Southern Alberta, Canada, is a good example of this (McNeil et al. 2005; Lucas et al. 2007). Not only does it contain multiple trackways of a number of extinct Pleistocene megafauna who co-occurred on this shore, but skeletal material and even archaeological artifacts are found. The tracks often overlap, providing a temporal sequence for movement. Tracks preserved include camel, horses, bison, musk ox, and mammoth, as well as caribou (McNeil et al. 2005). The abundance of mammoth tracks, in particular, has allowed quantitative analysis of population behavior and demographics. For example, the spacing of adult woolly mammoth tracks suggests that 4-5 mph was a typical walking speed (McNeil et al. 2005). Moreover, by measuring the sizes of the various footprints preserved, body size and rough age stages could be assigned (e.g., baby, subadults, adults, mature adults) and population demography investigated. A relatively low proportion of juveniles was found at Wally's Beach compared with that typically found in extant African elephant herds, leading the authors to conclude that this was a population in decline (McNeil et al. 2005).

One discovery in the late 1800s—of "devil's corkscrews," large elongated, twisted fossilized sediments in the Badlands of Nebraska—was a puzzle to science for a long time (Fig. 8.5). At first these structures, which could reach 3 m long, were identified as giant, freshwater sponges, and then later as giant root traces (Barbour 1892, 1895). This led to a lively debate in the literature. When the fossil bones of rodents were found within some of the spiral chambers, Fuchs (1892, as cited in Barbour 1895, p. 518) suggested, *"I think we have before us all the essential elements of Daemonelix, and that accordingly we are justified in viewing these strange fossils as nothing else in reality than the underground homes of Miocene rodents, apparently of the family Geomyidmae."* In a rather droll rebuttal to this criticism, Barbour (1895, p. 520) argued that the deposits were aqueous and, thus, *"Unless . . . all geologists [are] wrong, then Dr. Fuchs' gopher is left to burrow and build its nest of dry hay in one or two hundred fathoms of Miocene water."* As further study established, these were indeed trace burrow fossils but ones made by three species of Miocene beaver in the genus *Palaeocastor* (Martin 1994). Nonetheless, their ichnofossil designation, *Daemonelix*, still reflects their original colorful nickname.

What information can we glean about *Palaeocastor* from their burrows? Unlike other beavers, *Palaeocastor* did not live in or near water. Instead, analysis of sediments reveals that these medium-bodied rodents were strictly terrestrial and occupied semiarid upland environments, particularly in western Nebraska and eastern Wyoming (Martin and Bennett 1977). In some

Fig. 8.5. a, Trace burrow fossil commonly known as "devil's corkscrews." These elongated twisted fossilized sediments can reach 3 m long (see ranger standing next to the fossil for scale) and were constructed by the Miocene beaver *Palaeocastor*. **b**, Reconstruction of what *Palaeocastor* probably looked like; these beavers lived in grasslands and not streams. **c**, Skull of *Palaeocastor*. Circa late nineteenth or early twentieth century, courtesy of James St. John and the Agate Fossil Beds National Monument, Nebraska, USA; Nobu Tamura, CC BY 3.0; Kevin Walsh, CC BY 2.0.

areas, they occur in great abundance; more than two hundred separate burrows have been identified as part of one large *Palaeocastor* colony in Nebraska (Martin 1994). The morphology of these ichnofossils is also very interesting; while originally speculated as an adaption to reduce the ability of predators to enter the burrows, they are now considered to be an adaptive response to paleoclimate (Meyer 1999). The spiral twists help limit air circulation, which results in more consistent temperature and humidity within the burrow, suggestive of an environment with temperature extremes (Meyer 1999). There are side chambers on some, identified as nurseries, bedding areas, and/or food chambers (Martin and Bennett 1977). Some *Daemonelix* even contain the remains of possible predators, including a small-bodied ancient raccoon relative, *Zodiolestes daimonelixensis* (Martin and Bennett 1977; Martin 1994).

In addition to environmental reconstructions, analysis of paleoburrows like those of *Daemonelix* yields insights about the animal itself. Body size is related to the diameter of the burrow as well as the pattern of scratches and grooves on the interior of the structure, and branching patterns of the burrow system can help identify age stages of the animal and features of their general ecology (Martin and Bennett

1977; Gobetz and Martin 2006; Gobetz 2007). Empirical work on modern fossorial mammals helps with such interpretations. In one study, for example, rodents were introduced to large Plexiglas-sided boxes filled with substrate and burrow formation was followed (Gobetz 2007). The author found clear differentiation of digging strategy between nursing pups and mature adults; juveniles dug more frequently with their incisors, whereas adults used their claws (Gobetz 2007). Moreover, by manipulating the environment, it is possible to investigate the influence of moisture or sediment size and composition on the formation of burrows, allowing more refined interpretation of paleoburrow trace fossils.

The Importance of Animal Dung

If you have ever had a pet rabbit, you know that animals, particularly herbivores, produce dung as a byproduct of digestion. A lot of dung. Thus, it should come as no surprise that entire ecosystems revolve around recycling the nutrients in animal waste (Richardson 2001). These include a number of saprobic bacteria, slime molds, beetles, and fungi. Of particular interest here are *Sporormiella*, a genus of more than eighty ubiquitous and cosmopolitan fungi whose life cycle is closely tied to herbivores, particu-

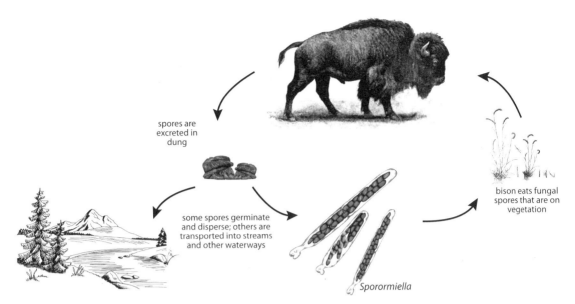

spores are
excreted in
dung

bison eats fungal
spores that are on
vegetation

some spores germinate
and disperse; others are
transported into streams
and other waterways

Sporormiella

Fig. 8.6. *Sporormiella* cycle. *Sporormiella* are found on the surface of leaves or stems and are eaten by herbivores as they forage. The journey through the digestive system triggers germination, which occurs in the animal dung after the *Sporormiella* are excreted. After maturation of the fruiting bodies in the dung, new spores are released into the air. Some land on plants, and the cycle repeats. Other spores may be transported into streams and waterways, where they are deposited in layers with other sediments and become part of the geologic record. Because of the obligate dependence on large-bodied herbivores for germination, the presence of *Sporormiella* is considered a reliable indicator of herbivore activity. Images drawn by author or courtesy of the public domain. Bison drawing from Beddard 1902.

larly small- and large-bodied mammals (Bell 1983; Krug et al. 2004).

Sporormiella spores are common on vegetation. Thus, herbivores inadvertently ingest them along with plant materials while foraging (Bell 1983; Richardson 2001; Krug et al. 2004). After passing through the digestive tract of the herbivore, where the chemical and physical environment can help trigger germination, they are excreted along with the partially digested plant material and microbial bodies (Bell 1983). It is in animal dung where they complete their life cycle and the fruiting bodies develop (Fig. 8.6). After maturation, the fruiting bodies rupture and new spores are discharged to the surrounding vegetation, and the cycle repeats (Bell 1983; Richardson 2001; Krug et al. 2004). This obligate life cycle is coprophilous, that is, "dung-loving," for fairly obvious reasons. While many genera of coprophilous fungi exist, just how obligate their relationship is with herbivore dung and their utility for paleoecology has been questioned (Baker et al. 2013; Newcombe et al. 2016). Hence, our

focus here on the *Sporormiella* type, which is considered one of the most reliable indicators of herbivore activity (Baker et al. 2013).

How Does *Sporormiella* Analysis Work?

Characterizing the abundance of *Sporormiella* spores over time employs the same approach employed in palynology (pollen analysis). Pollen grains (and spores) are produced in great quantities, and because they are small particles, they tend to be well mixed by the atmosphere and fall in a "rain." The majority fall to the ground, where they quickly decay. But pollen grains are resistant to decay in nonoxidizing conditions, and so if they happen to fall into an appropriate sedimentary location (like a lake or ocean), they can persist in the environment for long periods of time (Prentice 1988). Because pollens and spores can be identified with varying degrees of taxonomic precision, they record the vegetative history of a local to regional community over time. While pollen studies have been around for quite

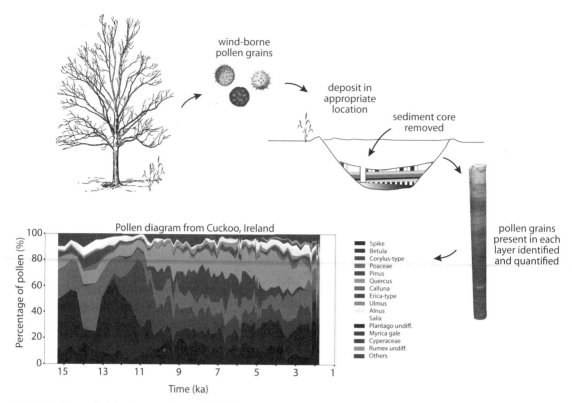

Fig. 8.7. Pollen analysis is a long-standing and reliable method for reconstructing past vegetation abundance and composition. As pollen is dispersed by wind, some falls into depositional environments where grains are preserved. When a sediment core is removed, the pollen grains present in each layer can be identified and quantified, revealing estimates of abundance over time. This technique does not record the abundance of vegetation dispersed by animals or other vectors. Further details about this study are further discussed in chapter 9. Pollen diagram created by Merikanto, CC BY-SA 4.0. Other images drawn by author or courtesy of the public domain

some time, it was Dr. Margaret Davis who developed quantitative methods to reconstruct late Quaternary vegetation history and dynamics (Davis 1963, 1969). Her work revolutionized palynology.

To conduct spore or pollen analysis, a set of sequential and/or overlapping cores are taken from lake sediments or other sources in the field and (normally) encased in PVC tubes (Fig. 8.7). Once in the lab, they are split longitudinally into two halves and generally imaged with high-resolution photography; other analyses may include magnetic susceptibility (Traverse 1988; Faegri et al. 1989). Typically, one half of each core is archived at a core repository, and the other portion undergoes further analysis. Cores are then subsampled, with the size of each "cookie" aliquot dependent on the investigator and questions of interest. From these, samples are removed for pollen or spore analysis at standardized intervals.

Samples are cleaned to remove carbonates, silicates, and other contaminants (e.g., Faegri et al. 1989) and then spores (or pollen) are mounted on microscope slides for identification (not always easy!) and counting. While this was traditionally done by hand using a light microscope, new automated techniques for palynology have been in development (Holt and Bennett 2014). Normally, investigators report the relative abundance since the actual numbers are influenced by the sampling methodology, including the size of the core as well as the sedimentation rate (Prentice 1988; Traverse 1988; Faegri et al. 1989). Given the historic importance of pollen analysis to late Quaternary environmental reconstructions, it is not surprising that an entire body of literature is devoted to the understanding, quantification, and taphonomy of pollen (e.g., Davis 1969; Prentice 1988; Traverse 1988; Bradley 1999;

Faegri et al. 1989). It is important to realize that the pollen spectrum produced does not include all plant taxa within the vegetational mosaic. A number of "silent" plant species are not represented, such as those that are animal dispersed, produce limited quantities of pollen, or have poor preservation (Prentice 1988). Similarly, the deposition of spores may be influenced by taphonomic factors (e.g., Raper and Bush 2009). However, given these caveats, palynology—and spore analysis—does provide a way to examine changes in relative abundance of wind-dispersed plants/spores over time.

Sporormiella and Paleoecological Inference

There are a few qualities of *Sporormiella* and other coprophilous fungi that make them particularly useful for paleoecology (Davis 1987). First, *Sporormiella*-type fungi have a broad distribution that encompasses much of the boreal and temperate regions of the globe (Kirk et al. 2008). Thus, they are ubiquitous in habitats where megafauna are, or were, found. Second, they have distinctive features that enable their detection in pollen samples. These include their color (dark brown) and morphological characteristics such as a pronounced sigmoid germination pore (Davis 1987; Davis and Shafer 2006; Gill et al. 2013). Third, fungal spores have high preservation potential and are commonly found in the fossil record, particularly in lake sediments (Davis 1987; Feranec et al. 2011). Fourth, unlike other fossil or paleoecological indica-

tors, cores are generally temporally continuous. Thus, the relative abundance of coprophilous fungi can be quantified over time as an indirect indicator of local herbivore biomass (Davis 1987; Gill et al. 2013).

Because large-bodied herbivores produce copious quantities of dung, an underlying assumption with coprophilous fungal analyses is that a high abundance of spores reflects a high abundance of megaherbivores in the environment (Davis 1987). And a recent empirical study demonstrated that abundance of *Sporormiella* was positively correlated with bison grazing intensity at local scales (Gill et al. 2013). Thus, stratigraphic changes in spore abundance within cores have been used to infer the dynamics of megafaunal population sizes, including their extirpation from local environments (Davis 1987; Burney et al. 2003; Robinson et al. 2005; Davis and Shafer 2006; Robinson and Burney 2008; van Geel et al. 2008; Gill et al. 2009; Kiage and Liu, 2009). Indeed, numerous studies have reported that *Sporormiella* spores were fairly common in Pleistocene lake sediments, were rare in mid-Holocene records, and became common again only after livestock were widespread (Fig. 8.8), which perfectly parallels the trajectory of megafaunal extinction in the Americas and the historic spread of domestic animals (e.g., Robinson et al. 2005; Davis and Shafer 2006; Robinson and Burney 2008; Gill et al. 2009).

Integrating *Sporormiella* abundance with other paleoecological proxies can provide even more

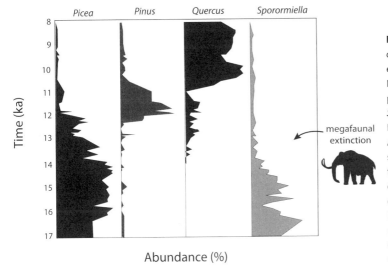

Fig. 8.8. *Sporormiella* abundance can be quantified along with pollen to examine ecological changes within communities. Here, the abundance of spruce (*Picea*), pine (*Pinus*), oak (*Quercus*), and *Sporormiella* was examined over the late Pleistocene and early Holocene at Applewood Lake in Ohio. The study showed a transition from a spruce forest to an oak-pine savanna. These changes were synchronous with the extirpation of megafauna from this area. Redrawn from Gill et al. 2009; mammoth silhouette from PhyloPic (http://phylopic.org) under Creative Commons license.

powerful inferences. For example, in the eastern United States, a dramatic late Pleistocene drop in *Sporormiella* abundance was followed by the development of novel non-analog vegetative communities in the Holocene and enhanced fire regimes (Gill et al. 2009, 2012). Indeed, palatable hardwood forest species increased in abundance; evidence these authors used to argue for the important engineering role of keystone megaherbivores on the environment (Gill et al. 2009, 2012). These authors combined pollen, charcoal, and *Sporormiella* records as well as employed x-ray fluorescence spectroscopy to characterize elemental ratios for species previously established to reflect paleoenvironmental conditions (e.g., Zr/Al ratios are a proxy for sedimentary input, Rb/Sr reflect weathering intensity, V/Cr indicate anoxic sediments, Mg/Ca are a proxy for moisture, etc.; Roy et al., 2009). Similarly, using both pollen and *Sporormiella* records, major shifts in the composition of vegetation communities have been documented after the anthropogenic extinction of many large-bodied herbivores in Australia (Rule et al. 2012) and Madagascar (Burney et al. 2003). In Australia, the greatly reduced herbivore pressure led to a wholescale ecosystem shift with the replacement of mixed rainforest by sclerophyll vegetation (Rule et al. 2012). Such integrative work allows the reconstruction of ancient ecosystems and also highlights the role that large-bodied herbivores play in ecosystems.

Potential Challenges with Coprophilous Fungi as a Paleoecological Proxy

A few issues complicate the use of *Sporormiella* and other coprophilous fungi to investigate the changes in megaherbivore abundance over the late Quaternary. First, while some coprophilous fungal species may have preferences for particular taxa or clades (Lundqvist 1972; Richardson 2001; Baker et al. 2013), it is not currently possible to reliably identify spores below the level of genus (Feranec et al. 2011; Baker et al. 2013). More refined identifications await the development of a comprehensive modern reference collection, which is currently lacking. Thus, it is not possible to link *Sporormiella* presence or abundance to specific species or even a clade of herbivores (e.g., bovids, lagomorphs, etc.), which complicates the interpretations of changes in abundance over time

(Raper and Bush 2009; Feranec et al. 2011; Baker et al. 2013). For example, while the volume of dung produced by individual small-bodied herbivores such as rabbits and hares is certainly tiny compared that of a mammoth, their populations can reach very high densities. When we observe a shift in spore abundance, to what extent does it reflect changes in the population dynamics of small- versus large-bodied mammals? Do animals of different sizes respond similarly to environmental perturbations? More refined spore identifications and calibration with modern studies of spore affinities may help address these issues. For now, this problem is largely ignored and abundance is typically considered to mostly reflect the large-bodied herbivores in the environment.

A second issue, although probably relatively minor, is that dung samples from birds and perhaps even reptiles apparently also support coprophilous fungi; *Sporormiella* were identified in all avian taxa tested in New Zealand (Wood et al. 2011). Indeed, these authors demonstrated the postsettlement decline of native avian herbivores (including the extinct moa) in New Zealand using spores recovered from a forest soil core. It may well be that many species of herbivorous birds and reptiles are also hosts for coprophilous fungi, but few studies have looked (Wood et al. 2011). On continents such as North America, where large-bodied mammals drastically outnumber large-bodied terrestrial herbivorous birds (or did in the late Quaternary before the rise of the poultry industry), this probably contributes little.

However, a more troubling issue is that recent evidence challenges the fundamental assumption that coprophilous fungi require herbivore dung to complete their life cycle (Newcombe et al. 2016). Is this relationship actually facultative? Is this true of all species or just certain clades? Is it influenced by environmental conditions? Should this finding prove to be the case more widely, what changes in spore abundance data need to be better investigated and quantified? It seems unlikely that herbivores don't contribute substantially to the spore burden, but perhaps environmental conditions play a bigger role than previously recognized. For example, spore abundance has been shown to vary with distance from the lakeshore (Raper and Bush 2009). Thus, if a core contains sediments laid down during fluctuations in environmental conditions, changes in spore abundance could also reflect variation in the core's proximity to the shoreline (Raper and Bush 2009).

For all of these reasons, *Sporormiella* analyses are currently best suited for evaluating ecosystem-level abundance dynamics of herbivores and responses to large-scale disturbance, especially when paired with other paleoecological proxies (e.g. Gill et al. 2009, 2012). Here, it can provide an additional tool helpful for exploring the past.

Nature's Furry Historians

A particularly useful source of information for reconstructing late Quaternary environments arises from the activities of woodrats (*Neotoma*) (Fig. 8.9). These herbivorous rodents have been referred to as "nature's furry historians" because of their habitat of collecting things from their environment. Gathered materials are used to construct or reinforce houses or dens around the base of a tree, cactus, or rock-shelter. When the houses and, especially, their

Fig. 8.9. The bushy-tailed woodrat, *Neotoma cinerea,* is found throughout much of the western United States. It preferentially lives in caves or rocky outcrops, where it creates large debris piles, or middens. These can be preserved for tens of thousands of years under the right conditions. The animal in this image had just been released from a live trap in Wyoming. Note the long, furry tail that gives the animal his name. In colder climates, the tail becomes almost squirrel-like. Photo by author.

associated debris piles are situated under the appropriate conditions, encapsulated materials can persist for tens of thousands of years, providing a unique window into past environmental history.

So, just how does this work?

Packrats

Woodrats are medium-sized rodents found today throughout much of North America (Fig. 8.10). They are most diverse in the southwestern portions of the United States, such as Colorado, New Mexico, and Texas, where as many as three species can occur in relatively close proximity, occupying only slightly different habitats (Hall 1981). Woodrats are even found in Death Valley. This is the hottest, driest place in the Western Hemisphere, with triple-digit temperatures common both day and night from May to September (Smith et al. 2009). However, surprisingly, woodrats are not physiologically well adapted to the arid and hot environments where they often live. For starters, they are highly sensitive to environmental temperature (Lee 1963; Brown 1968; Brown and Lee 1969; Smith et al. 1995; Smith and Betancourt 1998, 2003, 2006), which often leads to size-selective mortality (see Fig. 5.5). Indeed, woodrats are the poster child for Bergmann's rule (see chapter 5). Secondly, they lack physiological adaptations common to other desert rodents, such as estivation (the ability to concentrate urine) and specialized nasal passages, both of which help reduce water loss (MacMillen 1964; Tracy and Walsberg 2002). The lack of physiological adaptations to heat may reflect a relatively recent occupation of hot and arid habitats; genetic studies suggests that expansion into arid and warm interior deserts occurred only within the last 100,000 to 50,000 years (Patton et al. 2007).

So, given this apparent mismatch between physiology and ecology, how do woodrats successfully live in such hot, stressful environments? It turns out that like humans, woodrat survival in extreme environmental conditions is entirely dependent on the thermal buffer provided by a protective structure: woodrats construct elaborate dens or houses around cacti and trees, in rock outcrops or crevices, or even in human buildings (Betancourt et al. 1990). These can be quite large (more than 3 m tall and/or wide) freestanding conical structures composed of rocks

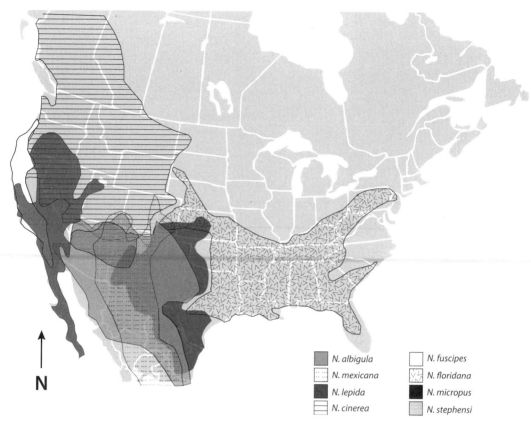

N

▨ N. albigula		☐ N. fuscipes	
⋯ N. mexicana		⋎ N. floridana	
▦ N. lepida		■ N. micropus	
▤ N. cinerea		▥ N. stephensi	

Fig. 8.10. Geographic distribution of eight common species of *Neotoma*. While *Neotoma* are found in many habitats across the continental United States, they reach their greatest diversity in the Southwest. Note the overlapping distributions of many of the species in New Mexico, Texas, and Colorado, which in fact differ in both body size and habitat preferences. For example, the white-throated woodrat, *N. albigula*, is typically found in arid habitats with abundant cactus, whereas the Mexican woodrat, *N. mexicana*, occupies arid woodlands, and the bushy-tailed woodrat, *N. cinerea*, boreal forests. All species construct dens or houses and produce midden piles. However, not all are located where environmental conditions favor preservation.

and vegetation such as branches, leaves, and twigs and/or cactus spines (Fig. 8.11). Because good sites are occupied by successive generations of woodrats, if protected from the weather, the piles get larger with each generation. Woodrat houses also often incorporate other found materials, such as potsherds, bits of handwoven baskets, shotgun shells, potato chip bags, cow patties, bottle caps, gum wrappers (i.e., the archeological artifacts of the future), or whatever else attracts the animal's attention while it is foraging. This eclectic collection of materials is what leads to the other common name for woodrat, the "packrat." Where available, woodrats preferentially live in rock crevices, caves, or outcrops (Smith 1995), and they orient the plant and other debris to block access to cracks in the stone. Woodrats even readily take

advantage of unoccupied houses or cars for shelter. For example, Ancestral Puebloans involuntarily shared their stone structures with woodrats; packrat paleomiddens have been found in ruins within Mesa Verde, Canyon de Chelly, and Chaco Canyon.

Houses are important refuges for woodrats. Depending on the construction, they can buffer against extreme environmental temperatures by up to 20°C (Murray and Smith 2012). Moreover, the cactus spines and joints as well as the maze-like interiors with multiple entrances also provide significant protection against predators (Vorhies 1945; Lee 1963; Brown 1968; Smith 1995). Thus, houses are an essential behavioral adaptation that allow woodrats to occupy what otherwise would be thermally stressful conditions.

Fig. 8.11. Examples of woodrat dens. **a**, A fairly exposed den of the desert woodrat, *N. lepida*, in Joshua Tree National Park. **b**, A bushy-tailed woodrat, *N. cinerea*, sitting on top of a den in a rocky crevice in Wyoming. **c**, A bushy-tailed woodrat watching the author investigate his den in Wyoming. **d**, An old image of a desert woodrat den in Furnace Creek, Death Valley National Park, where woodrats build their dens in large mesquite trees, which are about the only vegetation in the sand flats. Panel a: Photo by Robb Hannawacker, CC-BY-2.0; b and c: Photos by author; d: Photo by Joseph Grinnell circa 1919, courtesy of the Museum of Vertebrate Zoology at UC Berkeley.

Paleomiddens

Part way up we came to a high cliff and in its face were niches or cavities . . . in some of them, we found balls of a glistening substance looking like pieces of variegated candy stuck together . . . It was evidently food of some sort, and we found it sweet but sickish, and those who were hungry . . . making a good meal of it, were a little troubled with nausea afterwards.
—W.L. Manly, *Death Valley in '49*, 1894, p. 126

While from the outside it may look like a pile of wind-blown debris, woodrat houses contain multiple interconnecting chambers (Lee 1963; Brown 1968; Smith 1995). These include food caches, nests, and several debris piles or middens. The debris piles are where the animals deposit waste materials, including fecal pellets, plant fragments, animal bones, and other objects removed from the inner chambers. As woodrats urinate on their midden piles, the materials are coated and begin to adhere. If the midden happens to be located in an arid and protected location—such as a cave in the Southwest United States—the urine desiccates and cements the materials tightly together (Betancourt et al. 1990).

It turns out that desiccated urine is a good preservative; indeed, it is called *amberat* because of its similarity to fossil tree sap, or amber. And like amber, it can potentially preserve materials for tens of thousands of years, even fragile molecules like DNA (Hornsby et al., in prep.). Over time, the material consolidates into a hard, indurated or cemented asphalt-like block (Fig. 8.12). At this point, researchers often call refer to them as paleomiddens to differentiate these late Quaternary archives from human middens or other archeological collections. Because indurated

plant fragments
from indurated
midden

rocky crevice containing a
paleomidden (Grapevine
Mountains, Death Valley)

intrepid
graduate
student
shown for
scale (~6 ft)

Fig. 8.12. Woodrat paleomiddens in Death Valley. **a,** This dense midden with macrofossils exposed is covered with a layer of amberat, which is desiccated urine. **b,** Rocky crevice in the Grapevine Mountains where a midden was extracted; note the graduate student, Larisa Harding, shown for scale. Photos by author; drawing of bushy-tailed woodrat by Sue Simpson for the author.

paleomiddens are mostly composed of fecal pellets (or dung) containing the remains of undigestible plant material, they are not very tasty or nutritious—as the ill-fated '49ers found out when they became lost in Death Valley on their way to the gold fields of California. However, their documentary quality *is* quite high. So once again, animal dung can yield important information about past paleoecology.

The oldest paleomiddens (>40,000-50,000 years old) push the ability of radiocarbon dating to quantify, although most are Holocene in age (Betancourt et al. 1990). Some locations have exceptional preservation, especially when in arid desert environments. For example, my lab has worked extensively in Death Valley, where we have recovered over 120 paleomiddens from a transect along Titus Canyon on the east side of the valley—many of these well over 30,000 years old (Smith et al. 2009, 2014). This is a time when large-bodied mammals such as woolly mammoth, mastodons, horses, and camels still roamed across North America. Woodrats created

these debris piles in caves that might have well also housed a saber-tooth cat or some other prehistoric creature. Thus, woodrats have unintentionally become nature's furry historians. The unique historical records recorded by paleomiddens provides a valuable and unique insight into the local ecology and evolutionary history of ancient plant and animal communities.

Paleomidden Processing

Most paleomiddens are firmly attached to the walls or floors of rock crevices, often limestones (Betancourt et al. 1990). Because exposure to water results in dissolution of the amberat, older middens are fairly uncommon in areas with high humidity, such as the Pacific Northwest, or where rock has exfoliated. Ancient indurated middens are removed from the rock substrate using a geological pick and hammer, which can take considerable effort depending on the size and orientation of the deposit (Fig. 8.13a and Fig. 8.13b). We have occasionally spent

Fig. 8.13a. Hunting for woodrat paleomiddens along Titus Canyon, Death Valley National Park. My lab worked at this site for several years and recovered more than one hundred middens, about half of which were Pleistocene in age. **a**, Upper Titus Canyon. Virtually every crevice visible yielded a packrat midden, although some were hard to access. **b**, Former graduate student Ian Murray recording data while I was inside the cave hammering out the paleomidden. **c**, Author with former graduate student Ian Murray assessing a "messy" midden. This was difficult to sample because it had been reworked by generations of woodrats. **d**, Former graduate student Larisa Harding pointing to an Indian basket that had been collected by a woodrat and reworked into its midden. We found several such artifacts along this transect; all were left in situ, and the National Park Service was notified of their whereabouts. Such middens were obviously Holocene in age. All photos by author.

hours in a twisted position trying to hammer out a particularly refractory midden. Particular care needs to occur to make sure that discrete layers, potentially representing different depositional intervals, are not mixed when they are collected. Once the midden is removed, it is cleaned in the field to remove any midden rind—that is, areas on the surface that have potentially rehydrated and could be contaminated with more modern material. And of course, collection sites are georeferenced and permanently marked with metal tags, and topography, aspect ratio, relief, and nearby modern vegetation are recorded.

Back in the lab, processing of the paleomidden is relatively straightforward. The field sample is broken into several pieces; these include an aliquot archived for pollen analysis and another nonprocessed portion

retained as a voucher specimen. The remaining paleomidden is then disassociated by soaking it in a four-liter bucket of tap water for about a week. Lab studies have demonstrated that soaking does not significantly alter the nature of any of the materials present in the midden. Interestingly, the same amberat that preserved the contents as an asphalt-like chunk for many thousands of years quickly rehydrates into urine when exposed to moisture. Sampling and processing methods of middens is generally similar for different researchers (Spaulding et al. 1990), though there can be important variations, particularly in terms of the size of the collected sample. To some extent, this reflects the distance materials need to be carried after collection; when field vehicles are close, samples can be considerably larger.

Fig. 8.13b. Three massive middens, all of which filled entire caves. **e**, Examining a midden in Titus Canyon, Death Valley National Park. **f**, A huge Pleistocene midden also from Death Valley National Park that was exposed by the exfoliation of the surrounding rock; note the debris accumulating as the midden weathers. **g**, A massive, active woodrat debris pile, which covers older, indurated middens. While valuable information is contained within this cave, the possibility of mixing or reworking makes it difficult to sample. All photos by author.

After soaking, the bucket of organic material is power-washed and wet-sieved to remove nonorganic rock fragments from the disassociated midden matrix. Finally, the remaining biotic matter is dried for a few days in a forced-draft drying oven at approximately 60°C. At this point, all organic material is dry-sieved through a series of standard stainless-steel mesh geological sieves and subsequently hand-sorted to remove fossil pellets, plant macrofossils, and fossil teeth and bones from other matrix materials.

The age of the recovered midden is determined by direct radiocarbon dating of a sample of the organic fraction (chapter 4) using a tandem accelerator mass spectrometer. Generally, an outside lab does this, and preprocessing to a graphite target may or may not occur depending on the capabilities of the particular research group. Unfortunately, the high cost associated with obtaining radiocarbon dates means that many recovered middens are not dated, especially if they appear to contain the same vegetation types and/or abundance as nearby samples. Radiocarbon ages are converted to calendar years (years before 1950 AD) using a calibration curve such as the Cal-Pal program (www.calpal-online.de; see chapter 4).

The taphonomy of midden formation is complicated. While some middens are large and complex and clearly have had many generations of woodrats contributing to them, others are much more homogeneous and may have been formed rather quickly. Thus, midden deposition can be slow or fast, or even episodic. In 2006, my students and I found a midden in Titus Canyon in Death Valley that was well over

4 m in height. We rather colorfully dubbed this behemoth "Oozing Midden" because it was so huge and complex; it was a challenge to decide how to properly sample. We ended up removing and dating many different layers, which yielded radiocarbon dates more than 8,000 years apart. Yet, subsampling of other middens collected within Titus Canyon yielded dates within a few hundred years of each other (Smith et al. 2009). While the exact duration of a depositional episode cannot be easily resolved, validation studies on modern midden formation suggest multiple generations (~10-50) probably contribute in most instances (Smith and Betancourt 2006).

What happens next depends on the research questions being addressed. For example, there has been much interest in characterizing changes in the structure and composition of plant communities from the late Pleistocene to present; such work provides a window into the wholesale environmental perturbations that followed the retreat of glacial sheets at the terminal Pleistocene. This is straightforward to examine using the plant macrofossils contained within paleomiddens. Typically, aliquots of recovered plant fragments are identified under a dissecting scope using a reference collection of plant materials for the geographic area. In many cases, it is possible to identify the fragments down to the particular species of plant. However, for some taxa this degree of accuracy is difficult and macrofossils are identified only to the genus level (Fig. 8.14). Rarefaction techniques allow the determination of the minimum sampling effort necessary to accurately detect presence and proportion of common and rare macrofossils. Counts are converted and reported as relative abundance measures, which standardize for unequal sampling efforts. More details of plant macrofossil analyses are discussed in the bible of packrat middens by Betancourt et al. (1990). Other analyses on vegetation include characterizing stable isotope signatures of the vegetation (see chapter 6) or changes in the stomatal density of leaves, which is an adaptive response to low water availability (Van de Water et al. 1994). Similarly, it is possible to quantify and identify the insects within middens (Ashworth 1973; Elias 1990).

Recent studies have demonstrated that it is also possible to extract well-preserved aDNA from the plants, insects, or animal remains captured within the paleomiddens (Hornsby et al, in prep). Woodrat amberat significantly slows the degradation of DNA of materials encased within middens, although extracting and piecing together the fragmented strands remains tricky (Hornsby et al. in prep). Researchers have been able to document species turnover at a fine temporal scale using aDNA and have even been able to characterize the role of both abiotic and biotic factors in modulating the geographic range. However, to date, few studies have employed packrat paleomiddens as a source of aDNA (but see Hornsby et al, in prep). In the future, these sorts of studies may allow the characterization of the expansion and retraction of woodrat lineages through time. And, as my lab has demonstrated, the bones or even fecal remains within paleomiddens can be directly employed to determine the diet and/or body mass of ancient populations of woodrats (Smith et al. 1995; Smith and Betancourt 1998, 2003, 2006; Smith et al. 2009).

What's Hidden in a Midden?

The plant macrofossils collected by woodrats and preserved within their middens have allowed the documentation of changes in plant communities within the Southwest United States over the past 30,000 years (e.g., Wells and Jorgensen 1964; Wells 1966, 1976; Wells and Berger 1967; Van Devender 1977, 1986; Betancourt and Van Devender 1981; Van Devender et al. 1985; Betancourt et al. 1990; Van de Water et al. 1994; Lyford et al. 2003; Jackson et al. 2005). The first studies using middens were able to trace the movement, and subsequent extirpation, of juniper over the Great Basin Desert (Wells and Jorgensen 1964). But midden studies have also quantified the spread of ponderosa and piñon juniper forests across the Southwest in response to the changing climate of the late Quaternary. And such work has documented the migration of what we now consider to be the iconic plant of the Great Basin Desert—creosote—into North America from its origins in South America. Indeed, our work in Death Valley quantifies the late Pleistocene extirpation of juniper and the movement of creosote into Death Valley in the early Holocene in response to changing environmental conditions (Fig. 8.14). Much (although

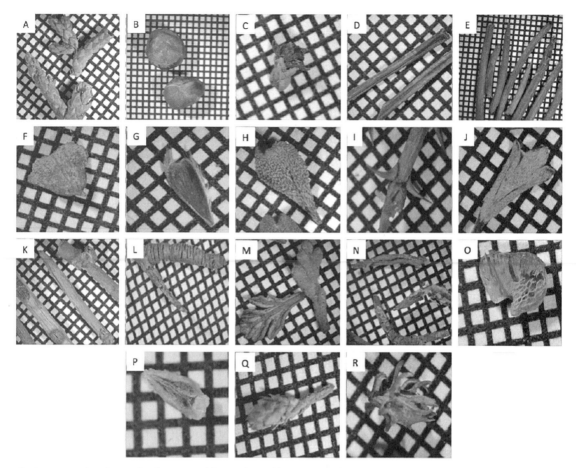

Fig. 8.14. Examples of macrofossils recovered from paleomiddens recovered in Titus Canyon, Death Valley National Park. Identifications are made by comparing fragments under a dissecting scope with known reference plant samples from the surrounding area. Background grids are 1 cm by 1 cm. Quantifying the amount of each within aliquots can assess relative abundances. Species as follows: (A) *Juniperus osteosperma* twigs, (B) *J. osteosperma* fruit, (C) *J. californica*?, (D) *Pinus flexilis*, (E) *Pinus monophylla*, (F) *Atriplex* sp., (G) *Encelia* sp., (H) *Ericameria* sp., (I) *Ribes* sp., (J) *Artemisia* sp., (K) *Ephedra* sp., (L) *Cercocarpus* sp., (M) *Purshia* sp., (N) *Eriogonum* sp., (O) *Sphaeralcea* sp., (P) *Cryptantha* sp., (Q) *Peucephyllum schottii*, and (R) *Ambrosia* sp.

not nearly all) of the raw data collected by scientists has been deposited into the USGS/NOAA North American Packrat Midden Database (https://geochange.er.usgs.gov/midden).

While most workers have focused on the abundant plant macrofossils middens contain, many other objects are found and studied as well—rocks, archaeological artifacts, insects (e.g., Elias 1990), bones, and especially woodrat fecal pellets. The latter are by far the most numerous constituent, but in the past they were generally discarded or, at best, used to date the material. However, in 1995, we published a paper that demonstrated that fossilized

fecal pellets could be used to characterize the body size of the woodrat that produced them; a lab study on captive animals had demonstrated that pellet width was significantly related to body mass (Fig. 8.15; Smith et al. 1995, Smith and Betancourt 2006). Further validation studies with modern woodrats confirmed that neither diet, habitat quality, nor other environmental influences appreciably influenced the pellet width to body size relationship (Smith et al. 1995; Smith and Betancourt 2006). For example, we blind-measured pellets from different species of woodrats trapped in different habitats, seasons, and locations and compared estimated to actual field

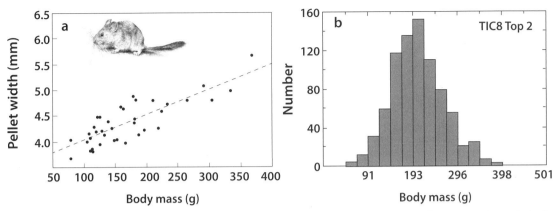

Fig. 8.15. Woodrat pellet width is a proxy for body mass. **a,** Measuring the width of fossil pellets allows us to estimate the body mass of the individual that produced it; this relationship between pellet width and body mass is not influenced appreciably by gender, species, or diet (Smith et al. 1995). **b,** Pellets within a typical paleomidden sample are normally distributed. We estimate that it takes around twenty years for a midden to be formed, thus generations of animals contribute to each recovered midden. The particular midden shown here was middle Holocene in age (about 3,700 years old). Both aDNA and statistical analysis of the pellets confirmed it was constructed by the desert woodrat, *Neotoma lepida*. Bushy-tailed woodrat by Sue Simpson for the author.

masses. Our values for percent predicted error (%PE; the difference between the predicted and actual mass divided by the predicted mass) were quite good: less than 21%PE for pellets 4 mm or larger. The %PE is a comparative index of predictive accuracy often used in paleontology (Van Valkenburgh 1990). The typical values found in the literature are about 18%PE–55%PE for various molar and bone measures. Thus, fossilized woodrat fecal pellets yield a better measure of body size than most cranial or postcranial elements.

Because discretely radiocarbon-dated middens often contain thousands of pellets, and a mountain range may contain dozens of distinct middens, we demonstrated that midden sequences could be employed to investigate population-level body size responses to late Quaternary climatic change across the western United States (Smith et al. 1995, 2009; Smith and Betancourt 1998, 2003, 2006; Balk et al. 2019; Figs. 5.5, 5.7; Fig. 8.16). Moreover, successive generations of woodrats often occupy the same cave and contribute to stratified middens, which also allows the characterization of chronological changes *within* a population.

Woodrats as Tools for Understanding the Response of Mammals to Climate Change

As discussed in chapter 5, woodrats are the poster child for Bergmann's rule. A robust relationship

exists between adult body mass and ambient environmental temperature, with smaller individuals found in hotter environments. This pattern holds at all scales: from individuals, populations, subspecies, and even species within the genus (Lee 1963; Brown 1968; Brown and Lee 1969; Smith et al. 1995, 1998; Smith and Betancourt 1998, 2003, 2006; Smith and Charnov 2001). We suspect that the underlying mechanism is probably a direct physiological response; lethal upper temperature scales as an inverse function of body mass (Brown 1968; Smith et al. 1995; Smith and Charnov 2001).

Our work—to date, my lab is the only one that has employed packrat middens in this fashion— illustrates the various ways that climate change influenced woodrats over the late Quaternary (see Fig. 5.7). Each woodrat species we have studied demonstrates remarkable resilience to climate shifts, often adapting largely in situ. In short, woodrats responded as predicted from Bergmann's rule: they were larger-bodied during colder periods and smaller-bodied during warmer conditions (Fig. 8.16). Thus, across the western United States, woodrat population body size decreased during the Pleistocene to Holocene transition as climate warmed, increased during the Younger Dryas cold episode, decreased during the warm conditions of the middle

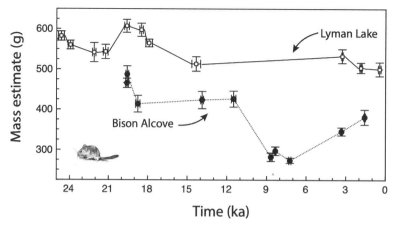

Fig. 8.16. Bushy-tailed woodrat (*Neotoma cinerea*) body size over time. These sequences represent population of woodrats living at a high elevation site near Lyman Lake, Arizona, (closed circles) and a lower elevation site near Bison Alcove, in Arches National Park, Utah, (open squares) over the past 25,000 years. Note the geographic variation in body size between these two sites; the larger animals occupy cooler environments. Both populations fluctuated in body mass over time, with shifts corresponding to climate changes. Redrawn from Smith and Betancourt 2003, 2006. Bushy-tailed woodrat by Sue Simpson for the author.

Holocene, and increased again during the Little Ice Age (Smith et al. 1995, 2009; Smith and Betancourt 1998, 2003, 2006; Balk et al. 2019). Only in sites at the margins of their geographic or elevational range were populations extirpated (Smith and Betancourt 2003, 2006; Balk et al. 2019); sometimes this was accompanied by a species replacement of a smaller- or larger-bodied species depending on the direction of temperature shift (Smith and Betancourt 2003, 2006). Moreover, within species there is remarkable congruence in the magnitude of the responses to the environmental perturbations of the late Quaternary; sites located closely together display similar morphological responses to climatic variation (Smith and Betancourt 2003). Regressions using various independent temperature proxies (tree rings, lakes, or ice cores, etc.) clearly demonstrate that temperature changes were the proximate driving force behind these morphological shifts (Smith et al. 1995, 1998, 2009; Smith and Betancourt 1998, 2003, 2006; Smith and Charnov 2001; Balk et al. 2019). We discuss the use of the woodrat paleomidden record in climate change studies further in chapter 11.

Owl Pellets

In addition to ancient dung, animal vomit is also a useful paleoecology proxy—at least when produced by birds of prey, or raptors. Such undigested and regurgitated food is typically referred to as a pellet.

Birds lack teeth. Thus, raptors use their feet, talons, and beaks to tear apart small mammals, birds, or reptiles into smaller pieces, which are then swallowed, along with their bones, feathers, teeth, fur, and/or scales. Because much of this is inedible, these parts need to come back up. Hence, indigestible bits are regurgitated as pellets at the roost, which in turn concentrates prey skeletal material in one area (Terry 2004). While all raptors—the Falconiformes (diurnal raptors, such as eagles, hawks, and falcons) and the Strigiformes (owls)—regurgitate pellets containing the indigestible bits of their prey, those produced by owls tend to contain much more bone than other raptors. This is because owls swallow most prey whole, although the head may first be removed for larger prey items such as lagomorphs (Andrews 1990; Yalden 2009). Indeed, owl pellets may contain up to 40%-50% bone, compared to the 5%-10% found for diurnal predators (Andrews 1990).

Pellets can readily be dissected and their contents analyzed to yield insights into the birds diet. Indeed, this has often been done in public schools to teach about food webs; hence, the success of the book *Owl Puke* (Hammerslough 2004) and others like it. And it explains the ready availability of owl pellets for sale on the internet: a small owl pellet currently sells for just over $1.00, although if you want one with verified mammal remains, the price jumps to $6.00 or more. However, the analysis of owl pellets is more than an engaging elementary or secondary school science activity. It is rather an important contributor to our late Quaternary fossil record.

Processing and Analysis of Owl Pellets

The analysis of owl pellets is relatively simple. Generally, intact pellets are disassociated by soaking them in water, or water and a diluted hydrogen peroxide solution (Yalden 2009; Terry 2010). They also may be manually separated after drying, using forceps. At that point the skeletal elements are sorted based on size and type and macroscopically identified to species and age class using a reference collection established for the area. Counting each unique element and orientation (i.e., left humeri, right femora) gives researchers a conservative determination of the minimum number of individuals for each stratum found. The percentage volume (i.e., a prey item's percentage of contribution to overall volume ingested), percentage occurrence (i.e., percentage of the number of pellets where a particular prey item is found), and importance value (i.e., percentage volume divided by percentage occurrence) can also be computed. Typically, species composition, evenness, and relative abundance measures can be obtained for each discreetly dated level (Yalden 2009).

While the actual processing of intact owl pellets is straightforward, the collection and interpretation of remains is more complicated (Andrews 1990; Avenant 2005; Yalden 2009). How they are deposited, the degree of disarticulation, and the extent of time averaging all depend on local conditions. For example, as pellets degrade over time they release skeletal material. In the best circumstances, this results in a layered buildup of skeletal remains, which can be constrained temporally. However, this doesn't always occur, especially if the roost deposits are disturbed by the activities of other animals. Moreover, the extent of disintegration of owl pellets is dependent not only on age but also temperature and moisture.

In general, validation studies confirm good agreement between the actual ecological community present as determined by livetrapping and that constructed based solely on owl pellet analysis (Terry 2004, 2010; Avenant 2005). Death assemblages do closely reflect the species composition and abundance patterns of the nearby living small mammal community and are not overly influenced by the dietary preferences of the predator (Andrews 1990; Avenant 2005; Terry 2010). This is true even of historical or late Holocene sites, although as expected, species richness increases with the degree of time averaging present (Terry 2010). However, there are some important caveats. Larger-bodied prey, such as lagomorphs, are too big for some owls and/or too big to be consumed whole (Andrews 1990; Yom-Tov and Wool 1997; Yalden 2009). Thus, generally, they are not present in high abundance in the record (but see Yom-Tov and Wool 1997). Additionally, since owls are nocturnal, most recovered prey within pellets are as well. This means that diurnal small mammals such as squirrels or chipmunks are not well represented in owl pellets (Andrews 1990; Yalden 2009).

Reconstruction of Ancient Environments

Scientists have long employed pellets to investigate the predatory habits of birds and to characterize seasonal or historical changes in the abundance and composition of small mammals within ecosystems (e.g., Davis 1909; Spiker 1933, Andrews 1990; Terry 2004, 2010; Yalden 2009; Terry and Rowe 2015). Indeed, Oliver "Paynie" Pearson, a pioneering mammalogist from the University of Berkeley Museum of Vertebrate Zoology, allegedly claimed that owls were his best field assistants when he was conducting field work in Patagonia. This was because they offered a longer-term and more complete window into the abundance and species composition of the area than did his trapline surveys.

Moreover, small mammals can be more diagnostic for paleohabitat reconstruction than larger-bodied ones because they respond to the environment in a more fine-grained fashion (Brown and Nicoletto 1991). However, their copious fossil accumulations in caves can be produced in many ways, which complicates ecological interpretations (Andrews 1990). Thus, cave sites with owl roosts, especially those with ongoing deposition, provide an excellent opportunity to employ standardized taphonomy to examine shifts in the historic or paleo-environment over time (Andrews 1990).

Owl pellet analysis can also be useful in conservation paleoecology, particularly in developing an appropriate baseline for management efforts. For example, work at the late Quaternary Homestead Cave site, near the Great Salt Lake in Utah, has uncovered a near continuous owl pellet fossil record that extends approximately 13,000 years (Terry and Rowe 2015). It consists of eighteen strata with over 180,000 identified skeletal elements to date. Terry and Rowe (2015) employed this fossil record to characterize the dynamics of small mammal communities in the Great Basin over time and to establish a baseline with which to compare historic anthropogenic effects in this area. They found that the overall energy flow within the small mammal community was relatively constant over the Pleistocene and into the Holocene. A decline in the abundance of large-bodied rodents as climate warmed was compensated for by an increase in medium-sized small mammals (Terry and Rowe 2015). However, the replacement of shrublands by invasive annual grasses around 150 years ago drastically changed both the energy flow and average body size within the community. Thus, the modern community is functionally distinct from the long-term average. Because small mammals play a critical role in desert ecosystems, the authors speculate this means an unraveling of the ecological function of this habitat.

Ancient DNA (aDNA) in Paleoecology

After some initial hiccups—including a series of ever-more sensational aDNA discoveries that embarrassingly turned out to be modern human contaminants and not actually widely reported dinosaur remains (e.g., Woodward et al. 1994; Zischler et al. 1995; Pääbo et al. 2004; Willerslev and Cooper 2005)—the use of genetic analysis in paleoecology has blossomed. This is because of a fortuitous combination of rapid advances in technology, including next-generation sequencing, the development of better DNA extraction techniques, as well as greatly reduced costs; sequencing is now ten to one hundred thousand times cheaper than it was a few decades ago. Concomitant with these technological innovations was the development of a series of

"authenticity criteria" among scientists, which set standards and improved the quality of the science (Cooper and Poinar 2000; Pääbo et al. 2004; Nicholls 2005; Willerslev and Cooper 2005). While most genetic studies target how species and populations evolve over time and the causative factors underlying such changes, the recent development of environmental DNA techniques potentially allows the characterization of large-scale biodiversity patterns over time (Thomsen and Willerslev 2015).

DNA and Fossils

DNA (deoxyribonucleic acid) is the hereditary material for most life on Earth. While most DNA is found within the nucleus, some is found in organelles, such as mitochondria (mtDNA) or, in the case of plants, choloropasts (cpDNA). The sequence of chemical bases of DNA, called nucleotides, codes the information necessary for building and maintaining organisms. Both normal metabolic activity and the environment constantly cause DNA damage, which means that repair mechanisms are constantly active (Lindahl 1993; Pääbo et al. 2004). After the death of an animal, when enzymatic repair no longer occurs, DNA degrades rapidly and there is an irreversible loss of nucleotide sequence information. Postmortem decay is accelerated by microorganisms (Lindahl 1993).

How quickly does this happen? Can fossils be a reasonable source of DNA for analysis? Using a temporally calibrated fossil assemblage normalized for ambient temperature, Allentoft et al. (2012) discovered that DNA disintegration in fossil bones is best fit by first-order decay kinetics. Moreover, taphonomic processes such as pH and temperature and bone diagenesis also influence postmortem DNA decay. For example, while the average half-life of a 500 base pair mtDNA sequence was estimated to be about 9,500 years at $-5°C$; it was 1,200 years at 5°C, 180 years at 15°C, and only 30 years at 25°C (Allentoft et al. 2012). Nuclear DNA degrades even faster—about twice as rapidly—as mtDNA (Allentoft et al. 2012). Thus, long-term preservation of DNA in the fossil record is relatively rare and requires exceptional conditions, such as freezing or rapid desiccation (Lindahl 1993; Pääbo et al. 2004; Nicholls 2005). Despite earlier wild claims (e.g., Woodward et al. 1994; Zischler et al. 1995), the oldest substantiated fossils still containing measurable DNA are about 400,000 years old; most are much younger (Willerslev et al. 2003). Often these come from biological remains found frozen in permafrost. Thus, aDNA is most useful when examining late Quaternary fossils. Fortunately, this turns out to be a particularly important time period for understanding evolutionary dynamics and responses to environmental change.

Analysis of aDNA

The basic steps in aDNA analyses involve: (1) the extraction of DNA, (2) cloning and amplification, (3) sequencing, and (4) analysis. While this list is straightforward, the steps are anything but that. For example, extraction is often quite difficult: nuclear DNA is often not well preserved, and when present, it is generally in low copy number. Thus, mitochondrial DNA, which is much more common in cells, has often been the focus of studies, although even this too can be difficult to obtain from fossils. And, of course, the maternal inheritance of mtDNA complicates the analytical interpretation of results. Most importantly, DNA extracted from fossil specimens

tends to be truncated into short segments of less than 300 base pairs, which may be damaged and are, moreover, quite low in number compared to contaminants (Pääbo et al. 2004; Poinar et al. 2006). The actual extraction techniques involve grinding the fossil bone sample into powder, digesting the sample in a lysis buffer to release DNA from the sample, and then purifying and binding the DNA through use of a silica spin column and centrifugation (Rohland and Hofreiter 2007; Rohland et al. 2018).

The cloning, amplification, and sequencing of these tiny fragments of DNA have also been a problem. Indeed, it wasn't until the invention of the polymerase chain reaction (PCR), that it even became possible to routinely amplify the few DNA copies that had survived in ancient specimens (Pääbo et al. 2004). PCR became possible after a heat resistant enzyme (*Taq* DNA polymerase) was isolated from thermophilic bacterium found in the Lower Geyser Basin of Yellowstone National Park (Fig. 8.17). PCR is a sort of molecular photocopy that uses primers— complementary sequences to the segment of DNA being targeted, free nucleotides, and *Taq* polymerase—

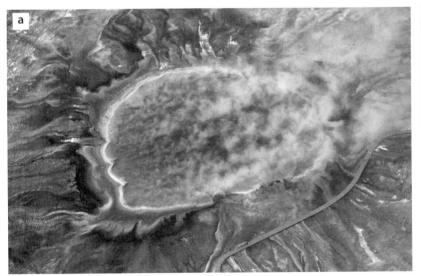
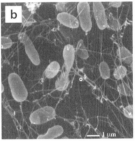
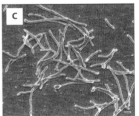

Fig. 8.17. Thermophilic bacteria are found in the hot springs of Yellowstone National Park. **a,** Aerial view of Grand Prismatic Spring in Yellowstone National Park. The spring is approximately 75 m wide by 91 m long. Note the steam rising from the extremely hot water; the center is surrounded by huge mats of brilliant orange algae, bacteria, and archaea. **b,** The famous thermophilic bacterium *Thermus aquaticus*, whose discovery led to the development of *Taq* and rapid advances in PCR. **c,** Thermophilic bacteria are also found in deep-sea vent fluids. Panel a: Photo by Jim Peaco, National Park Service, courtesy of Wikimedia Commons. b: Photo by Mark Amend, NOAA Photo Library, courtesy of Wikimedia Commons. c: Photo by Diane Montpetit, Food Research and Development Centre, Agriculture and Agri-Food Canada, courtesy of Wikimedia Commons.

as reagents. Briefly, the sample is heated so the DNA separates into two pieces of single-stranded DNA, and as they are cooled, the primers bind to the appropriate sites and the enzyme then synthesizes two new strands using the original as a template; each new molecule thus contains one old and one new strand of DNA. This cycle of denaturing and synthesizing new DNA is repeated more than thirty times, exponentially amplifying the original concentration of DNA. PCR targets DNA sequences from specific genes using primers unique to the group under investigation. The development of primers specific enough can be a roadblock to successful results, particularly if studying an extinct species with no close modern relatives (Woods et al. 2017). Often the mtDNA cytochrome *b* gene is used when looking at phylogenetic relationships within mammals (Irwin et al. 1991; Johns and Avise 1998). Its high sequence variability makes it particularly useful for comparison of species in the same genus or family.

The enormous amplifying power of PCR also increases the concentration of contaminants present, which leads to problems with the reliability of sequences obtained (Pääbo et al. 2004). This has led researchers to propose a set of best practices, including dedicated laboratories for aDNA work that are physically separated from buildings where contemporary DNA analyses are performed and, for particularly important discoveries, confirmation by reproduction of the analysis and results in a second, independent laboratory (Pääbo et al. 2004; Willerslev and Cooper 2005). Contamination problems have been compounded with newer, less specific sequencing techniques.

New Methodologies

Over the past several decades, enormous effort and money has led to the development of a wide variety of new methods. For example, recent breakthroughs in rapid, large-scale next-generation sequencing has transformed molecular analyses, and the field of aDNA in particular (Knapp and Hofreiter 2010; Woods et al. 2017). These techniques sequence millions of small fragments of DNA in parallel and then rely on sophisticated bioinformatics analyses to reconstruct the fragments (Knapp and Hofreiter 2010; Woods et al. 2017). This allows the amalgamation of

enormous amounts of data, yielding much more genetic information than does traditional PCR. For example, a metagenomic approach allowed the sequencing of 28 million base pairs of woolly mammoth (*Mammuthus primigenius*) DNA from Siberia (Poinar et al. 2006), which would have been unthinkable even fifteen years ago. DNA metabarcoding is another recent large-scale method that holds promise. Here, next-generation sequencing of a mass collection of fossil remains is conducted using universal PCR primers. The sequence data generated is then identified to species using a database of existing DNA data (Guimaraes et al. 2016; Woods et al. 2017). Such an approach potentially could be used to do armchair paleontology without having to individually identify fossil remains, assuming, of course, that all potential species were represented in the database.

Another exciting recent development is that of sedimentary ancient DNA (sedaDNA), or environmental DNA (eDNA), which follows the discovery that genetic material of plants, animals, and microbes can be extracted from sediments (Willerslev et al. 2003; Thomsen and Willerslev 2015; Tyler et al. 2019). For mammals, these materials likely come from waste materials (yes, dung is useful once again!), antlers, nails, or skin cells (Lydolph et al. 2005). These techniques may allow the characterization or tracing of community composition over time, somewhat like pollen analysis (see the earlier discussion on *Sporormiella* in this chapter). And, it may reveal the presence of species for which there is no fossil record. For example, analysis of sedaDNA from interior Alaska suggested a longer persistence of horses and mammoths in this region than does the fossil record (Haile et al. 2009). However, some methodological issues plague interpretation of these results, such as the influence of water movement and potential biases in the deposition or survival of DNA (Tyler et al. 2019).

aDNA Studies

The discovery that DNA could be extracted and sequenced from ancient remains promoted the development of a new and exciting field of evolution. The first demonstration of the potential of aDNA studies was the sequencing of 221 base pairs of DNA extracted from a 140-year-old museum specimen of the extinct quagga, a curious partially stripped relative of the plains zebra (Higuchi et al. 1984). This led to a whole host of firsts, such as the sequencing of dodo, woolly mammoth, and even ancient humans (Pääbo et al. 2004). The idea of being able to characterize evolution in action was irresistible.

Since then, ancient DNA has been used to resolve the evolutionary history and biogeography of many extant and extinct mammals (e.g., Debruyne et al. 2008; Wyatt et al. 2008; Chang et al. 2017; Froese et al. 2017; Presslee et al. 2019). For example, the inclusion of aDNA from extinct mastodon and mammoth clarified elephantid genetic relationships and demonstrated the validity of separate African forest and savanna elephant species (Palkopoulou et al. 2018). These two taxa apparently split around 500,000 years ago, but a phylogenetic analysis omitting these two taxa might not have been able to resolve this ancient divergence. And a recent analysis of sloth mitogenomes not only resolved the phylogenetic affiliations of the clade but also confirmed a previously discredited hypothesis of a biogeographic connection between northern South America and the Greater Antilles during the Oligocene (Delsuc et al. 2019).

Some aDNA studies yield information that has significant conservation value. For example, an analysis of late Pleistocene brown (grizzly) bears in Alaska has radically altered our view of their population dynamics and the connectivity of populations (Leonard et al. 2000; Barnes et al. 2002). Today, mtDNA lineages are segregated geographically, but they coexisted around 35,000 years ago. This dispels the idea that brown bear lineages represent distinct subspecies, which are adapted only to local conditions. Interestingly, Barnes et al. (2002) demonstrated that there was little interchange of females between these populations, which has implications for population structure. And studies of North American and Eurasian bison over the late Pleistocene demonstrate that mtDNA genetic

diversity is correlated with periods of climate change (Shapiro et al. 2004).

A recent publication used high-throughput DNA sequencing of fossil collections of bison, bears, and mammoths to demonstrate the surprising finding that fossil collections have a strong male bias (Gower et al. 2019). This was quite unexpected. The authors proposed a number of factors that might drive this result. For example, this bias might arise as an artifact of taphonomy if male bones are larger and denser and thus better preserved in the fossil record. Or, the greater home range of male animals compared to that of smaller-bodied females might increase the probability of fossil preservation. This latter interpretation was supported by a latitudinal difference in the bias of collections; the male bias in collections decreased at more northernly latitudes where female bear ranges were larger (Gower et al. 2019). Nonetheless, these findings suggest that sexual differences in preservation potential may complicate the interpretation of fossils.

FURTHER READING

Bell, A. 1983. *Dung Fungi: An Illustrated Guide to Coprophilous Fungi in New Zealand*. Wellington: Victoria University Press.

Betancourt, J.L., et al. 1990. *Packrat Middens: The Last 40,000 Years of Biotic Change.* Tucson: University of Arizona Press.

Bromley, R.G. 1996. *Trace Fossils: Biology, Taphonomy, and Applications*. London: Chapman and Hall.

Davis, M.B. 1963. On the theory of pollen analysis. *American Journal of Science* 261:897-912.

Smith, F.A., and J.L. Betancourt. 2006. Predicting woodrat (*Neotoma*) responses to anthropogenic warming from studies of the palaeomidden record. *Journal of Biogeography* 33:2061-2076.

Willerslev, E., and A. Cooper 2005. Ancient DNA. *Proceeding of the Royal Society B* 272:3-16.

Woods, R., et al. 2017. The small and the dead: a review of ancient DNA studies analyzing micromammal species. *Genes* 8:312-326.

9 Reconstructing Past Climate

Many of us routinely check the weather report in the morning as we are getting ready for school or work. And, as we often find out later, the weather report can be wrong. Sometimes very wrong. This is especially true in places like Santa Fe in northern New Mexico, where the high elevation and surrounding mountains mean there is significant spatial and temporal variability in weather conditions. Indeed, the quip Mark Twain reputedly made, *"If you don't like the weather now, just wait a few minutes,"* is as apt for New Mexico as it was for New England. In winter, I keep an extra down coat in the car just in case.

Predicting weather is tricky because it is the day-to-day state of the atmosphere—the temperature, humidity, cloudiness, wind, and precipitation. Climate is easier because it is the weather *averaged* over time. While forecasters might not be able to tell me if the high today will be 2°C or 8°C with a high level of precision, they can tell me that the average for January 2 in Santa Fe is 5.8°C (elevation ~2,150m; longitude: −105.975, latitude: 35.6194; https://www.usclimatedata.com/climate/santa-fe/new-mexico/united-states/usnm0292). The World Meteorological Organization, an agency of the United Nations, is the scientific body responsible for managing these modern climate records (https://public.wmo.int/en).

But, imagine that you wanted to know about climate in the past? How does one go about reconstructing past climate or climatic fluctuations in Earth history? After all, to understand the ecology and evolution of long-dead mammals, we need to understand something about the environment in which they lived. This includes an understanding of the climate of the time. Fortunately, it turns out that there are quite a lot of historical and paleontological proxies and records; the choice of which to use depends on the temporal interval of interest (Fig. 9.1). Here we review these, and in chapter 10 we employ them to discuss mammalian responses to climate change.

The Historical Record of Climate

For the last 170 years, we can use the instrument record (Brohan et al. 2006) to determine past climate. This is derived from piecing together the historical record of in situ measurements made by numerous meteorological stations, weather balloons, buoys, ships, and other sources. We have sufficient observation stations to recreate global temperature from about 1856 on (National Research Council 2006). For some regions, however, the instrumental record extends centuries; the longest is the Central England temperature series, which began in 1659 (Fig. 9.2). This unique record contains monthly (and daily after 1772) mean surface air temperatures for the Midlands region of England (Manley 1974; Parker et al. 1992). Thus, we have a unique direct window into near-time climatic events for a particular region. These

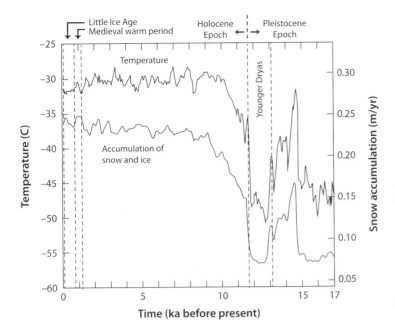

Fig. 9.1. Paleoclimate over the Pleistocene and Holocene.

include the Little Ice Age—an extended period of mountain glacier expansion in the Northern Hemisphere over the sixteenth through nineteenth centuries—and 1816, the year without a summer (Mann 2003). The latter was caused by the massive eruption of Mount Tambora in what is now Indonesia that injected vast amounts of dust and aerosols into the atmosphere (Oppenheimer 2003).

Historical climatic events can also be recreated using a number of other documentary techniques, such as the comparison of oil paintings of glaciers in the French and Swiss Alps with modern photographs, logs, or diaries of when harvests began or trees blossomed, and even the sampling of ancient gas trapped within fishing floats (Pfister et al. 1999; National Research Council 2006; Fagan 2007; Mann et al. 2008; Brázdil et al. 2010). For example, Benjamin Franklin kept a detailed diary when he was the American ambassador to Paris in the 1780s. He wrote about the weather, including a description of a constant and dry fog on which *"the rays of the sun seem'd to have little effect"* ("Meteorological Imaginations and Conjectures, [May 1784]," *Founders Online*, National Archives, https://founders.archives.gov /documents/Franklin/01-42-02-0184). This weather pattern has since been attributed to a volcanic eruption in Iceland (National Research Council

2006). Similarly, the onset of spring flowering of cherry trees (*Prunus jamasakura*), a major cultural event, has been recorded in Korea and Japan since the fourteenth and ninth centuries, respectively (Aono and Omoto 1993; Aono and Kazui 2008). Because the flowering dates of cherry trees are closely related to the monthly mean temperature, the phenological record from diaries kept by emperors, aristocrats, monks, and merchants provides a temperature chronology for this region.

However, most historical documentary methods are generally limited to the past few hundred years; they also tend to be discontinuous in time (National Research Council 2006; Mann et al. 2008).

Temperature Proxies

Thus, the historical record takes us only so far. To examine the climate of the late Quaternary or earlier in the Cenozoic, we need other proxies. Fortunately, these exist, although they vary in their spatial and temporal resolution (Fig. 9.3). The National Oceanic and Atmospheric Administration (NOAA) maintains the National Centers for Environmental Information (NCEI), databases of curated paleoclimate proxies spanning the last few millennia to just over 100 Ma (www.ncdc.noaa.gov/data-access/paleoclimatology -data/datasets).

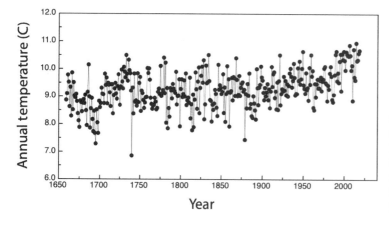

Fig. 9.2. Central England temperature series, which began in 1659. This is the longest instrumental record of monthly temperature for any site on the Earth. Daily temperatures were recorded from 1772 onward. Notice the clear upward trajectory in temperature since 1925.

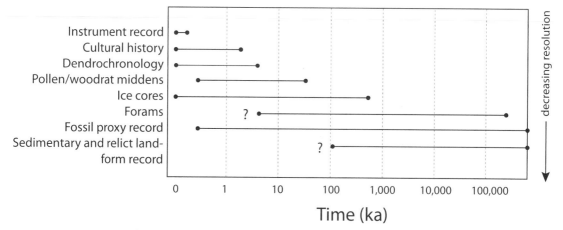

Fig. 9.3. Various climate records or proxies and the time frame over which they provide information. As these proxies extend over longer time frames, they become less precise and temporally constrained.

Dendroclimatology

Dendroclimatology, or tree-ring analysis, is a robust proxy for determining local environmental conditions over the past few thousand years. Trees grow in a predictable way in response to both temperature and moisture, which is reflected in the variation in growth rings or layers of the vascular cambium. If you examine the cross section of a trunk, the thickness of these growth rings reflects local environmental conditions (Martinez 1996; Hughes et al. 2010). Thus, a thick growth ring means that conditions were favorable for growth, whereas a thin one (or in some unusual cases, no ring at all) suggest the opposite. Because the rings near the core of the tree are the oldest and the ones near the bark are the youngest (Fig. 9.4), one can recreate the history of

environmental conditions over time. As seems to be true of so much in science, this pattern was first commented on by the artist and visionary Leonardo da Vinci long ago: "*Li circuli delli rami degli alberi segati mostrano il numero delli suoi anni, e quali furono più umidi o più secchi la maggiore o minore loro grossezza,*" that is, "*The rings around the branches of trees that have been sawn show the number of its years and which [years] were the wetter or drier [according to] the more or less their thickness*" (Da Vinci 1817).

The scientific study of tree rings started with the pioneering work of Andrew Ellicott Douglass. In the early 1900s, Douglass was an astronomer at the Lowell Observatory in Flagstaff, Arizona. He was particularly interested in solar variability and its potential effects on Earth's climate. Faced with a lack

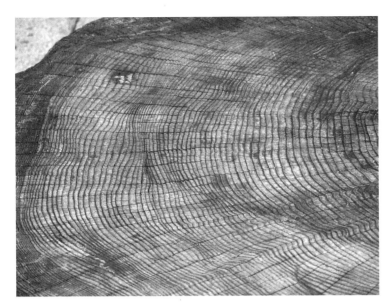

Fig 9.4. The growth rings of a tree. This trunk is about one meter in diameter; note that rings are not uniform in width. Differences reflect variation in growing conditions. Photo courtesy of Adrian Pingstone.

of weather records—and surrounded by forests—he hit upon the idea of using growth layers in the long-lived pines as a proxy for climatic conditions. This became the basis for the new discipline of dendroclimatology (Hughes et al. 2010). Douglass moved to the University of Arizona in Tucson in 1906 and founded the first, and arguably most famous, laboratory of tree-ring research there in 1937 (see Laboratory of Tree-Ring Research at ltrr.arizona.edu).

The mechanics of dendroclimatology are relatively straightforward. A pencil-width core is extracted from the trunk of a living tree using an increment borer. Once removed, a core is generally stored temporarily in a paper straw, where it is allowed to dry. At this point, the core is mounted in a core mount with the wood fibers perpendicular to the field of view and sanded to expose the cell structure of the rings. Analyses of the rings are conducted using a high-powered dissecting microscope. (See Rocky Mountain Tree-Ring Research at www.rmtrr .org/basics.html for more details.)

However, counting tree rings is not quite as straight-forward as it seems. For example, dendroclimatology is most effective in temperate regions where there is seasonality. And its efficacy varies with tree species; Douglas fir (*Pseudotsuga menziesii*) is a preferred species because of its typically well-defined and circular rings (Martinez 1996). Local conditions can lead to the formation of false rings, or even no rings, which can be

hard to interpret (Hughes et al. 2010). Thus, the most common method employed is a cross-dating technique called the skeleton plot. The growth characteristics of cores from multiple trees are collected from an area and matched to build a chronology (Hughes et al. 2010). And it is possible to extend these records backward in time using downed tree samples from old buildings or archeological ruins. NOAA archives paleoclimatic data from tree rings, including the International Tree-Ring Data Bank (www.ncdc.noaa.gov/data-access /paleoclimatology-data/datasets/tree-ring), the largest public archive of tree ring data.

The oldest tree-ring chronologies span the entire Holocene. For example, crossmatching has resulted in a combined oak and pine tree-ring record that extends from the present to approximately 12,460 years ago for Central Europe (Friedrich et al. 2004). This includes particularly useful thermal events for climate research such as the warming and cooling periods of the Younger Dryas. In North America, the oldest chronologies have been built using bristlecone pine, *Pinus longaeva*, from the White Mountains of California (Fig. 9.5; Ferguson and Graybill 1983; Salzer et al. 2019). Bristlecone pine are found at high elevations where they are exceptionally long-lived and slow growing. Indeed, the aptly named Methuselah Walk is a four-mile hiking trail that takes you through the oldest known individual living trees. This includes Methuselah, a bristlecone pine that as

Fig. 9.5. Bristlecone pine. This is a long exposure taken after sunset in the Bristlecone Pine Forest of the White Mountains in California. This particular grove of bristlecones are the oldest living things on the planet, many are over 4,000 years old and the oldest date to more than 4,700 years. Photo by Rick Goldwaser, CC BY 2.0.

of 2020 was more than 4,850 years old according to OldList, a database of ancient trees from Rocky Mountain Tree-Ring Research (www.rmtrr.org/oldlist .htm). Some bristlecone pines have been reported as old as 5,062 years, although these dates have not yet been confirmed.

Using a combination of live bristlecone pine trees and dead/downed wood along the Methuselah Walk, a fully anchored chronology was compiled that goes back more than 8,600 years (Ferguson and Graybill 1983). Recent work linking several "floating chronologies" has extended this record to more than 10,300 years (Salzer et al. 2019). Interestingly, growth rates of bristlecone pine have increased since the mid-nineteenth century, perhaps reflecting shifts in temperature and precipitation (Salzer et al. 2009).

Ice Cores

The gold standard for reconstructing paleoclimate over the Pleistocene are ice cores. Data for these are archived at the World Data Center for Paleoclimatology (www.ncdc.noaa.gov/data-access /paleoclimatology-data/datasets/ice-core). An ice core is just what it sounds like—a vertical column of ice extracted with considerable effort from glaciers, much like an oversized soda straw. As ice builds up in cold climates, it forms distinctive visible layers that follow the law of superposition (chapter 4); the lower layers are older than the upper ones. Moreover, the interconnected air spaces filled with

atmospheric gases get pinched off into small bubbles within the layers as the weight of the upper snow compresses them (Alley 2002). Thus, by extracting a vertical core from a glacier, we can analyze the physical properties and concentrations of the ice and various gases (e.g., CO_2, CH_4, $\delta^{18}O$, etc.) trapped within the air bubbles. The characterization of atmospheric gas compositions over time allows the reconstruction of paleoclimate. For example, recall from chapter 7 that the ratio of ^{18}O to ^{16}O varies in a predictable way with environmental temperature. Dating of the cores is done with a combination of techniques, including counting visible layers; identifying geologic events such as major volcanic eruptions; electrical conductivity; and the use of isotope ratios (Alley 2002; Talalay 2016).

Ice cores have been drilled using mechanical or thermal drills in glaciers and/or ice sheets on virtually all continents (Fig. 9.6; Talalay 2016). After all, 10% of Earth's land surface is covered in ice (Alley 2002). Hand augers are sometimes still used for short (<30 m) sequences, especially when prospecting for good deeper sites. Cores are typically extracted in sections of 1-6 m in length, which means that many repeated drill cycles are necessary for constructing a long sequence. Once removed, the cores must be kept cold—well below freezing (Alley 2002). In the beginning, core analyses were done on site in snow-covered trenches at −20°F; now, samples are generally

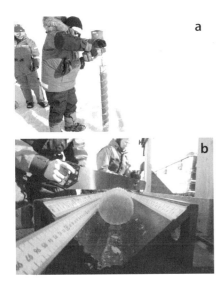

Fig. 9.6. Ice cores. **a**, Hand auger being used by Los Alamos National Laboratory scientist Dr. Scott Elliott near Barrow, Alaska, in 2018. **b**, Core being sectioned. **c**, Ice core storage area at National Ice Core Laboratory in Denver, Colorado. **d**, A section of the GISP2 ice core from 1,837 to 1,838 m deep. The annual layers are clearly visible; these result from differences in the size of snow crystals deposited in winter versus summer and variations in the abundance and size of air bubbles trapped in the ice. Counting layers is how the age of the ice is determined. This section of ice was formed about 16,250 years ago and represents a span of about 38 years. By analyzing the ice and the gases trapped within, scientists are able to learn about past climate conditions. Photos courtesy of United States National Ice Core Laboratory.

quickly transported off site; quick and climate-mediated transport is essential for their integrity (Alley 2000). The National Science Foundation Ice Core Facility (NSF-ICF, icecores.org) in Denver, Colorado, is a physical depository for the storage and curation of meteoric ice cores obtained from glaciated regions of the world (Fig. 9.6).

While shallow cores had been drilled prior to the 1950s, it was only in 1957 or 1958 that the advent of deep drilling efforts began (Jouzel 2013). Working in polar regions was difficult with challenging environmental conditions, but the long-term ice stability in these areas allowed the development of exceptionally well-preserved and detailed climate records (Jouzel 2013). Notable deep cores were recovered from Camp Century, Greenland, and Byrd Station, Antarctica, by the US Army's Cold Regions Research and Engineering Laboratory (Alley 2002). Since then, a number of nations have been involved in coring efforts at the poles, including among others the United States, Soviet Union, Denmark, Norway, Sweden, Netherlands, Switzerland, Australia, and France. For example, international efforts by various consortiums led to the drilling of the Greenland Ice Sheet Project (GISP), Greenland Ice Core Project (GRIP), North Greenland Ice Core Project (NGRIP), and European Project for Ice Coring in Antarctica (EPICA). However, the complexity and high cost of these polar expeditions means that such long-term cores are still fairly limited. The oldest and deepest cores still come from Greenland and Antarctica and extend to more than 3 km in depth. Until recently, the NGRIP core, which stretches back approximately 130,000 years, was the longest from the Northern Hemisphere; the Antarctic cores stretch back an amazing 800,000 years, which covers eight of the approximately twenty-one glacial-interglacial cycles of the Pleistocene (Jouzel 2013).

Amazing as the EPICA core is, it has recently been displaced as the longest ice core record. In 2019, researchers published their description of the world's oldest ice core to date (Yan et al. 2019). This amazing core records over 2.7 million years of Earth history, including the only direct evidence for *all* the glacial-interglacial cycles of the Pleistocene. And it reveals surprising insights. For example, analyses of the gases suggest that the duration and magnitude of ice ages changed significantly around 1 Ma. While glacial-interglacial cycles in the early to middle Pleistocene occurred about every 40,000 years, this changed to about 100,000 years at the middle Pleistocene (Yan et al. 2019).

Pollen Analysis

Pollen analysis, or palynology, can also provide indirect evidence of past environmental temperature (see NCEI's collection of pollen datasets at www.ncdc.noaa.gov/data-access/paleoclimatology-data/datasets/pollen). As with *Sporormiella* (chapter 8), it involves characterizing the composition and abundance of microscopic spores and/or pollen; in this case, the reproductive materials produced by stamens and anthers of seed plants (Prentice 1988). Wind-dispersed plants produce enormous amounts of tiny (15-100 micron) pollen grains and spores (Fig. 9.7), which are released, travel widely, and mix with the atmosphere (Davis 2000). If they happen to rain out in an appropriate sedimentary location, such as a pond, lake, or ocean, they can be buried and preserved for long periods of time (Prentice 1988; Davis 2000). Packrat middens have also been used as a source for pollen analysis in the western United States (Betancourt et al. 1990; see chapter 8).

Because pollen and spores can be identified with varying degrees of taxonomic precision, quantifying the taxa within a pollen core yields the history of local to regional vegetation over time (e.g., Davis 1963, 1969, 2000; Davis ad Shaw 2001; Davis et al. 2005; Grimm et al. 2013). The process is fairly straightforward, although tedious. A pollen sample/core is extracted from the sediment, the pollen and spores are isolated from the matrix through a combination of chemical and physical techniques, cleaned, and then mounted on microscope slides. A relative abundance diagram is generated by microscopically identifying the grains and counting

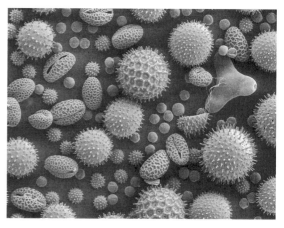

Fig. 9.7. Pollen grains from a variety of common plants: sunflower (*Helianthus annuus*), morning glory (*Ipomoea purpurea*), hollyhock (*Sildalcea malviflora*), lily (*Lilium auratum*), primrose (*Oenothera fruticosa*) and castor bean (*Ricinus communis*). The image is magnified x500; the bean-shaped grain in the bottom left corner is about 50 μm long. Image courtesy of the Dartmouth Electron Microscope Facility.

their abundance (Fig. 9.8). This allows characterization of the changes in vegetation for thousands to millions of years. Because the geographic distribution of plants is strongly mediated by temperature and precipitation, pollen analysis also allows the reconstruction of paleoclimate. This is done by characterizing the overlap in abiotic tolerances of the plants found together, which reflects the environmental conditions at the time the grains were deposited (Davis 2000).

Pollen analysis has been widely employed around the globe to examine the vegetative history of the Quaternary (Grimm et al. 2013). Impressively, palynologists have had a long tradition of making data freely available to both the wider scientific community and the public at large. In addition to numerous pollen archives within the United States, other rich sources of pollen data include Austria's PalDat (Palynological Database, www.paldat.org), the Canadian Pollen Database via University of Ottawa's Laboratory of Paleoclimatology and Climatology (www.lpc.uottawa.ca/data/cpd/) and the European Pollen Database (www.europeanpollendatabase.net).

Fossil Foraminifera and Cenozoic Climate

Thus far, none of the climate proxies we have discussed go back in time more than about 1-2 Ma. Yet, we might well want to know how mammal evolution

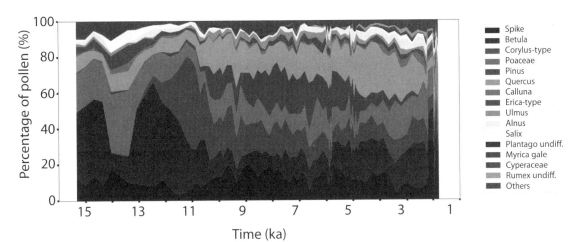

Fig. 9.9. Foraminifera in the Indian Ocean, South Coast, East Bali (field width = 5.5 mm). Foraminifera are a gold mine for reconstructing Cenozoic climate. Image by Alain Couettee, CC BY-SA 3.0.

was influenced by climate earlier in the Cenozoic. To examine climate at this scale—66 Ma—requires a different approach. This is where fossil foraminifera come in.

Forams are single-celled protists with shells or "tests" (internal shells) usually largely made of calcium carbonate ($CaCO_3$) or aragonite (Fig. 9.9). There are about 10,000 species alive today and perhaps another 40,000 fossil taxa (Mikhalevich 2013). Most live, or lived, in the sediments at the bottom of the ocean, although some are planktonic and some may even be

found in soils (Lejzerowicz et al. 2010). Forams have been around since about Cambrian (~540 Ma), and their fossils are abundant throughout the Phanerozoic (Mikhalevich 2013). Although many foram species are tiny (about the size of the period at the end of this sentence), they are among the most abundant shelly species in many marine environments.

Importantly, the chemistry of a foram shell reflects the chemistry of the seawater where it lived and grew. Because temperature influences the isotopic fractionation of oxygen in the oceans—warmer water evaporates off more of the lighter isotope (^{16}O; chapter 7)—shells that grew in warmer conditions have different isotopic signatures than those shells that grew in colder environments (Zachos et al. 2001, 2008). Thus, shells with enriched $\delta^{18}O$ levels were grown when oceans were cooler.

The long duration of forams in the fossil record, their widespread distribution across different spaces and depths, and their high abundance in ecosystems, means they can provide a critical window into variations of stable oxygen isotopic composition over time, space, and depth. Measuring the oxygen isotope ratio in foraminifera found in deep-sea sediment cores has allowed the reconstruction of ocean temperatures across the last 100 million years of Earth history (Fig. 9.10; Zachos et al. 2001, 2008). Moreover, differences in the relative abundance of foram species common to specific environments

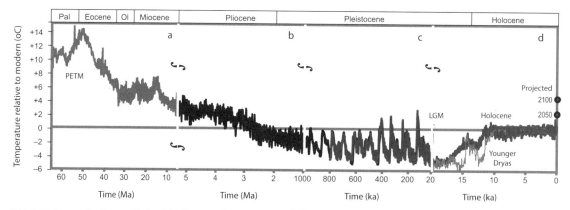

Fig. 9.10. Cenozoic climate. Each of the four panels represents a different temporal scale. **a**, Early to late Cenozoic (65 to 5.3 Ma). Note the high temperatures of the Paleocene-Eocene Thermal Maximum (PETM). Temperature estimates were derived from analysis of oxygen isotopes extracted from microscopic marine organisms (Zachos et al. 2001; Hansen et al. 2013). **b**, Pliocene to Pleistocene (5.3 to 1 Ma). These data are also based on oxygen isotope analysis of microscopic marine organisms (Hansen et al. 2013). Note the onset of glaciations in the Pleistocene. **c**, Latter part of the Pleistocene (1 Ma to 20 ka). Temperature here is derived from the EPICA Dome C core in central Antarctica (Jouzel et al. 2007). Note the strong 100,000-year-dominated glaciation cycles. This panel ends at the Last Glacial Maximum (LGM) around 21,000 years ago. **d**, Terminal Pleistocene and Holocene (20 ka to present). Here, multiple datasets are plotted, including the EPICA Dome C and North Greenland (NGRIP) ice cores, and the instrumental record. Finally, projected temperatures in 2050 and 2100 are indicated by dots (IPCC 2017). All temperatures are shown as relative to modern, with positive values indicating higher, and negative values indicating lower, global temperature. Figure modified slightly after Glen Fergus, CC BY-SA 3.0.

also yield information about ocean chemistry, currents, and surface winds. The use of the foram record has allowed the quantification of mammalian evolutionary patterns with paleoclimate (e.g. Smith et al. 2010a; Secord et al. 2012); we talk more about this in chapter 10.

FURTHER READING

Alley, R.B. 2002. *The Two-Mile Time Machine*. Princeton, NJ: Princeton University Press.

Fagan, B. 2007. *The Little Ice Age: How Climate Made History 1300-1850*. New York: Basic Books.

Grimm, E.C., et al. 2013. Pollen databases and their application. In *Encyclopaedia of Quaternary Sciences*, edited by S.A. Elias, 831-838. Amsterdam: Elsevier.

Hughes, M.K., et al. 2010. *Dendroclimatology: Progress and Prospects*. Berlin: Springer-Verlag.

IPCC (Intergovernmental Panel on Climate Change). 2007. *Climate Change 2007: The Physical Science Basis*. Contribution of Working Group I to the Fourth Assessment Report of the Intergovernmental Panel on Climate Change. Cambridge: Cambridge University Press.

Zachos, J., et al. 2001. Trends, rhythms, and aberrations in global climate 65 Ma to present. *Science* 292:686–693.

Part Three

Using Paleoecology to Understand the Present

10 The Past as Prologue

The Importance of a Deeper Temporal Perspective in Climate Change Research

The Earth's climate is changing. Not only is it getting warmer virtually every year, but the frequency of extreme events—hurricanes, tornados, storms—is increasing (IPCC 2014; Fig. 10.1). This is of obvious concern for all of us, especially given the high proportion of cities that are threatened by rising sea levels and storm surges (IPCC 2014). Yet, climate change itself is not new to the Earth system. From the "snowball" Earth of the Proterozoic, where our globe came close to freezing over several times, to the extreme glacial periods of the Ordovician and end-Carboniferous, or the more recent twenty-one or so ice ages of the Pleistocene, climate has changed many times over the Earth record (see Fig. 9.1). Indeed, changes in climate are inevitable and often drastic. Even temperature changes as rapid as those expected over the next one hundred years are not novel; after all, the warming at the end of the Younger Dryas some 11.5 ka led to a 5°C to 6°C increase in just a few decades (Dansgaard et al. 1989; Alley et al. 1993; Alley 2000; Kobashia et al. 2008).

What is new this time around is the driver—humans. The evidence is now unequivocal that ongoing shifts in Earth's climate are the result of the enormous amounts of carbon dioxide, methane, and other molecules humans have released into the atmosphere by the combustion of fossil carbon over the past century (IPCC 2014; Fig. 10.2). The most recent estimates suggest our activities already have caused about 1°C of global warming above pre-industrial levels. This may not sound like much, but in the past a one- to two-degree drop was enough to plunge us into the Little Ice Age. And it's not going to get better any time soon; global warming is currently increasing at 0.2°C per decade because of these past and ongoing emissions (IPCC 2014). Moreover, these are global averages, which mask the severity of the problem. In the Arctic, for example, the rates are two to three times higher (IPCC 2014). On June 20, 2020, Verkhoyansk, a Siberian town about 3,000 miles east of Moscow, Russia, reached 38°C (100.4°F)—the highest temperature ever recorded in the Arctic Circle and Siberia ("Heat and Fire Scorches Siberia," NASA Earth Observatory, June 2020, https://earthobservatory.nasa.gov/images/146879/heat-and-fire-scorches-siberia). These changes are of particular concern because melting permafrost will likely release sequestered greenhouse gases, which may exacerbate our climate problem.

As we try to grapple with the consequences of our changing climate—or head off the worst of it—policy makers require robust and high-quality scientific data. Yet, there are few analogs in human-recorded history. This is where the fossil record comes in: the increasing availability of fine-scale paleoclimate data has led to greater appreciation for the rapidity and frequency of past shifts in the Earth climate system and focused attention on the historical record as a

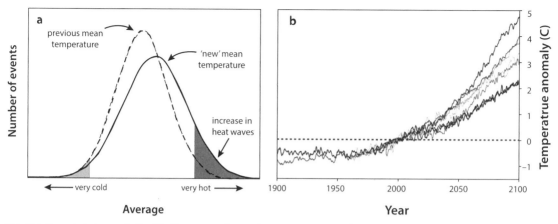

Fig. 10.1. Expected changes in Earth's climate over the coming century. **a**, How changes in mean temperature can lead to higher numbers of extreme events. As climate warms, what would have been extreme heat waves in the past become more common. **b**, Intergovernmental Panel on Climate Change (IPCC) projections for the next century. All climate models project at least a 2°C increase in global temperature; some suggest this could be as great as 5°C

means of assessing likely biotic responses. Here, we discuss a variety of studies examining both ecological and evolutionary responses of mammals to climate shifts, especially those at the Paleocene-Eocene Thermal Maximum and late Quaternary.

The focus on the late Quaternary is appropriate not only because of the well-preserved fossil record, but because a number of climatic events have been well characterized for this time frame. These include the Younger Dryas (12.8-11.5 ka), which contained *both* cooling and warming episodes. It terminated in a particularly abrupt 5°C-6°C temperature increase over just a few decades; perhaps the closest analog for current anthropogenic climate change we have (IPCC 2014). Other episodes were the cold event at 8.2 ka that led to an abrupt 3°C decrease in global temperature and persisted for several centuries, and the mid-Holocene warm period (~7-5 ka), sometimes referred to as the Climatic Optimum, a period of elevated temperature that may have largely been restricted to the Northern Hemisphere (Alley 2000; Rohling and Pälike 2005). Importantly, no mammal species were driven to extinction by any of these events despite their rapidity or severity. This suggests some combination of distributional, morphological, and/or phenological responses were sufficient to cope with the environmental perturbations that occurred.

How Have Animals Responded to Past Climate Changes?

Animals have a limited range of possibilities when coping with a changing environment: they may become locally extinct, they can move to more favorable conditions, or they can adapt physiologically, behaviorally, or morphologically to the new conditions (e.g., Smith et al. 1995; Graham et al. 1996; Smith and Betancourt 1998, 2003, 2006; Hughes 2000; McCarty 2001; Lyons 2003; Parmesan and Yohe 2003; Davis et al. 2005; Blois and Hadly 2010; Beever et al. 2017). Of course, they can also do some combination of all three. How animals respond is strongly mediated by their biology and life history, as well as their particular physical and biological environment. For example, mammals with long generation times are unlikely to be able to effectively adapt in situ to climate change occurring over short time periods. However, because a long generation time is associated with larger body size, a larger home range, and greater dispersal capabilities, such animals are more likely to find suitable new habitats within their geographic distribution. Of course, the type and magnitude of the response to climate change is dependent on the temporal scale (Barnosky et al. 2003). Most paleoecological work has focused on the late Quaternary mammal record, which documents the entire gamut of responses to environmental

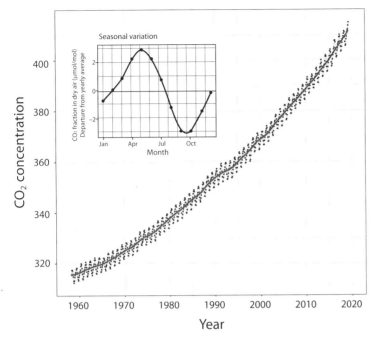

Fig. 10.2. The Keeling curve. This figure depicts atmospheric carbon dioxide concentrations measured at Mauna Loa, Hawaii, since 1958. While the longest continuous record is from Mauna Loa, measurements have been independently confirmed at many other sites. The inset illustrates the annual fluctuation in carbon dioxide, which is the result of seasonal variation in carbon dioxide uptake by land plants. Data from Dr. Pieter Tans, NOAA/ESRL, and Dr. Ralph Keeling, Scripps Institution of Oceanography, CC-BY 4.0.

shifts, including tolerance and local extirpation as well as adaptive changes in genetics or morphology.

Movement: The "Easiest" Response?

Mammals are mobile. This is reflected in their fairly large geographic ranges, even for the smallest of species (Brown 1995). Consequently, mammals may encounter a wide variety of environmental conditions within portions of their distribution. Indeed, the range of temperatures experienced within the geographic distribution of terrestrial mammals is about 10°C–60°C, with an average of about 21°C (Fig. 10.3; Smith et al., in prep.). Thus, when faced with changing temperatures, mammals have the option to shift, expand, or contract their geographic distribution, although urbanization and fragmentation of habitats by humans now hampers this ability. During major climate cooling events, mammals have even expanded their distributions to include new continents; the opening of the Thulean and Beringian land bridges led to the interchange of fauna between North America and Europe multiple times during the early Cenozoic (Webb and Opdyke 1995; Woodburne 2004; Smith et al. 2006).

Over the past few decades, considerable effort has gone to characterizing how climate change will, or

has, led to shifts in the distribution and abundance of species (see Hughes 2000; Jackson and Overpeck 2000; Lyons 2003; Parmesan and Yohe 2003, Root et al. 2003, Moritz et al. 2008; Ackerly et al. 2010; Ordonez and Williams 2013). Detailed, spatially explicit models have been developed to quantify the velocity of climate change and assess whether species can keep pace (Loarie et al. 2009; see also Table 10.1 and Table 10.2 herein). Such modeling efforts are generally based on characterizing the environmental niche space now occupied by species and projecting forward (Berry et al. 2002; Peterson et al. 2002; Thuiller et al. 2004; Araújo et al. 2006; Araújo and Luoto 2007). While much work has focused on past distributional shifts of plants, mostly because of their excellent late Quaternary pollen record (e.g., Davis and Shaw 2001; Williams et al. 2002, 2009, 2010; Blois et al. 2012), until recently generally lacking was a consideration of how mammals responded in the past (but see Blois and Hadly 2010 and references therein).

FAUNMAP

The first quantitative studies of how mammal geographic distributions altered in response to late Quaternary climate shifts were conducted by Russ

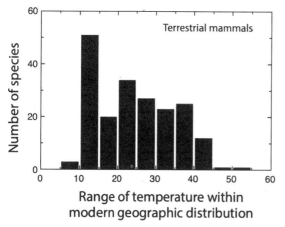

Fig. 10.3. The range of temperature contained within a mammal's geographic range. Data for more than 3,000 terrestrial, nonvolant mammals. Note that for the average mammal, their range encompasses an approximate 21°C variation in temperature. From Smith et al., in prep.

Graham and his colleagues (Graham 1986; Graham and Mead 1987; Graham and Grimm 1990; Graham et al. 1996). These early efforts led to a working group codirected by Graham and Ernie Lundelius, Jr., called FAUNMAP I. The aim of this collaboration was to construct an electronic relational database linking a geographic information system (GIS) with mammal faunal lists for thousands of paleontological sites across the contiguous United States. The temporal window was 50,000 to 500 years ago, a span that corresponded with the effective range of radiocarbon dating. The time span was later extended to include the entire Pleistocene (FAUNMAP II), and the spatial extent expanded to North America. FAUNMAP I and II were ultimately integrated into a broader effort—the Neotoma Paleoecology Database (www.neotomadb.org)—where today it provides the foundational information for mammals (Goring et al. 2015). The original intent of FAUNMAP was to address a long-standing question in ecology about the individualistic versus symbiotic nature of communities (e.g., Clements versus Gleasonian; Box 10.1). Thus, the group was initially interested in whether mammal communities tracked late Quaternary climate shifts together or individualistically according to their specific thermal tolerances.

Analyses by the FAUNMAP group suggested that many, if not most, mammals responded individualistically to environmental perturbations over the late Quaternary (Graham et al. 1996). Although many species of mammals did shift their range in apparent response to climate change, they did so at varying speeds, in varying directions, and to varying extents (Fig. 10.4). While many taxa moved north, likely tracking the movement of plant communities or favorable thermal conditions, some mammals stayed put or, like the northern pocket gopher, shifted their geographic distribution to the west (Graham et al. 1996). These species clearly were responding to other fluctuations in the environment, such as alterations of soil composition and/or structure (Graham et al. 1996).

Table 10.1. Computed velocity of net distributional shifts of mammals over the late Quaternary

Temporal interval	Midpoint, elapsed time (yrs)	Average centroid shift (km)	Maximum centroid shift (km)	Net rate at midpoint (km yr⁻¹)	Net rate for fastest taxon at midpoint (km yr⁻¹)
Full Glacial to late Glacial	15,000	1,190	2,957	0.08	0.20
Late Glacial to Holocene	10,250	1,400	5,786	0.14	**0.56**
Holocene to Modern	5,375	1,310	3,184	0.24	**0.59**

Source: Data from Lyons 2003.

Note: Values in bold exceed the estimated global mean velocity (see Table 10.2) required to adapt to projected anthropogenic climate change over the next one hundred years (e.g., Loarie et al. 2009); however, climate velocity for many temperate habitats is less than the global rate. For example, expected climate velocity for grasslands and tropical and temperate forests is about 0.08 kilometers per year (km yr⁻¹). The net rate of range displacement met or exceeded this in all time periods during the late Quaternary. These values are conservative in two ways. First, they are net directional displacements and, thus, underestimate actual range movement. Second, the time interval is defined as the difference from one midpoint to another; many of the sites cluster more tightly in space, especially those of the full Glacial to late Glacial. Because rates are influenced by the time interval, we have probably underestimated the actual rate of range shifts.

Table 10.2. Some terms used in biodiversity research

Term	Description
Allopatry	Without overlapping geographic ranges; that is, not co-occurring at the same place and time.
Climate velocity	The speed and direction a species must move to remain within its present climate space; also described as the speed at which species must migrate to maintain the same environmental niche space.
Environmental niche space	The combination of abiotic factors that make up a species niche.
Niche	Here we refer to the Hutchinsonian niche. That is, the n-dimensional hypervolume that defines the requirements of a species. These may be abiotic environmental conditions and/or biotic interactions and resources.
Non-analog climate	Climate with no modern counterpart; that is, a unique set of climatic conditions that occurred in the past.
Non-analog community	An assemblage of species that do not occur together today; often associated with times of non-analog climate.
Range centroid	The centroid is the geometric center of a two-dimensional figure. Here we refer to the center of the geographic distribution.
Species evenness	How similar the various species in a community are in terms of their abundance; when communities are not even, it indicates dominance by one or more species. Often measured mathematically by indices.
Species richness	The number of different species in a particular ecological community; often referred to as alpha diversity.
Species turnover	Changes in the species richness of a community over time and/or space; sometimes referred to as beta diversity.
Sympatry	Species that occur at the same place at the same time.

Moreover, co-occurring species did not always move together. The consequences of individual range shifts were the disassociation of species assemblages; mammals that previously had overlapping ranges in some cases became allopatric. Thus, some Holocene and/or modern mammal communities were different from Pleistocene ones. Indeed, these non-analog (Table 10.2) Pleistocene mammal assemblages likely reflected a unique abiotic environment present at that time (Graham et al. 1996). A remarkable finding from these analyses was that modern community structure, often taken as the status quo by modern biologists, only emerged in the late Holocene (Graham et al. 1996).

Later research used FAUNMAP data to quantify the magnitude and direction of range shifts of terrestrial mammals in the continental United States over the last 40,000 years (Lyons 2003, 2005). Data were compiled into four time periods: the Wisconsin or pre-Glacial (40,000-20,000 ybp), the late Glacial (20,000-10,000), the Holocene (10,000-500), and modern (500-present), which represented times of considerable climate differences; the temporal windows chosen varied a bit between papers (e.g., Lyons 2003, 2005). The geographic range of each species was characterized from fossil localities using an Alber's equal area projection; the area enclosed was computed and the centroid determined using a Cartesian coordinate system. Ingeniously, by projecting geographic ranges from one time period to another, Lyons (2003) was able to map the dynamics of mammal range shifts in relationship to environmental change. Importantly, she quantified shifts in the location of the range, the range size, and direction of movement from one time period to another (Fig. 10.5). In later work, she conducted simulation modeling to further explore the overall pattern of mammal range shifts over both space and time (Lyons 2005) and explored the ecological traits of species and their range shifts (Lyons et al. 2010).

Overall, these studies found overwhelming evidence of the dynamic nature of mammal geographic ranges over the past 40,000 years (Lyons 2003, 2005). The geographic distribution of North American mammals moved, shrank, and expanded, sometimes in surprising ways (Fig. 10.6). Nor was movement driven solely by species moving north to track receding glaciers (Lyons 2003). Although many taxa had conservative ranges that changed only

Box 10.1

Are Communities Superorganisms or Assemblages of Co-occurring Species?

Communities are populations that coexist in space and time. How they are structured and interact, and whether they have emergent properties beyond those of their constituent organisms, has been an area of scientific dissent. Two scientists, Fredric Clements and Henry Gleason (left image and right image, respectively), are famous for their diametrically opposing views of communities and the importance of ecological succession, which led to a vigorous and heated debate in ecological circles in the early to mid-1900s (Tansley 1935; Kingsolver and Paine 1991).

In 1916, Fredric Clements, a botanist with the Carnegie Institution, published a foundational paper where he proposed ecological communities were a sort of "superorganism" largely structured by the tight ecological interactions among the constituent species. He argued that communities had emergent properties transcending those of the species themselves. In the same way that animal organs work together to make a body function, he felt that the species associations within in a plant community were essential for energy to flow "properly" (Clements 1916, 1936). Moreover, Clements argued that communities evolve together and, thus, can be tracked over both time and space as they respond to environmental perturbations. Hence, communities were both deterministic and repeatable, and perturbations from a "climax" state of a community would be followed by a deterministic succession leading back to the original state. He stated, *"As an organism the formation arises, grows, matures, and dies . . . Furthermore, each climax formation is able to reproduce itself, repeating with essential fidelity the stages of its development"* (Clements 1916, p. 3).

Gleason (1926), who was a botanist at the New York Botanical Garden, agreed that communities are composed of interacting species, but he strongly disagreed that biotic interactions were the glue that held them together. Instead he argued that communities change over time as species respond individualistically to environmental fluctuations. In his view, communities were not deterministic but ephemeral, composed of loose associations that changed continuously in space and time rather than discrete, precisely defined units that stayed intact over evolutionary time scales. Thus, the predictability of species composition within a community decreased inversely with time. As he put it, *"an association is not an organism, scarcely even a vegetational unit, but merely a coincidence"* (Gleason 1926, p. 16). Moreover, he argued for an important role for the physical environment in determining the composition of a community (Gleason 1926).

Like so much in science, today most ecologists take a middle ground between these two extreme views. Work by Robert Whittaker, Thompson Webb, and others led to a rethinking of this lively debate (Whittaker 1953; Webb 1974a, 1974b). An important contribution was the use of pollen records to recreate the response of plant communities to late Quaternary climate change. This, and subsequent work by Russ Graham and colleagues with fossil mammal assemblages, established the individual responses of organisms to changing environments (Graham 1986; Graham and Mead 1987; Graham and Grimm 1990). Today there is wide agreement that species are distributed individualistically and that community composition typically changes along environmental gradients. However, there is some predictability to species composition in communities; to the extent that species share the same life history tolerances, they sometimes respond similarly to environmental perturbations. While some successional processes do occur, they are not deterministic but heavily dependent on historical contingency.

Box Fig 10.1 Left Frederic Clements. Reprinted from Oberg 2019, DigitalCommons@University of Nebraska-Lincoln.

Box Fig 10.1 Right Henry Allen Gleason holding three folders of herbarium specimens. Courtesy of the Vertical Files of the LuEsther T. Mertz Library of the New York Botanical Garden, NYBG Staff Photographer, CC BY-SA 4.0.

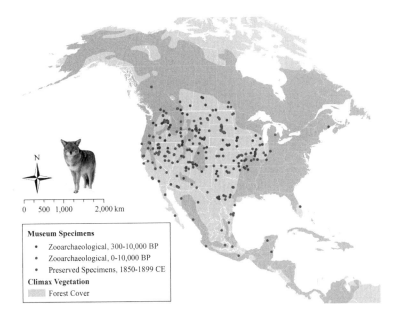

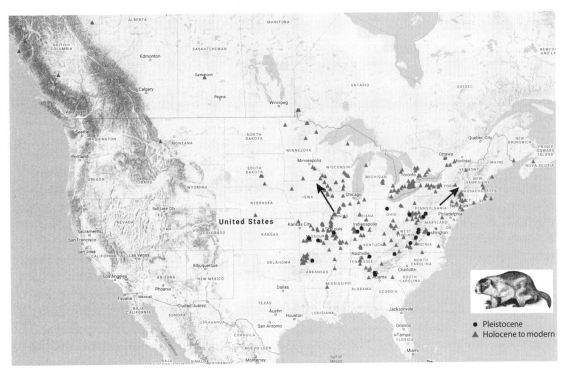

Fig. 10.4. FAUNMAP has been used to examine the distributional response of mammals to late Quaternary climate shifts. **a**, Expansion of coyote range (*Canis latrans*) across both North and Central America since the Holocene. Coyotes have dramatically expanded their range over this time range—most strikingly since 1900. This is likely due to the extirpation/ extinction of larger-bodied competitors by humans. **b**, Changes in the geographic distribution of marmots (*Marmota monax*) from the late Pleistocene (filled circles) to the Holocene (triangles). Overall, there has been a vast expansion of the geographic range in this taxa. Reprinted from Hody and Kays 2018, CC-BY 4.0; Generated by the author.

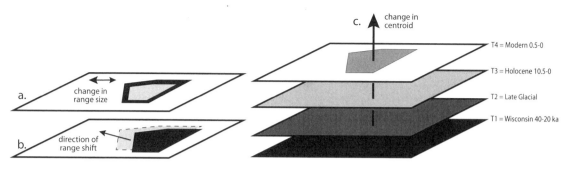

Fig. 10.5. In a series of classic papers, Dr. Kate Lyons examined the shifts in geographic range of terrestrial nonvolant mammals in North America over the late Quaternary. Her work characterized the (a) change in range size, (b) cardinal direction of range shift, and (c) change in the range centroid over four time periods.

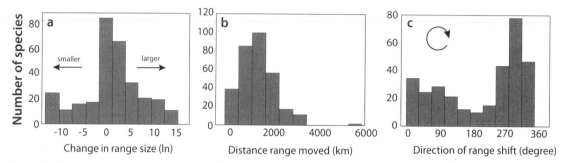

Fig. 10.6. Changes in mammal geographic range from the glacial to Holocene time periods. **a**, Lyons found that most mammals increased their overall geographic range over this time, although a few showed declines. **b**, Ranges were quite dynamic and a few species shifted their range many thousands of kilometers. **c**, During this time interval, species mostly shifted their ranges to the north, although there was movement in virtually all directions. Other temporal transitions yielded different patterns. Redrawn from Lyons 2003, 2005.

slightly over time, others displayed large differences between each temporal interval. For example, from the pre-Glacial to the late Glacial time periods, the centroid of the geographic range moved by an average of about 1,190 km, while the area encompassed expanded (Lyons 2003). There was a trend for movement toward the western portion of the continent, although species moved in all directions. Moreover, the extent of the northern range shifted more than did the southern edge, but these changes were independent and not simply the result of the entire range tracking more favorable thermal environments. Interestingly, trophic affiliations played some part in range dynamics; large shifts toward the north were driven by carnivores and artiodactyls, and southern shifts were primarily carnivores (Lyons 2003). Larger species tended to shift their range more than smaller ones, although the latter were more variable in their response (Lyons et al. 2010). Spatial

location was also important as a predictor for range movement, leading to more similar than expected communities centered in the middle of the continent and non-analog communities found around the area of glaciation (Lyons 2005).

The Glacial to Holocene time interval covered the contraction and melting of the large ice sheets as well as the Younger Dryas cold episode and other periods of rapid climate change. Consistent with these large-scale environmental perturbations, Lyons (2003) found the greatest expansion and movement of mammal geographic ranges. For example, the mean range size expanded and the mean centroid was displaced by about 1,400 km (Lyons 2003). While distributional shifts occurred in all directions, slightly more ranges were displaced toward the south and southeast than in other cardinal directions (Lyons 2003). Moreover, the northern edge of the range shifted more than did the southern edge (1,274 vs. 981 km), but again, these shifts were independent of one another. Tropic affiliation also played a role, with carnivores and rodents driving the large shifts in latitudinal limits (Lyons 2003). Such changes meant much of the western half of the continent and the area near the glaciation were composed of non-analog communities (Lyons 2005), probably reflecting non-analog climate conditions (Williams et al. 2002).

In contrast, Lyons (2003) found that mammal geographic range size decreased from the Holocene to Modern. Moreover, while there was still considerable movement in the location of the range within the continent—the average shift in the centroid was 1,310 km—there was no apparent directionality to this displacement. Both northern and southern range shifts were equally likely and of approximate equal magnitude (1,042 vs. 1,072 km, respectively). Intriguingly, large shifts in the northern distributional edge were mostly rodent species (Lyons 2003). Spatially, this led to more similarity in communities across the continent than expected on the basis of individualistic responses to environmental perturbations (Lyons 2005).

Predicting Future Distributional Shifts

So, what does all this tell us about the ability of modern mammals to adapt to ongoing anthropogenic change? Modern studies have estimated the velocity of expected climate change—defined as the movement along the Earth's surface needed to maintain a constant temperature. This varies spatially and with habitat. Velocity estimates range from 0.08 km yr[1] in mountainous biomes such as tropical and subtropical coniferous forests, temperate coniferous forest, and montane grasslands to as high as 1.26 km yr[1] in mangrove forests and deserts (Loarie et al. 2009). The overall global mean velocity along Earth's surface needed to maintain constant temperatures is 0.42 km yr[1] (Loarie et al. 2009). This is the rate mammals must shift their range if they are to successfully track anthropogenic climate change.

Using conservative values of range displacement from Lyons (2003), which reflect net change and hence underestimate the actual range movement over time, it is clear that during most climate transitions over the late Quaternary, mammals were able to cope simply by shifting their distributions (Table 10.1). Indeed, the fastest moving taxa achieved

net velocities of more than 0.5 km yr[1]. However, in some locations, and for some species, additional morphological or behavioral adaptations may have been necessary. For example, the net rate of movement during the Holocene to modern transition was 0.24 km yr[1]; a value high enough to cope with the extent of climatic change expected for many, but not all, habitats in the future. Because no species went extinct during this time interval, it suggests that further adaptations were employed. Further work comparing the magnitude and direction of distributional shifts in the late Quaternary, with spatially explicit models of anthropogenic climate change and habitat fragmentation, may well identify areas where contemporary species will have problems coping.

To date, the Lyons studies (2003, 2005) remain the most synoptic examination of the distributional response of mammals to an important environmental transition. This work demonstrated that how mammals respond to climate and environmental change was clearly complex. For example, because many co-occurring species shared habitat or environmental requirements, they responded to climate change in similar ways, leading to more "stability" of communities over time and space than previously appreciated (Lyons 2005). Moreover, the direction of range shifts did not simply track climate change. More recent work has built on this temporal approach. For example, employing species distribution models of a subset of North America mammals and using both hindcasting and forecasting models, Williams and Blois (2018) found that species traits and life histories were more important in predicting range shifts than either temperature or precipitation velocity over the past 16,000 years. Indeed, as reported by Lyons (2003, 2005), the direction of range shifts was not strongly influenced by climate change (Williams and Blois 2018). And, surprisingly, many mammal species displayed very little distributional response to climate shifts over the late Quaternary (Lyons 2003). For these species, other modes of adaptation to environmental change may be possible. We discuss these next.

Morphological Adaptation

A number of well-established ecogeographic rules exist for mammals (chapter 5). These describe variations in the morphological traits of animals that occur over physiogeographic gradients, often thermal clines (Mayr 1956). Because the fundamental driver for these is probably energy—the ability to acquire and use energy or to reduce energy use—they tend to be strong selective forces on animals. Of these, Allen's and Bergmann's rules are the best supported for mammals (Millien et al. 2006). And, to the extent these rules reflect morphological adaptations to conserve or dissipate body temperature, which is of profound importance for endothermic animals such as mammals, they are clearly relevant in our era of changing climate.

How universal morphological change is among taxa as a response to ongoing climate change is debated (e.g., Millien et al. 2006; Parmesan 2006; Gardner et al. 2011; Boutin and Lane 2014; Teplitsky and Millien 2014). Factors correlated with temperature, such as precipitation or productivity, may be more important drivers of body size changes for some species (e.g., Rosenzweig 1968; James 1970; Yom-Tov and Nix 1986; Blois et al. 2008; Yom-Tov et al. 2008). And the extent to which morphological shifts are phenotypic or genetically based is also unclear for most organisms, although body size is highly heritable for most mammals (Falconer 1953, 1973; Leamy 1988; Smith et al. 2004, Smith and Betancourt 2006). Further, where a population is located within the geographic range may influence the likelihood that rapid adaptation can occur; genetic constraints may be greater near range boundaries (Hoffmann and Blows 1994; Garcia-Ramos and Kirkpatrick 1997). This is an ongoing and topical area of research. What the fossil record can provide to this debate is an indication of how often morphological change has occurred in the past, what the likely drivers and limits to this adaptive response were, and a quantification of the rate of change over time.

Morphological Adaptation in "Real" Time

As discussed in chapter 5, morphological change can occur quickly (Carroll 2008; Hendry et al. 2008; Hoffmann and Sgrò 2011). Classic examples include the significant shifts in body size and color of house sparrows (*Passer domesticus*) across the United States, which occurred over less than a century (Johnston and Selander 1964), as well as the changes in body

size and beak morphology among Darwin's finches on the Galapagos island of Daphne Major. The latter were driven by drought, changes in food availability, and temperature swings (Grant and Grant 2002). Body size decreases have been observed among British passerines during the last thirty years (Yom-Tov et al. 2006), as well as in passerines on the east coast of Australia over the last one hundred years (Gardner et al. 2009). And previous work by my group has documented a rapid dwarfing of the average body mass of rodents in response to climate warming within a single decade (Smith et al. 1998). Indeed, in a study of modern rodents living on the floor of Death Valley—one of the hottest and driest places on Earth—we found that the mean body mass of the population closely tracked environmental temperature over the course of the year, becoming larger during the winter and falling precipitously in late spring and summer (Murray and Smith 2012; Smith et al. 2014). Closer examination suggested the pattern was driven by size-selective mortality of animals during the winter and summer months. Moreover, daily activity and foraging patterns closely tracked ambient temperature as well (Fig. 10.7); animals did not emerge from their dens if it was too cold or too warm (Murray and Smith 2012; Smith et al. 2014). Thus, environmental temperature directly influenced the acquisition of energy and reproductive activities. The relationship we documented between animals and temperature over the course of a day translated into tangible morphological change of the woodrat population at the scale of months, years, decades, centuries, and even millennia (Smith et al. 2009, 2014).

Morphological Adaptation over the Late Quaternary

Palaeontologists have long noted that climate has been associated with changes in the body size of mammals over evolutionary time (e.g., Guilday 1971; Kurtén 1973; Lundelius et al. 1983). For example, fossils indicated that white-tailed deer (*Odocoileus virginianus*) in the Midwest were small during the middle Holocene but became larger during the late Holocene (Purdue 1989). This was attributed to a roughly 2°C increase in summer temperatures in the middle Holocene. Similar shifts in body mass were

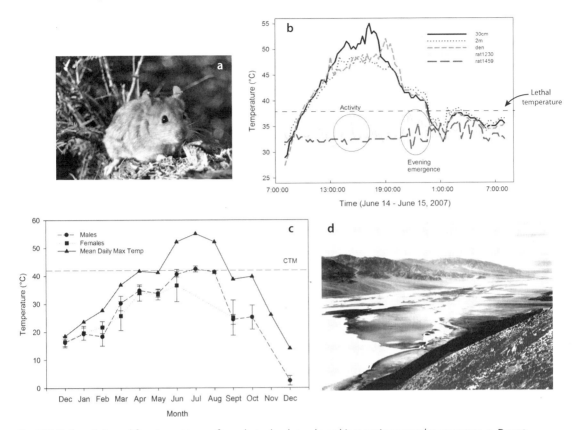

Fig. 10.7. Daily activity and foraging patterns of woodrats closely tracks ambient environmental temperature. **a,** Desert woodrat, *Neotoma lepida*, sitting on a mesquite branch in Furnace Creek, Death Valley National Park. **b,** Data from digital temperature loggers (i-Buttons brand) at Furnace Creek, Death Valley. These devices were set with a five-minute recording time and attached to the fur of several woodrats and to the dens at several elevations. Note the striking difference between the den temperature (where the animals were during the day) and the outside shade temperature. While temperatures outside approached 55°C (131°F), they stayed 15°C-20°C cooler within the den. The animals stayed inside their respective dens until ambient temperature dropped below lethal at 43°C (109°F); at this point they took several foraging bouts. The foraging bouts are indicated by spikes in the animals' temperature records as they leave the den. Most bouts lasted only five to fifteen minutes. **c,** Mean temperature for each month of the year (averaged across the five years of the study), with the mean temperature when animals leave their den recorded for males (solid circles) and females (solid squares). Note that foraging or mating activities do not initiate until temperature drops below lethal. Males emerge before females, which reflects their greater investment in mating. As climate change increases the temperature at this site, animals will have a shorter window of opportunity to forage or mate. **d,** "Looking across desert toward mountains, Death Valley National Monument, California." This 1941 photograph was one of a series commissioned by the National Park Service for the headquarters of the Department of the Interior in Washington, DC. The theme was *"nature as exemplified and protected in the U.S. National Parks."* Sadly, World War II intervened and the project was never completed. Note the unusual presence of water on the valley floor; both 1940 and 1941 were exceptionally wet years, with rainfall more than 50%–75% above the long-term average (Western Regional Climate Center, https://wrcc.dri.edu). Panels a, b: Photo and figure by the author. c: Redrawn from Murray and Smith 2012; Smith et al. 2014. d: Ansel Adams, courtesy of the National Archives and Records Administration.

observed in several squirrel species over the Holocene (Purdue 1980). However, temperature has not always been the driver for size changes over time; the mean body size of the ground squirrel (*Spermophilus beecheyi*) in California increased from the Pleistocene to modern, apparently owing to

shifts in precipitation but not temperature (Blois et al. 2008).

A confounding factor when dealing with the fossil record is whether the shifts in size at a location reflect in situ adaptation or "cline translocation"—that is, displacement of the geographic range driven by a changing environment (Koch 1986). This is where integration of modern and fossil records can be particularly useful. By characterizing patterns of spatial variation for modern animals, one can determine how far animals would have to shift their range to achieve the observed morphological changes over time. For most taxa, a combination of both adaptation and movement likely explains morphological patterns over time (Koch 1986; Smith and Betancourt 2006; Balk et al. 2019). For example, for bushy-tailed woodrats (*Neotoma cinerea*) averaging 350 g, to cope with a 1°C warming they can either migrate 100 km in latitude, decrease in body size by 1%, move upslope by 200 m, or some combination of these responses (Balk et al. 2019). Clearly character-izing the relative contributions of each would be exceptionally useful for predicting the ability to respond to anthropogenic climate change, although this is not always possible.

The Usefulness of Debris

Woodrat middens have allowed some of the most fine-grained examinations of how a small rodent adapted to the many temperature fluctuations of the late Quaternary. As mentioned earlier, *Neotoma* are the poster children for Bergmann's rule, and for this genus at least we have a mechanistic understanding of how and why they are sensitive to temperature (e.g., Figs. 5.5, 5.7; chapter 8). Midden sequences have been collected in mountainous areas across much of the western United States, particularly in the more arid and southern regions (Betancourt et al. 1991). They show that woodrat population body mass closely tracked climate over the late Quaternary in a predictable fashion, becoming larger during cooler periods and smaller during warming events (Smith and Betancourt 1993, 1998, 2006; Smith et al. 1995; chapters 5 and 8 herein). Of all the middens collected to date, the exceptional preservation conditions in Death Valley National Park led to the recovery of the most complete and fine-grained chronosequence (Smith et al. 2009, 2014). More than 130 middens were collected from Titus Canyon and the surrounding areas within the Amargosa Range on the eastern boundary of the national park (Fig. 10.8). Of these, fifty-three were later radiocarbon-dated; unfortu-nately, funding constraints limited the number. Titus Canyon is a narrow east-to-west-situated gorge with

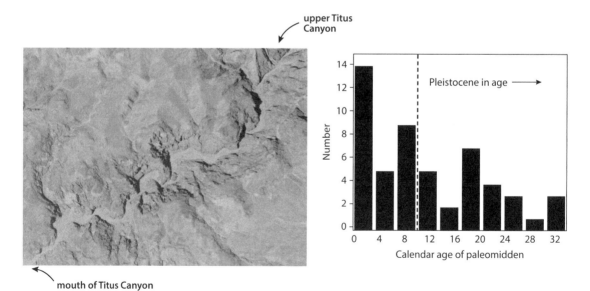

Fig. 10.8. A highly resolved chronosequence of paleomiddens collected by my lab group from Titus Canyon, Death Valley National Park. **a,** Aerial view of Titus Canyon; the lower part is a slot canyon that is extremely narrow. Middens are uncommon here because floods wash out deposits. Most middens were found in the rocks and caves in the upper reaches of the canyon. **b,** About 130 middens were collected from Titus Canyon and the surrounding areas within the Amargosa Range on the eastern boundary of the national park; ages in thousands of years; radiocarbon dating revealed that more than half were Pleistocene in age.

considerable elevation gain as it broadens out. While the western mouth of the slot canyon is about 200 m in elevation, the upper drainage climbs steeply to over 2,000 m. Most paleomiddens were collected in dolomite and limestone caves and crevices facing the canyon (Smith et al. 2009).

The oldest middens recovered from Titus Canyon were more than 33,000 years old. While about half the samples were Pleistocene in age, the entire Holocene was also well represented. Moreover, studies revealed that not one, but *two* different species of woodrat (*N. cinerea*, the bushy-tailed woodrat, and *N. lepida*, the desert woodrat) once lived in Titus Canyon and the Amargosa Range (Smith et al. 2009, 2014; Hornsby et al., in prep.). This was surprising because these two species differ significantly in body size as well as habitat requirements and diet. Further, *N. cinerea* is not found anywhere on the east side of Death Valley today. Species identifications were originally conducted using

the body size range of fecal pellets found in conjunction with fossil bones, and the original results were largely confirmed by ancient DNA analysis on the fecal pellets (Smith et al. 2009, 2014; Hornsby et al., in prep.). Thus, in one location, we were able to examine all the ways that animals potentially respond to late Quaternary climate shifts: adaptation, extinction and/or replacement, and movement.

These studies suggested that bushy-tailed woodrats (*N. cinerea*) were found originally in allopatry within the canyon (Smith et al. 2009, 2014). For the first 20,000 years of the paleomidden record (e.g., the late to terminal Pleistocene, ~33 ka to ~13 ka), *N. cinerea* populations responded to temperature changes with rapid and substantial shifts in body size (Fig. 10.9), likely combined with some limited movement up and downslope. The desert woodrat, *N. lepida*, expanded into the lower elevations of Titus Canyon around the latest Pleistocene. Although all

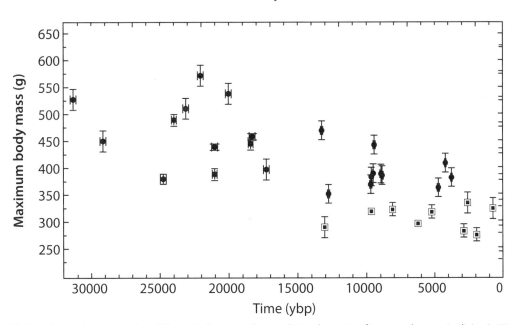

Fig. 10.9. Woodrat body size over time. This graph illustrates the population dynamics of two woodrat species living in Titus Canyon, Death Valley National Park, for over 30,000 years. The larger species (closed circles, *Neotoma cinerea*) was present at the onset, but by the early Holocene, climate warming at this site led to its extirpation from the canyon. A smaller-bodied species (open squares with dots, *Neotoma lepida*) was then able to colonize and remains in the canyon today. Note that both species fluctuated in size over time; these shifts reflected climate changes over this time frame. Thus, in one area, we see all the possible responses of animals to climate change: adaptation, range contractions or expansions, and extirpation. Redrawn from Smith et al. 2009.

midden localities were within the thermal niche of desert woodrats by 21 ka (Hornsby et al., in prep.), modern studies suggest that the much larger and more dominant bushy-tailed woodrats likely restricted activity and space use when both were present (Smith et al. 2009). Indeed, overlapping radiocarbon dates from paleomiddens suggest the two species were always displaced from each other by at least 300 m elevation within the canyon, with bushy-tailed woodrats in caves and alcoves at the higher elevations (Smith et al. 2009, 2014). Rapidly warming climates not only selected for smaller body size but also forced the larger bushy-tailed woodrats farther up in elevation. At about 13 ka—near the end of the warm Bølling-Allerød interstadial—*N. cinerea* was extirpated from Titus Canyon and the Amargosa Range entirely (Smith et al. 2009, 2014; Hornsby et al., in prep). This abrupt turnover coincided with climate warming above the thermal heat stress onset threshold of around 30°C for *N. cinerea* (Hornsby et al., in prep.). Desert woodrats migrated upslope as the larger and more dominant bushy-tailed woodrats vacated habitable spaces. Ultimately, they reached an elevation where they could no longer tolerate the cooler conditions of the winters. Today, bushy-tailed woodrats are extirpated from the Amargosa Range, and desert woodrats are found in Titus Canyon, but only up to about 1,800-1,900 m; no woodrat species now occupy the highest elevations. The fine-grained scale of this work, combined with our knowledge of the thermal niche of *Neotoma* from laboratory and modern fieldwork, allows the quantification of the adaptive envelope as well as the temperature thresholds lead to extirpation.

Morphological Adaptation over Deep Time

The common descriptor for the Pleistocene epoch is "the ice ages." And, indeed, more than twenty glacial-interglacial cycles occurred over the last 2.5 million years. However, the late Quaternary is not the only time within the Cenozoic when climate fluctuated abruptly or dramatically. The early Cenozoic experienced a number of what are known as hyperthermals: sudden and brief time intervals of extreme global warming and massive carbon input that abruptly altered Earth's climate (Lourens et al. 2005; Nicolo et al. 2007; Zachos et al. 2001, 2008). The

most prominent of these events was the Paleocene-Eocene Thermal Maximum (PETM, or ETM-1), a hyperthermal episode that transpired around 55.5 Ma. While the underlying mechanism for the PETM is still debated, it is recognized globally by a large negative $\delta^{13}C$ carbon isotope excursion and warming of the globe by about 5°C-6°C (Pagani et al. 2006; Zachos et al. 2001, 2008). The warming climate led to substantial changes in aquatic and terrestrial ecosystems (McInerney and Wing 2011; Figueirido et al. 2012). For example, the immigration of Perissodactyla and Artiodactyla from Eurasia to North America at this time profoundly altered the diversity and structure of mammal communities (Woodburne et al. 2009; Figueirido et al. 2012). But there is also evidence of adaptation; in particular, the dwarfing of numerous archaic and modern mammalian lineages, including ungulates, primates, omnivores, and others (Gingerich 1989, 2003; Clyde and Gingerich 1998, Chew 2015).

Ross Secord and his colleagues conducted a detailed examination of mammalian morphological responses to the PETM in the Clarks Fork Basin of northwestern Wyoming (Secord et al. 2012). The well-excavated site has proved rich in mammal fossils (Gingerich 1989; Clyde and Gingerich 1998). Moreover, it has a continuous and well-resolved stratigraphic section that spans the PETM interval (Gingerich 1989). Focusing on the small early horse *Sifrhippus* (*Hyracotherium* in earlier literature) that first appeared in North America and Europe during the PETM, Secord et al. (2012) quantified the morphological response of this equid to the abrupt and severe shifts in climate. This work was particularly noteworthy (and briefly discussed in chapter 7) because the authors were able to explicitly test between two main causal hypotheses proposed for Bergmann's rule—temperature and productivity.

As in many other paleontological studies, body size changes in *Sifrhippus* were quantified using the area of the first molar (M1; chapter 5). The local paleoclimate was reconstructed using $\delta^{13}C$ and $\delta^{18}O$ isotopic signatures extracted from preserved apatite in fossil tooth enamel; recall that the structurally bound carbonates in enamel undergo minimal diagenetic alteration and can persist for millions of years (chapter 7; Lee-Thorp and Sponheimer 2003; Koch 2007). Secord et al. (2012) found that the carbon signature clearly mirrored the abrupt negative carbon excursion associated with the PETM in other locations, whereas the oxygen isotopes reflected changes in air temperature. Moreover, these changes in temperature *were* significantly related to shifts in the body size of *Sifrhippus*. The morphological changes they documented were impressive: dwarfing of around 30% in *Sifrhippus* at the onset of the PETM and an abrupt body size increase of more than 75% at the ending of the episode as climate cooled (Secord et al. 2012).

To address whether the morphological shifts in *Sifrhippus* were driven by productivity rather than temperature, the authors needed some sort of proxy for environmental productivity. But this is challenging to characterize in the fossil record. Thus, Secord et al. (2012) took a creative approach by comparing the geochemical signatures recorded in tooth enamel for mammals varying in their habitat affinities. As discussed earlier (chapter 7), quantifying the flux of oxygen through an animal can be difficult. For herbivores, it is heavily influenced by local surface and plant water (Kohn 1996; Vander Zanden et al. 2016). Because drinking water varies in $\delta^{18}O$ both spatially and temporally (i.e., higher values in warm conditions, and lower values in colder conditions), enrichment tends to be positively related to aridity of the habitat for evaporation-sensitive species (Levin et al. 2006). Taking advantage of this difference, Secord et al. (2012) compared the isotopic signatures of two mammals: the evaporation-sensitive *Sifrhippus* and the herbivorous ungulate *Coryphodon*, also present at Clarks Fork Basin and generally considered to be heavily dependent on water habitats. The mean difference between the two oxygen isotope values was computed as a proxy for aridity and assumed to be inversely related to productivity. Interestingly, Secord et al. (2012) found good correspondence between this metric and mean annual precipitation estimated from analysis of paleosols in a nearby section, suggesting they had indeed captured something akin to productivity with this metric.

Overall, the authors found no relationship between changes in productivity and body size. Indeed, they found the opposite trend: body size decreased during wetter conditions. Thus, they concluded that morphological changes observed over the Paleocene-Eocene Thermal Maximum were driven by temperature, consistent with expectations from Bergmann's rule.

While the PETM is recognized as the largest of the hyperthermals occurring during the early Cenozoic, a number of lower-magnitude transient events have been discovered (Lourens et al. 2005; Stap et al. 2010; US Geological Survey's Eocene Hyperthermals Project at www.usgs.gov/centers/fbgc/science/eocene-hyperthermals-project). These include the Eocene Thermal Maximum 2 (ETM2, or H1), which occurred approximately 53.7 Ma and led to an almost 3°C warming of terrestrial environments, and the H2, at 53.6 Ma, that led to an estimated 2°C warming (Lourens et al. 2005; Nicolo et al. 2007; Stap et al. 2010). Recent studies have begun examining potential morphological responses to these abrupt climate shifts as well (Chew 2015; D'Ambrosia et al. 2017). For example, the stratigraphic record of the Willwood Formation in the Bighorn Basin of Wyoming includes both the ETM2 and H2 hyperthermal events. Here, significant decreases in body size have been observed for several mammalian lineages, including an earlier equid, *Arenahippus pernix* (D'Ambrosia et al. 2017). Intriguingly, this taxon dwarfed by about 14% during the 3°C warming of the ETM2 and then increased by about 20% at the terminus; note that this response is about half that reported by Secord et al. (2012) for the approximate 6°C warming of the PETM. This leads to an interesting question: does a predictive relationship exist between the magnitude of morphological change of equids and the degree of temperature change? With only two hyperthermals investigated to date, it is hard to say, but ongoing work may clarify the limits, degree, and directionality of potential adaptive responses (D'Ambrosia et al. 2017).

Limits to Morphological Adaptation as a Response to Climate Change

Modern ecologists have generally dismissed the possibility of adaptation as a response to anthropogenic climate change (but see Hoffmann and Sgrò 2011 and references therein). It is generally assumed that adaptation operates at time scales too slow to allow this to be a viable way for animals to cope with rapid, abrupt shifts in temperature (Bradshaw and Holzapfel 2001; Huntley 2007; Hoffmann and Sgrò 2011). However, as we have discussed, animals did successfully respond to late Quaternary climate change by a combination of local adaptation and slight shifts in distribution or elevation (e.g., Smith and Betancourt 1998, 2003, 2006; Smith et al. 2009). Thus, contemporary scientists may be underestimating the ability of animals to adapt to ongoing environmental change (Hoffmann and Sgrò 2011). Nonetheless, it stands to reason that there are limits to in situ morphological adaptation. There are undoubtedly rates of temperature change too rapid for mammals to cope with and/or thresholds where temperature exceeds the thermal niche of an animal (Balk et al. 2019). Characterization of such limits, and the potential asymmetry of response to warming versus cooling climate, is essential for predicting or mitigating ongoing anthropogenic environmental change.

Under what circumstances should we anticipate animal populations could adapt to environmental change? Based on evolutionary theory, we can make some broad predictions (Mayr 1956). First, animals must have reasonably short generation times so that the response to selection can occur on a time frame consistent with the scale of environmental shifts. For example, mammals with generation times much greater than a year—say, elephants—may be less likely to be able to adapt in situ. Second, large well-connected populations tend to have a greater ability to adapt than small isolated ones. This makes sense: by definition, larger populations contain more individuals. Hence, they are more likely to include animals with traits favored by selection. Third, it is probable that generalists are more adaptable than specialists. Generalists tend to have a broader range of traits that might be favored by natural selection. Finally, species that inhabit areas with significant topography are more likely to be able to cope with rapid climate change. This is because combining vertical movement with local adaptation provides more scope to respond to environmental change. Movement up- and downslope is easier for species than latitudinal movement; that is, lapse rates are higher for elevational gradients than latitudinal ones (Meyer 1992). For example, in the western United States, temperature decreases by 1°C for every 1° latitude (or ~100 km) as one moves toward the North Pole, but decreases by approximately 6.2°C per kilometer as one moves upslope (Meyer 1992).

All of these criteria are met by *Neotoma*, which may explain their apparent ability to cope with late Quaternary climate shifts (Smith et al. 1995; Smith and Betancourt 1998, 2003, 2006; Smith et al. 2009). Given this, one can ask whether evolutionary rates observed in the paleontological record were ever high enough to adapt to the predicted rate of anthropogenic warming over the next one hundred years.

Quantifying the Adaptive Capability of *Neotoma*

We have characterized the adaptive capability of *Neotoma* in several ways (Smith and Betancourt 2006; Balk et al. 2019). First, we examined the spatial and temporal distribution of *Neotoma* paleomiddens over the late Quaternary in relationship to temperature to determine whether there was any pattern to their absence or presence. Finding biases in the paleomidden record would suggest that at some geographic locations or under some temperature conditions, local populations were extirpated. We predicted that animals at the distributional limits were more likely to be extirpated as temperature warmed or cooled (Balk et al. 2019); it is generally assumed that animals at the perimeters of their geographic range are close to the boundaries of their thermal niche and may have less adaptive capacity (Brown 1995). Second, we directly estimated the rate of morphological evolution over the late Quaternary using paleomidden chronosequences across their entire geographic range. Here, too, we looked for a geographic signature in adaptive ability. Thus, we computed the rate of morphological change in darwins for each individual chronosequence (see Box 5.2). As a reminder,

$$darwins = \frac{\ln \frac{x_2}{x_1}}{\Delta T}$$

where x_1 and x_2 represent the initial and final character state (in this instance, pellet width measured from paleomiddens), respectively at two adjacent time intervals, and ΔT is the difference in time in millions of years (Haldane 1949). We were able to circumvent mathematical issues inherent with rate comparisons by employing standardized measurement intervals (Haldane 1949; Gingerich 1983; Gould 1984).

Finally, we also computed the rate of body size change required to adapt in situ to predicted anthropogenic warming. For example, if woodrats were able to adapt to an estimated 5.8°C of warming, which is the current estimate for the next hundred years (IPCC 2014), the phenotypic change required translates into a phenomenal rate of 2,412 darwins. This represents a shift in mean body mass of a woodrat population from 350 g to 275 g. Did we ever encounter rates this high in the woodrat paleomidden record?

To estimate paleoclimate during the deposition interval of each midden and for the late Quaternary overall, we used oxygen isotopes measured from the Summit ice cores in central Greenland. These included the Greenland Ice Sheet Project 2, the Greenland Ice Core Project, and the North Greenland Ice Core Project, which all provide highly resolved and generally concordant paleoenvironmental records for the Northern Hemisphere over the last 100,000-plus years (Dahl-Jensen et al. 1998; NGRIP

2004; Jouzel et al. 2007). Temperature records were averaged into 100-year bins, and we then computed temperature change between adjacent bins in darwins, as we had done for middens. This allowed us to compare the overall distributions of mean temperature and rate shifts over time with the distribution of temperature and rates when paleomiddens were recovered (Figs. 10.10, 10.11). These approaches allowed us to ask whether the direction, magnitude, or rate of late Quaternary climate change were ever too great or rapid for woodrats to cope with.

These studies demonstrated that woodrats displayed remarkable resilience to late Quaternary temperature shifts (Smith and Betancourt 2006; Balk et al. 2019). There was no apparent asymmetry in the ability of animals to persist to warming or cooling climates; indeed, they responded equally well to environmental shifts in either direction

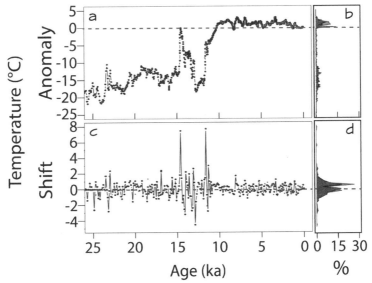

Fig. 10.10. Are woodrats more sensitive to temperature warming or cooling at a site? To address this question, we compiled data on chronosequences of middens across the range of bushy-tailed woodrats (*Neotoma cinerea*). If warming climates were harder to adapt to than cooling—or vice versa—we anticipated that we would see gaps in the chronosequences that corresponded to temperature shifts beyond what they were capable of responding to. Thus, we binned temperature derived from the Greenland Ice Sheet Project 2 (GISP2) into 100-year intervals and computed the mean temperature within an interval as well as the difference in temperature between bins. These data were compared to the middens found in that area. **a**, Comparison of binned temperature record in 100-year binned temperatures and the midden-centered binned temperatures (i.e., presence of middens). We found that middens were recovered throughout the range of temperatures estimated using the GISP2 ice-core temperature data over the late Quaternary. **b**, Frequency distribution of 100-year binned temperatures and midden-centered binned temperatures for all temperatures experienced. **c**, Comparison of 100-year binned temperature shifts and midden-centered binned temperature shifts. **d**, Frequency distribution of 100-year binned temperature shifts and midden-centered binned temperature shifts. We found no evidence of an asymmetric response to climate. In all instances, animals adapted in situ to the temperature shifts seen in the region. Redrawn from Balk et al. 2019.

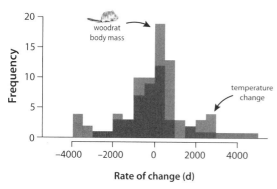

Fig. 10.11. Evolutionary change of woodrat body size change and climate. Rates of temperature change and morphological change were calculated in darwins (d). The distribution of rates of population mean body-mass change (in light gray) was not significantly different from the distribution of rates of temperature change (in dark gray) over the late Quaternary. Binning smoothed out eight extreme rates of temperature change: four decreasing temperature rates of −12,381 d; −7,457 d; −5,586 d; and −10,643 d; and four increasing temperature rates of 5,598 d; 7,482 d; 6,786 d; and 11,004 d. Redrawn from Balk et al. 2019.

(Fig. 10.10; Balk et al. 2019). This was a bit surprising because over evolutionary history, decreases in body size were quicker and/or easier to achieve than increases (Evans et al. 2012). However, paleomidden sequences were recovered over environmental conditions that ranged from 22°C cooler than modern to 3°C warmer (Balk et al. 2019). Neither the magnitude of temperature change nor the overall temperature influenced the likelihood that paleomiddens were recovered (Fig. 10.10). Simply put, temperatures when middens were formed were not different from the overall distribution of temperatures over the late Quaternary.

While climate was highly variable over the past 25,000 years, most shifts in temperature between adjacent time intervals were relatively small. Only a few intervals experienced rates exceeding 500 d, and many of these were concentrated at the late Pleistocene as ice sheets retreated and again in the late Holocene (Fig. 10.11). Importantly, the distribution of rates of woodrat morphological change did not differ significantly from that of the rate of climate change, suggesting that within most of their geographic range animals successfully coped with climate challenges over the late Quaternary. Indeed,

animals persisted even during temperature shifts as abrupt as 8°C over 100 years (Balk et al. 2019). The highest evolutionary rates of body size change observed were over 2,500 d, higher than the 2,412 d necessary to adapt to predicted anthropogenic climate change. Interestingly, we did recover a few climate transitions of more than 4,000 d, which indicated temperature fluctuations of greater magnitude than that expected in the near future (IPCC 2014). Unfortunately, we did not have paleomiddens spanning these time intervals that would have allowed us to examine the woodrat response—or lack thereof (Balk et al. 2019).

We did find evidence of range movement at the geographic range boundaries. While *Neotoma cinerea* were found as far south as northern Mexico during the full Glacial, the southern edge of their distribution shifted during the late Pleistocene and early Holocene. Today, the modern southern limit runs through northern New Mexico and Arizona (Smith 1997). And expansion into what is now the northern limit of their modern range could only occur after ice sheets retreated (Balk et al. 2019). However, for most locations, animals largely adapted in situ to the climate change of the late Quaternary. Overall, these studies confirm that morphological adaptation may indeed be possible for some mammals when confronted with anthropogenic climate change, especially when coupled with some distributional shifts (Smith and Betancourt 2006; Balk et al. 2019).

While morphological shifts in the body size of populations allow mammals to adapt to ongoing climate change, it is important to recognize that there are ecosystem consequences. Because size plays a fundamental role in mediating most fundamental ecological, life history, and physiological rates and processes (chapter 5; Peters 1983; Calder 1984), thermal selection can result in the restructuring of entire ecological communities.

Genetic Changes in Response to Climate Fluctuations

Thus far we have ignored the issue of whether traits actually evolve (i.e., determined by genetics) or are plastic (i.e., determined by the environment), or lie somewhere in between, in our discussion of adaptation to climate change. Many traits influencing behavior, morphology, and/or physiology are quantitative. This simply means that the variation between individuals in a population is the result of the combined effect of differences in genotype and the environment. The common way of calculating adaptive potential is through the Breeder's equation (Falconer 1989). This equation posits that the response to selection R, or the change in the mean of a trait over a generation, is given by

$$R = h^2 S$$

where h^2 is narrow sense heritability, and S is the selection differential (Falconer 1989). The selection differential is a measure of association between trait values and fitness, generally taken as the *within-population* change in the mean of the trait. If the values are known, this equation can be used to predict the adaptive responses of a population across a generation and, moreover, assess whether the response is sufficient to cope with environmental challenges (Falconer 1989).

Genotypic or Phenotypic Selection?

As paleontologists, we typically sidestep the question of phenotypic versus genetic evolution for several reasons. First, it is nigh impossible to estimate heritability with the fossil record. Commonly, h^2 is computed by measuring the phenotypic similarity of parents to their offspring, or better yet, of siblings to each other (Falconer 1989). While we can measure morphological changes in a trait or the overall size of an animal with the fossil record, we generally have no way of actually estimating heritability, which is essential for differentiating between the phenotypic and genetic underpinnings of a trait.

Second, some commonly measured traits in the fossil record probably *are* largely genetically based (Fig. 10.12). Empirical studies indicate that body size in modern mammals tends to be quite heritable, suggesting that morphological change is likely the result of both phenotypic and genetic selection (Falconer 1953, 1973; Rutledge et al. 1973; Leamy 1988; Smith et al. 2004; Smith and Betancourt 2006). For example, we estimated the broad sense heritability of body size in woodrats using field-caught pregnant woodrats that were brought into a laboratory to give birth. The offspring were measured as they grew and subsequent sib-sib analyses were conducted, yielding a very high broad sense heritability value of 0.86 (Smith and Betancourt 2006). This is consistent with

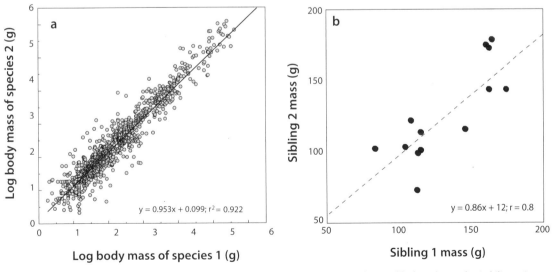

Fig. 10.12. Body size is extremely heritable in mammals. **a,** Results from a 'sib-sib' analysis yield a broad sense heritability estimate of 0.953 (the slope of the regression). This was done by randomly picking two species from each mammal genus and running a linear regression through the resulting data. The pattern is strikingly strong: species within a genus tend to be very close in body size. **b,** Results from a traditional sib-sib analysis performed on woodrats wherein field-caught pregnant females were brought into a laboratory and allowed to give birth; the ensuing offspring were measured. Plotted are pairs of siblings. Note that the broad sense heritability estimate is also extremely high at 0.86. Redrawn from Smith et al. 2004 and Smith et al. 2006.

the value obtained in a large-scale macroecological analysis ($H^2 = 0.95$; Smith et al. 2004). Note that while broad sense heritability (H^2) is the overall ratio of total genetic to phenotypic variance and not simply the additive variance (h^2), it does indicate the proportion of the phenotypic variance that results from genetic factors (Falconer 1989). Values this high strongly suggest that morphological adaptations to climate shifts are more than plastic responses; indeed, they are likely accompanied by substantial genetic changes in animal populations. Considerable ongoing research is focused on how anthropogenic environmental changes will, or are, influencing the genetic structure of populations (e.g., Hoffmann and Sgrò 2011 and references therein).

While environmental perturbations can directly select for adaptive genes, they can also influence the genetic structure of populations indirectly. For example, reduction in gene flow and/or effective population size often lead to random drift and genetic differentiation (Falconer 1989). Thus, climatic factors influencing the ability of animals to disperse or the connectivity of populations may well lead to changes in genetic structure and diversity. Phylogenetic studies have revealed an important role of past climate on gene flow in animals (e.g., Barnes et al. 2002, 2007; Orlando et al. 2002; Hadly et al. 1998, 2004).

The Influence of Past Climate on Genetic Structure and Diversity

The historical record reveals just how sensitive the genome is to environmental change. Over the past half century, climate change has led to documented genetic changes in a number of taxa (e.g., Hoffmann and Sgrò 2011). For example, Umina et al. (2005) and Balanyá et al. (2006) reported a shift in the frequency of particular alleles and chromosomal arrangements in the common fruit fly (*Drosophila melanogaster*) over the course of several decades. These traits were associated with adaptations to warm environments. Similar results have been seen in other taxa (Hoffmann and Sgrò 2011). Thus, we have every reason to expect that late Quaternary climate change also led to genetic changes in animal populations; directly measuring the genetic response of long-dead populations to environmental change is now sometimes possible through the use of ancient DNA (chapter 8). However, the prohibi-

tive cost, lack of appropriate fossil material, and necessity of clean rooms in physically separate laboratories (or even separate buildings) for aDNA analysis hampered the wide application of this technology until fairly recently. With recent advances, many new exciting aDNA studies are ongoing that track how the genetic structure and diversity of animals changed in response to past environmental perturbations (Blois and Hadly 2010).

For example, using aDNA extracted from dehydrated, preserved remains of southern elephant seals (*Mirounga leonine*) found along the coast of Antarctica, de Bruyn et al. (2009) explored founder-extinction dynamics over the Holocene. Elephant seals are crucially dependent on open beaches for breeding; the availability of suitable habitat is dependent on the extent of sea ice. Earlier work had demonstrated that elephant seal populations had expanded well south of their modern core populations during the warmer conditions of the mid to late Holocene (Hall et al. 2006). Clearly understanding how elephant seals respond to the gain and loss of critical breeding habitat is essential for modern conservation efforts. Using coalescent modeling of aDNA, de Bruyn et al. (2009) characterized genetic changes as shifting climate led to the exposure of new breeding habitat. They found that the rapid expansion of the population some 8,000 years ago led to the establishment of a highly diverse and genetically differentiated population over a relatively short time (de Bruyn et al. 2009). Interestingly, much of this genetic diversity was lost when further climate change led to the expansion of sea ice and the loss of the newly colonized breeding habitat. Thus, while elephant seals demonstrated high adaptive potential, they also show how sensitive populations can be to changes in environmental conditions (de Bruyn et al. 2009).

Paleontological studies focusing on small mammals have generally found that Holocene climate change led to decreases in population size and often reductions in genetic diversity (Blois and Hadly 2010). However, the magnitude and type of response were tied to the particular life history traits of the animals. For example, numerous studies employing the continuous late Holocene fossil record of Lamar Cave in Yellowstone National Park, Wyoming, have revealed important insights into the response of small mammals to climate change (Hadly 1996, 1997, 1999; Hadly et al. 2004). This site, which spans the last 3,000 years, has an excellent faunal record with over 10,000 fossil elements recovered to date, representing upwards of 80% of the species in the local mammal community (Hadly 1999). As expected, small mammals responded to late Holocene climate change through shifts in morphology, distribution, and abundance (Hadly 1996, 1997, 1999; Hadly et al. 2004). But such adaptations have not always been accompanied by reductions in genetic diversity. Thus, while populations of both montane vole (*Microtus montanus*) and northern pocket gopher (*Thomomys talpoides*) decreased sharply during the Medieval Warm Period (~1,438 to 470 years ago), only for the latter did the decreased abundance result in lowered genetic diversity (Hadly et al. 2004). This suggests that the montane vole maintained a higher ecological effective population size, perhaps because of more dispersal and higher connectivity between populations. Here, life history traits mediated the genetic response to climate changes over the late Holocene.

Past climate change has even perhaps influenced the evolution of novel traits. For example, the genetic diversity of social tuco-tucos (*Ctenomys sociabilis*) decreased sharply over the late Holocene, apparently in response to a volcanic eruption in the Andes and warming climates (Hadly et al. 2003; Chan et al. 2005, 2006). Today, this species lives in colonies in remote highland savannah areas of southern Argentina. Unlike other species in the genus, *C. sociabilis* are highly social, sharing burrows and perhaps even lactation responsibilities (Lacey and Wieczorek 2003, 2004; Tammone et al. 2012). Comparing modern and aDNA extracted from social tuco-tuco teeth, Chan et al. (2005, 2006) demonstrated a population decline around 2.6 ka, which was accompanied by a precipitous loss of genetic diversity. While numerous haplotypes were present at the beginning of the Holocene, diversity decreased sharply over time and especially after the middle Holocene. The authors hypothesized that this genetic bottleneck led to the development of sociality: because remaining tuco-tucos were all closely related, there was strong selection for individuals to develop cooperative behaviors to increase their inclusive fitness (Chan et al. 2005, 2006).

Future Directions in Genetic Studies of Climate Change

Increasingly, studies integrate different methodologies to obtain a more nuanced understanding of how animals responded to climate in the past. For example, integrating morphological analyses, ancient DNA work, and ecological niche modeling, Prost et al. (2013) examined the adaptation of Central European Palearctic shrews (*Sorex* sp.) to past climate. The use of aDNA allowed these workers to partition the genetic and nongenetic components of the response. These authors showed that shrew morphology, distribution, abundance, and genetic diversity were all heavily influenced by fluctuating late Quaternary climate. During the Pleistocene, a giant shrew was present in Central Europe. While this was originally described as an extinct Pleistocene species (*S. macrognathus*), aDNA analysis revealed that it was actually a large ecomorph of the common shrew (*S. araneus*), suggesting a high level of morphological plasticity in the clade. The large morphospecies was replaced in the warmer Holocene by smaller-bodied animals, but from a different genetic population (Prost et al. 2013). Interestingly, selection for larger (or smaller) body size in *Sorex* appeared to be mediated by shifts in food availability as well as temperature (Prost et al. 2013). Thus, dynamic population-level shifts in abundance, distribution, and morphology were accompanied by shifts in genetic diversity and structure.

As newer methods for obtaining and analyzing aDNA continue to be developed, we can look forward to increasingly fine-scaled studies using currently untapped paleontological resources. For example, sedimentary aDNA and bulk-bone metabarcoding (chapter 8) are two newly developed techniques that may allow for the more precise characterization of fossil assemblages; their use may allow us to examine how entire mammal communities responded to past environmental perturbations (Willerslev et al. 2003; Thomsen and Willerslev 2015; Swift et al. 2019; Tyler et al. 2019).

Phenological and Behavioral Changes

My cat, Electra, tends to sit under my desk lamp as I work in the winter. While I would like to believe she does this because of her undying love for me, in

Fig. 10.13. Cats readily find their preferred microhabitat. My cat, Electra, not only prefers sitting under my desk lamp on a chilly morning but also on top of my laptop, which gives off residual heat. Photo by author.

actuality the (non-LED) bulb gives off heat that creates a warmer microhabitat (Fig. 10.13). This behavior is not unusual. Mammals actively respond to stimuli in their environments, particularly temperature gradients. Indeed, endotherms like Electra spend about 90% of their metabolic energy maintaining a set internal temperature. Thus, over the course of a day, animals tend to adjust their thermal exposure to maximize time spent in the most favorable conditions and minimize the energy needed for thermoregulation. For example, they may move from the sun to shade as thermal conditions become stressful, or curtail foraging unless environmental temperatures are within a particular zone (Smith 1974; Murray and Smith 2012).

Consequently, the most immediate and direct way for animals to cope with climate change is through behavioral flexibility (Parmesan and Yohe 2003). Woodrats in Death Valley do not begin foraging until evening temperatures fall below lethal; this daily decision by animals is leading to a shorter window for foraging and mating activities as climate warms (Murray and Smith 2012). Even larger shifts in habitat use, emergence from hibernation, mating activity, and/or foraging patterns can easily occur over the lifetime of an animal. And to the extent behavior is

heritable and has fitness consequences, such traits can be selected upon (Falconer 1989). For example, the timing of parturition and rut in red deer (*Cervus elaphus*) on the Isle of Rum, Scotland, has shifted by five to twelve days over the past twenty-eight years; this was significantly associated with changes in the growing season (Moyes et al. 2011). Similarly, calving dates of caribou (*Rangifer tarandus*) on Greenland and Chillingham cattle (*Bos taurus*) in the United Kingdom have advanced, again coinciding with an earlier onset of the growing season in these populations (Post and Forchhammer 2008; Burthe et al. 2011). Meta-analyses of behavioral responses to ongoing climate change suggest that across all taxa examined thus far, biological rates (i.e., phenology) have advanced an average of 2.3 days per decade (Parmesan and Yohe 2003; Parmesan 2006).

However, behavioral flexibility is likely to be insufficient to adjust to wholescale changes in environmental conditions without concomitant change in other traits, some of which may be maladaptive. Moreover, some behavioral changes may lead to altered or mismatched synchrony between animals and habitat or food resources (Visser et al. 1998; Parmesan 2006). Mammals also differ in their ability to cope with environmental changes. For example, behavioral responses are more likely for those species that live multiple years and more likely when they are exposed to chronic stimuli, such as an increase in average warm season temperatures (Beever et al. 2017). Only recently have studies begun examining behavior as an adaptive response to climate shifts (Beever et al. 2017). Thus, we still lack a synoptic understanding of the limits or constraints on behavioral and/or phenological shifts as an effective response to climate fluctuations.

Studying behavior or phenology in the fossil record is difficult. While ichnofossils can sometimes provide insights into movement or ecological interactions between species (chapter 8), we are unlikely at present to be able to employ the fossil record to assess behavioral responses to late Quaternary climate change.

Multitaxon Response to Climate Change

Thus far we have largely focused on the individualistic responses of taxa to climate change. While the response of a species is contingent on its life history, ecology, and general biology, mammals are embedded in entire ecosystems. Thus, changes in the abundance or morphology of one species can influence others, leading to wholescale changes in community structure and composition. Studies of how entire mammal communities respond to climate perturbations are consequently an ongoing major emphasis in modern ecology (e.g., Parmesan and Yohe 2003; Parmesan 2006 and references therein).

To date, synoptic examinations of entire mammal communities to Quaternary climate change are fairly limited. For example, a recent study examined the influence of climate perturbations on the long-term dynamics and properties of mammal food webs over the last half of the Quaternary (Nenzén et al. 2014). They found that while turnover rates were highly correlated with glacial cycles and climate change over the past 850,000 years, ancient mammal food webs were fundamentally dynamically stable. Local extinctions were followed by immigrations of phylogenetically similar species without major structural changes to the food web (Nenzén et al. 2014). Thus, climate change led to changes in community dynamics, but without major changes in food web properties. While this study was suggestive, it did not include medium- or small-bodied mammals in the analysis.

A study of the Pleistocene-Holocene transition in Northern California revealed shifts in the evenness and richness of the small mammal community (Blois et al. 2010). These changes were significantly negatively correlated with climatic warming over this interval (Blois et al. 2010). The authors found that Pleistocene communities were significantly more even than Holocene communities, with fewer rare taxa present. This was important for several reasons. First, evenness metrics incorporate a consideration of abundance; thus, they are sensitive indicators of community change. Second, evenness may relate to community persistence under environmental stress (Blois et al. 2010); thus, decreases suggest less resilience. The decrease in evenness and richness of

small mammals may likely lead to other important shifts within the community: small-bodied mammals aid in soil aeration, facilitate seed and nutrient dispersal, and serve as a food source for other larger-bodied mammals within the community. However, this study did not incorporate the medium- or larger-bodied mammals present, probably because of the difficulty of obtaining adequate sample sizes of fossils.

Thus, a challenge for contemporary paleoecologists is to examine entire mammal communities in the context of environmental shifts over the late Quaternary. Synoptic studies, especially those that include all the potential responses of mammals to climate change, could be invaluable as we try to predict, and potentially mediate against, ongoing anthropogenic environmental perturbations.

FURTHER READING

Graham, R.W. 1986. Response of mammalian communities to environmental changes during the Late Quaternary. In *Community Ecology*, edited by J. Diamond and T.J. Case, 300-313. New York: Harper and Row.

Graham, R.W., and E.C. Grimm. 1990. Effects of global climate change on the patterns of terrestrial biological communities. *Trends in Ecology and Evolution* 5:289-292.

Millien, V. et al. 2006. Ecotypic variation in the context of global climate change: revisiting the rules. *Ecology Letters* 9:853-869.

Pardi, M.I., and F.A. Smith. 2012. Paleoecology in an era of climate change: how the past can provide insights into the future. In *Palaeontology in Ecology and Conservation*, edited by J. Louys, 93-116. New York: Springer-Verlag.

Parmesan, C. 2006. Ecological and evolutionary responses to recent climate change. *Annual Review of Ecology Evolution and Systematics* 37:637-669.

Parmesan, C., and G. Yohe. 2003. A globally coherent fingerprint of climate change impacts across natural systems. *Nature* 421:37-42.

Smith, F.A., and J.L. Betancourt. 2006. Predicting woodrat (*Neotoma*) responses to anthropogenic warming from studies of the palaeomidden record. *Journal of Biogeography* 33:2061-2076.

Zachos, J.C., et al. 2008. An early Cenozoic perspective on greenhouse warming and carbon-cycle dynamics. *Nature* 451:279-283.

11 Biodiversity on Earth

> We live in a zoologically impoverished world, from which all the hugest, and fiercest, and strangest forms have recently disappeared.
> —Alfred Russel Wallace, *Geographic Distributions of Animals* (1876, p. 150)

Ask a biologist or an informed layperson what the most pressing environmental problems are that we face today, and they would most likely respond with climate change and the rapid loss of biodiversity. Both of these have been the source of much international discussion and debate in the public sector, within government bodies, and in the scientific literature. To a paleontologist, species diversity (S) is the difference between origination (O) and extinction (E); that is, $S = O - E$. Diversity decreases when origination rates decrease, extinction rates rise, or both. While all of these are seen in the paleontological record, over modern times the decrease in diversity has been driven solely by rising extinction rates.

In the previous chapter, we discussed how the late Quaternary fossil record can help inform the scientific discussion on anthropogenic climate change. Here, we turn to the ongoing biodiversity crisis. Just how bad is this? And what can the extinction dynamics of the past tell us about the consequences of a major loss of biodiversity? Does the current crisis rise to the level of past mass extinctions?

How Serious Is the Loss of Biodiversity Today?

In a word—frightening. In less than fifty years, the human population has doubled to over 7.7 billion people, which has driven an ever-increasing demand for energy and resources (IPBES 2019). As popula-tions urbanize, resource extraction has led to substantial habitat degradation, shifts in climate and biogeochemical cycling, and reductions in the abundance and distribution of taxa (Vitousek et al. 1997; Decker et al. 2000; Thomas et al. 2004; Barnosky 2008; Burger et al. 2012; Burnside et al. 2012; Dirzo et al. 2014; Kolbert 2014; Boivin et al. 2016). On average, when the entire ecological footprint is included, each modern industrial human now "metabolizes" more than forty times the energy used by our ancestral hunter-gatherers (Decker et al. 2000). This has led to an ecological deficit; according to the Global Footprint Network (www.footprintnetwork .org/our-work/ecological-footprint), the annual anthropogenic demand for energy and ecological resources now far outstrips what the Earth regener-ates (see also Burger et al. 2012). For example, we cut down trees faster than they regrow, harvest fish faster than they are replenished, use more water than rivers and aquifers receive in rainfall, extract more mineral resources than are produced, and add carbon dioxide and other gases to the atmosphere faster than they can be sequestered by natural processes. Indeed, we need 1.75 Earths to sustain our modern average global rate of consumption; this rises to 4 Earths if all humans were at the living standard of the average citizen in the United States (Global Footprint Network; IPBES 2019). Earth "Overshoot Day"—the day where our demands for

---

Box 11.1
What Is Biodiversity?

Simply put, biological biodiversity refers to all aspects of the sum of the vast variety of life on our planet—the type, form, and function of living species and their interactions with the environment. Biodiversity includes important ecosystem services such as nutrient storage, recycling and cycling, soil formation, and climate. Studies demonstrate that greater species diversity ensures natural sustainability and environmental resilience.

There are several recognized levels of biodiversity. Loss at any one of these levels can result in ecological collapse:

Genetic diversity: the total number of genes within individuals, populations, species, and clades.
Species diversity: alpha diversity, or the number of species in a defined area.
Functional diversity: the range of things that species do as well as their ecological role within communities.
Ecosystem and community diversity: a combination of species richness and evenness; the diversity and uniqueness of biological communities and/or ecosystems.

---

ecological resources and services exceed that produced in a year—has moved significantly over time: from late September 2000 to July 29, 2019 (Global Footprint Network). Given such grim statistics, it should come as no surprise that these intense resource demands have led to modification of about 75% of the terrestrial surface of our planet thus far (IPBES 2019; Lovejoy 2019). Some projections suggest that only about 10% of the Earth will remain "substantially free" of the impacts of human activities by 2050 (WWF 2018).

Severe habitat alteration and degradation have led to dramatic reductions in wild animal abundance and distribution over the past century (Fig. 11.1). In particular, we have seen a fundamental shift from a world replete with wild mammals to one dominated by humans and our over 4.5 billion domesticated livestock (Barnosky 2008; Smith et al. 2016b, 2016d, 2018). Sadly, today somewhat less than 10% of terrestrial mammal biomass is made up of wild animals (Smith et al. 2016a, 2016d). This transformation resulted from anthropogenic extinction coupled with sharp declines in the abundance of native species in most major terrestrial biomes; these processes are ongoing and accelerating (Fig. 11.2; Ceballos and Ehrlich 2002; Dirzo et al. 2014; WWF 2018; IPBES 2019). For example, since 1970, there has been a 60% decline in vertebrate wildlife populations (WWF 2018). African lions have declined by 43% since 1993—and by much more than that if you consider their prehistoric range (Henschel et al.

2014). Giraffes are in even worse shape, particularly as we now appreciate that what we thought was one species is actually four genetically distinct ones (Fennessy et al. 2016). Even mammal taxa so common that they are considered of "least concern" by the International Union for Conservation of Nature (IUCN) are decreasing in abundance and showing contractions in their geographic range (Ceballos et al. 2017). Dwindling populations of common species is an alarming sign of the seriousness of biodiversity loss; after all, it is the first step along the trajectory toward extinction (Ceballos and Ehrlich 2002; Ceballos et al. 2017). And of course, functional extinction precedes species extinction and can occur more rapidly (Dirzo et al. 2014). For example, the death of Sudan, the last male Northern white rhino (*Ceratotherium simum cottoni*), in March 2018 left only two individuals—both female—making the species functionally extinct ("Last Male Northern White Rhino Is Put Down," Hannah Ellis-Petersen, *The Guardian*, March 20, 2018).

In sum, the current status of biodiversity on Earth is dire. More than 1 million species of plants and animals face extinction in the near term (Fig. 11.2), including most of the remaining large-bodied mammals on Earth (WWF 2018; IUCN 2020). Given our current trajectories, the largest mammal on Earth in two hundred years may well be a domesticated cow (Fig. 11.1; Smith et al. 2018). As the famous conservation biologist Thomas Lovejoy presciently noted some forty years ago, "*Reduction of 10 to*

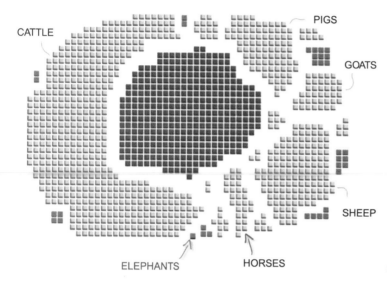

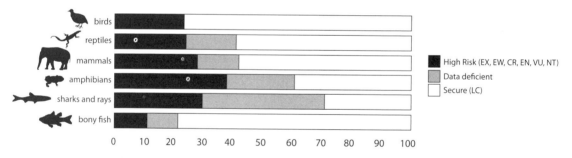

Fig. 11.1. The current biomass of Earth's nonvolant land mammals. Over the past few hundred years, in particular, most of the wildlife around the world has been replaced with humans and our domesticated livestock, the latter provisioned with supplemental food. Indeed, when livestock are included, the biomass of mammals is now greater than even in the terminal Pleistocene. This has led to widespread habitat modification around the globe. Modified with permission from xkcd .com (http://imgs.xkcd.com/comics/ land_mammals.png) and redrawn from Smith et al. 2016b.

Percentage of species in each IUCN risk category

Fig. 11.2. Global extinction risk for different vertebrate clades as determined by the International Union for Conservation of Nature (IUCN). Black bars indicate the highest risk categories: extinct (EX), extinct in the wild (EW), critically endangered (CR), endangered (EN), vulnerable (VU), and near threatened (NT). White bars measure those that are secure (or of least concern, LC). However, the risk is likely even greater than shown because many of the data-deficient (DD) mammals (in light gray) are probably also at risk, but we know too little about their biology or population dynamics to make informed assessments. Redrawn from IPBES 2019; animal silhouettes from PhyloPic (http://phylopic.org) under Creative Commons license.

20 percent of the Earth's biota will occur in about half a human life span. . . . This reduction of the biological diversity of the planet is the most basic issue of our time" (from Foreword, *Conservation Biology*, edited by Michael Soulé and Bruce Wilcox, 1980). Sadly, this may prove to have been an underestimate of the actual degree of the anthropogenic erosion of biodiversity.

Is the Current Rate of Biodiversity Loss Unusual in Earth History?

Things go extinct. As paleontologists are fond of saying, extinction is the ultimate fate of all organisms. Thus, in the evolutionary record of life, there is always a level of "background extinction," or species turnover. This background level fluctuates over geologic time (Raup and Sepkoski 1982). However, interwoven with the normal pulses of turnover over the past 540 million years have been a series of punctuated events identified as an abrupt and synchronous loss of at least 75% of taxa (Fig. 11.3); these are generally referred to as mass extinctions and result in the unraveling of Earth ecosystems (Raup and Sepkoski 1982; Sepkoski 1984). Some have proposed that the ongoing biodiversity crisis is the Earth's sixth mass extinction and that it will have similar effects (Myers 1990; Leaky and Lewin 1992; Kolbert 2014; Ceballos et al. 2015).

Mass Extinctions in the Fossil Record

Paleontologists generally recognize five instances over the past 540 million years when extinction rates were exceptionally high for geologically brief time intervals (Fig. 11.3). These occurred near or at the end of the Ordovician, Devonian, Permian, Triassic, and Cretaceous periods (Raup and Sepkoski 1982; Sepkoski 1984). While the exact magnitude of the loss of species varies among these events, and also depends on the analytic technique employed to characterize them, all of these episodes had extinc-

tion rates in excess of 75%. While the most famous extinction is undoubtedly the end-Cretaceous when the charismatic *T. rex* and other non-avian dinosaurs met their sad fate, this event probably ranks only fourth or fifth in terms of lethality. Far more significant was the end-Permian (~252 Ma), which is often called the Great Dying. It is hard to fathom just how severe this extinction was—within just a few hundred thousand years, more than 95%–98% of species on Earth were lost (Benton and Twitchett 2003; Erwin 2006). It indiscriminately wiped out both terrestrial and marine life. Indeed, life on Earth came perilously close to complete annihilation (Benton and Twitchett 2003; Erwin 2006).

These five mass extinctions led to major reorganizations of life on our planet. They sparked evolutionary innovations in their aftermath and profoundly shaped the trajectory and history of life on Earth (Sepkoski 1981; Raup and Sepkoski 1982; Sepkoski 1984; Raup 1991; Bambach et al. 2002, 2007; Erwin 2006; Payne and Clapham 2012). Indeed, all life today descends from the fortunate 2%–5% of organisms that survived the Great Dying—this is even more impressive when you consider some of these lineages may not have survived the two subsequent mass extinctions at the end-Triassic and end-Cretaceous.

What Drove Previous Mass Extinctions?

There are many postulated causes for mass extinctions. Some have argued they are cyclic, thus suggesting an extraterrestrial driver, such as cosmic bombardment caused by cyclic proximity to an asteroid belt or another planet (Raup and Sepkoski 1984; Rohde and Muller 2005). This idea is controversial, however, and it is not clear that there is any regularity to pulses in extinction. Bolides have been implicated, especially for the well-studied end-Cretaceous event approximately 66 Ma, where the impact site was identified on the Yucatan peninsula (Alvarez et al. 1984; Hildebrand et al. 1991). But it appears unlikely that all mass extinctions were due to impact events. Large igneous provinces (LIPs), which are massive simultaneous eruptions of multiple volcanoes that flood hundreds of thousands of square kilometers with lava, have been hypothesized to drive some events, such as the end-Permian (Benton and Twitchett 2003; Erwin 2006; Payne and Clapham 2012). Other causal mechanisms

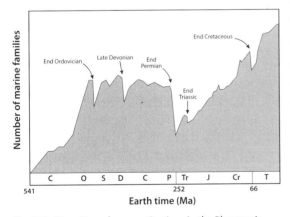

Fig. 11.3. Diversity and mass extinctions in the Phanerzoic fossil record. The curve is drawn after the famous Sepkoski curve (1981). Data are compiled at the family level and are shown for marine taxa that have a well-resolved record that extends throughout the Phanerozoic and are also common fossils. Note the five "big" mass extinctions at the end of the Ordovician (O), Devonian (D), Permian (P), Triassic (Tr), and Cretaceous (Cr). While these varied in their intensity, all resulted in the loss of more than 50% of life on Earth. The end-Permian, in particular, is known as the Great Dying because so much biodiversity was lost. Some estimates suggest as many as 97% of all species perished.

may include gamma ray bursts (pulses of intense radiation that can occur when a high-mass star collapses to form a neutral star or black hole), which have been proposed for causing the Ordovician extinction (Melott and Thomas 2009). The onset of glaciation, perhaps because of the rapid release of the potent greenhouse gas methane from methane clathrates, can trigger widespread extinction (Brand et al. 2016). And, as it turns out, biological innovations may even have led to the Devonian mass extinction (Algeo et al. 1995).

While the Devonian is sometimes called the age of fish, on land significant evolutionary events were also occurring. Although plants had colonized land by the Ordovician, they were fairly primitive, relatively small, and lacked roots, stems, or leaves (Thomas and Spicer 1987). By the early Devonian, early vascular plants had evolved, although they were still tied to water to complete their reproductive cycle. The evolution of the seed changed things. It allowed plants to reduce or eliminate their dependence on water for fertilization, freeing plants from a reliance on moist, lowland habitat (Thomas and Spicer 1987). Thus, the combination of vascular structure and seeds allowed plants to rapidly colonize terrestrial surfaces and develop much larger in size. Indeed, within a few million years, the maximum height of trees had increased from about 1 m to over 30 m and multistoried forests had arisen (Algeo and Scheckler 1998). As trees increased in size, they evolved ever larger, deeper, and more complex root structures. These penetrated deep into the ground, increasing the rate of rock weathering and pedogenesis (soil formation) as well as the release of minerals bound in rocks (Algeo et al. 1995; Berner 1997; Algeo and Scheckler 1998). Dissolved minerals flowed from the land into the oceans and likely triggered algae growth. As algae died, their decay may have caused ocean anoxia and the creation of dead zones (Algeo and Scheckler 1998). Moreover, a long-term effect of the rapid spread of plants across continents was a drawdown in carbon dioxide from the atmosphere and widespread global cooling, which likely led to a brief glaciation (Algeo et al. 1995; Berner 1997; Algeo and Scheckler 1998). Thus, by mediating weathering processes, plants influenced the geochemical linkage between the lithosphere and atmosphere.

Proximate Drivers of Extinction

What all of these events have in common is the destabilization of ecosystems around the Earth and the consequential extinction of species (Fig. 11.4). While the ultimate cause might be a bolide or LIP, the proximate mechanisms are changes in air and water chemistry and in environmental temperature, which kill species directly or indirectly through disruption of ecosystem function (Fig. 11.4; Roopnarine 2006; Payne and Clapham 2012). For example, the injection of sulfur dioxide (SO_2) into the atmosphere by volcanic eruptions can lead to short-term volcanic darkness, cooling, and photosynthetic shutdown; the infamous 1816 "year without a summer" was caused by the 1815 eruption of Mount Tambora in the Dutch East Indies. Sulfur dioxide was rapidly dispersed around the hemisphere and its conversion to sunlight-blocking sulphate aerosols drove short-term planetary cooling (Oppenheimer 2003). In the case of Mount Tambora, it led to global cooling of up to 0.7°C and widespread food shortages (Stothers 1984; Oppenheimer 2003). Parts of the eastern United States and Europe were particularly hard hit, experiencing snowfalls as late as August and massive crop failures. This is a common effect of volcanic eruptions. The 1991 Mount Pinatubo eruption injected 20 megatons of SO_2 into the stratosphere and led to a decrease in global temperatures of around 0.5°C for three years, while the 1980 eruption of Mount St. Helens lowered global temperature much less, by about 0.1°C ("The Cataclysmic 1991 Eruption of Mount Pinatubo, Philippines," USGS Fact Sheet 113-97, https://pubs.usgs.gov/fs/1997/fs113-97; "Volcanoes Can Affect Climate," USGS Volcano Hazards Program, https://volcanoes.usgs.gov/vhp/gas_climate.html). The release of SO_2, chlorine, and fluorine by eruptions also contributes to ozone depletion and acid rain, which not only kill life but also increase continental rock weathering and the subsequent flow of nutrients into aquatic environments (Fig. 11.4). And volcanoes release toxic metals, which can poison both marine and terrestrial ecosystems. The lethality of LIPs is influenced by where they occur, the rock substrate, and the type of event. Basalts, for example, are rich in sulfur dioxide. Volcanic darkness and cooling rarely last more than a few years after an eruption.

Eruption of giant volcanic province

Release of SO₂

Release of CO₂

Minor acidification
of ocean surface

Minor global
warming

Release of CO₂

Release of CH₄
from clathrates

Rapid global
warming

Ocean anoxia

Thermal expansion of
oceans, leading to
transgression

Changes in
ocean nutrients

Phytoplankton
extinction

Mass Extinction

Fig. 11.4. Both bolides and large igneous provinces can have catastrophic influences on the ecosystem. Shown here is part of the cycle of devastation that can follow the eruption of a giant volcanic province. While the ash and lava flows themselves have impacts—which depend on the extent and type of underlying rock—these lead to more regional effects than the atmospheric alterations. Several of the big-five mass extinctions were accompanied by large igneous provinces; whether they were the primarily underlying cause of the mass extinction is still a subject of some debate.

The long-term and potentially most catastrophic effect of massive eruptions is in the release of greenhouse gases into the troposphere and stratosphere (Fig. 11.4). The buildup of carbon dioxide in the atmosphere leads to a warming climate and rising sea levels, which can flood fragile nearshore habitats. For example, taxa dwelling in shallow water habitats were more likely to become extinct in the Ordovician mass extinction than those in the deep benthos (Finnegan et al. 2016). Changes in the ocean-atmosphere interface or in ocean temperatures also influence the amount of dissolved oxygen in ocean waters, creating hot and/or anoxic conditions that suffocate marine life. The Siberian Traps, an immense volcanic complex thought to be responsible for the end-Permian extinction, is estimated to have released 30,000 gross tons (1 GT = 10^{15} g) of carbon into the atmosphere (Svensen et al. 2009); ocean and soil temperatures may have risen by more than 6°C (Benton and Twitchett 2003). Changes in ocean circulation, especially of deep-oceanic currents that aerate the deep oceans, can also lead to the upwelling of nutrients and toxic material from oceanic depths.

Volcanism can eventually trigger glaciation. For example, silicate rocks deposited by volcanism draw CO_2 out of the air as they weather. During the late Ordovician, climate warming driven by greenhouse gas emissions from major volcanism was likely balanced by weathering of the uplifting Appalachian Mountains, which sequestered CO_2. After volcanism ceased, the continued weathering caused a significant and rapid drawdown of CO_2 in the atmosphere leading to glaciation. Glaciation can be especially detrimental if biota were previously adapted to a greenhouse world. Sequestering of water in ice sheets leads to a decline in sea level, which reduces habitat available for nearshore organisms. And shifts in ocean circulation or sea level can result in the upwelling of anoxic water from deep oceans, which deoxygenates upper levels of the water column. In short, the effects of LIPs are complicated.

The consequences of impacts are also complicated. Short-term effects include tsunamis, blast damage, earthquakes, and wildfires (Toon et al. 1997; Robertson et al. 2013). These can lead to widespread destruction, especially near the impact site. Even small bolides can have significant impacts. For example, in 2013 a small meteor entered Earth's atmosphere near Chelyabinsk, Russia. While only about 20 m in diameter, it was traveling at around 19 km/sec (Popova et al. 2013; Kring and Boslough 2014). Although atmospheric drag was sufficient to stop it well before it reached the ground—it exploded about 29.7 km above the surface—the resulting airburst still caused significant damage. The asteroid was moving faster than air molecules could be displaced, thus, as it descended into the atmosphere, a rapidly thickening compressed layer of high-temperature plasma developed (Kring and Boslough 2014). When the dynamic pressure became too great, the meteor exploded, generating a bright flash seen over 100 km away. This intense light was momentarily about thirty times brighter than the sun; many people watching it developed temporary flash blindness and/or ultraviolet skin burns (Popova et al. 2013; Kring and Boslough 2014). The asteroid also produced a hot cloud of dust and gas that extended

26.2 km from the impact and a large shock wave from the airburst. The shock wave was estimated to be the equivalent of around 500 kilotons of TNT—about twenty to thirty times the energy released from the Hiroshima atomic bomb—but fortunately most of this was absorbed by the atmosphere (Popova et al. 2013; Kring and Boslough 2014). Nonetheless, some 7,200 buildings in six cities across the region were damaged and an estimated 1,500 people injured. And this was a very small bolide that didn't actually even impact the surface of the planet.

Over geologic time, our planet has been bombarded by bolides many times (Fig. 11.5). As of 2020, about 190 terrestrial impact craters have been identified around the planet (Earth Impact Database, Planetary and Space Science Centre, University of New Brunswick, www.passc.net/EarthImpact Database/New%20website_05-2018/Index.html). However, this is certainly an underestimate of how often impacts have occurred, since craters are continually erased by erosion and other resurfacing activities, and moreover, most bolides probably land in the ocean. The extent and duration of the damage caused depends on the size of the bolide, impact

Bolide events 1994-2013
(Small asteroids that disintegrated in the Earth's atmosphere)

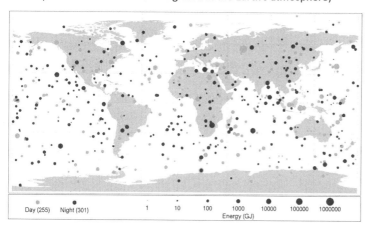

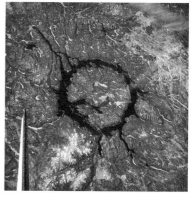

Fig. 11.5. The Earth has been hit by objects throughout its long history. *Left*: Shown here are 556 bolides that collided with Earth from 1994 to 2013; with the exception of the Chelyabinsk event, most asteroids were small enough to disintegrate in the atmosphere. The sizes of the dots are proportional to the energy released as measured in billions of joules (GJ) of energy; they represent objects ranging in size from about 1 to almost 20 m. The light gray are impacts that occurred during the day and the dark gray, at night. *Right*: Manicouagan Reservoir, Canada. This site is an impact crater some 100 km wide that was formed approximately 212 Ma. As might be expected, the crater has been worn down by glacier activity and erosion. Map courtesy of NASA Near Earth Object Program; Photo by NASA, courtesy Wikipedia Commons.

angle, and speed when it enters the atmosphere (Toon et al. 1997). Bolides 6 km or more in diameter are large enough to produce sufficient dust and sulfate to reduce light levels so much that vision would be nigh impossible and the fundamental process of photosynthesis would be halted (Toon et al. 1997). Land temperatures would drop below freezing, particularly in the interior of continents, and drought might ensue (Toon et al. 1997). The blast damage, earthquakes, and fires caused by ballistic ejecta would cover at least 10,000,000 km² of the surface and disproportionately more with larger bolides (Toon et al. 1997). For example, the ubiquity and thickness of the soot in the end-Cretaceous global debris layer suggests that much, if not all, of the terrestrial biosphere burned (Ivany and Salawitch 1993; Toon et al. 1997; Schulte et al. 2010; Robertson et al. 2013); the K-Pg bolide was about 10 km in diameter. If this wasn't enough disruption, depending on where the impact occurs, shock waves from a 6 km bolide could generate giant tsunamis 100 m or more tall, which can surge some 20 km inland. The K-Pg tsunami was estimated to have been 50-100 m tall when it reached Texas, a considerable distance from the impact site (Toon et al. 1997). But, these short- to medium-term effects are again dwarfed by the long-term impacts on Earth chemistry and climate (Toon et al. 1997; Schulte et al. 2010). And many of these are similar to the effects of massive volcanism just discussed, making it challenging to differentiate causal mechanisms for mass extinctions in the fossil record.

Are We In the Midst of a Sixth Mass Extinction?

The Earth hasn't been whacked by a large bolide lately, nor have massive volcanic provinces erupted. But, over the past one hundred years, humans have urbanized more than 75% of the planet, drastically reduced the populations of wild animals by harvest or usurping their habitats, pumped enormous amounts of greenhouse gases into the atmosphere, and released toxic compounds into terrestrial and marine environments. Do these ongoing activities, and their current and future impact on Earth's biota, rise to the level of a sixth mass extinction, as some have argued (e.g., Myers 1990; Leaky and Lewin 1992; Kolbert 2014; Ceballos et al. 2015)? Are current levels

of biodiversity loss comparable to mass extinctions in the fossil record?

The answers to these questions lie firmly in the realm of paleontology. We typically characterize past extinctions by the magnitude of the event (i.e., percentage of species lost) and, less frequently, by the rate of extinction (i.e., number of species lost per unit of time). The first is difficult to calculate for current biodiversity loss as the loss itself is ongoing, so we turn toward the latter approach. Whereas a precise answer is both complicated and taxon-specific (Jablonski 1994), it is not difficult to approximate a generalized species turnover rate. We start with the observation that the typical duration of a species in the fossil record is about 1-3 million years, although there is substantial heterogeneity (Raup 1981, 1991; Foote and Raup 1996; Alroy 2000; Vrba and DeGusta 2004) and the record may be missing many species (Plotnick et al. 2016). Using an average "life-span" for a taxon of 2 million years, that means that in any given year it has a 2×10^{-6} chance of "dying." Thus given, say, the 5,500 species of mammals found on Earth today, we'd expect about 0.00275 species would be lost due to background extinctions in a year (i.e., $5,500/2 \times 10^6$). However, paleontologists don't really think about time scales as short as a year. Translating into more reasonable durations, this is equivalent to 2.75 extinctions in 1,000 years, or 27.5 extinct species in 10,000 years. It should be noted that this computation is a bit dodgy because mean rates are sensitive to the interval over which they are measured, and the mean durations of species can vary substantially even within a clade (Raup 1981; Gardner et al. 1987; Gingerich 1993; Foote 1994; Jablonski 1994).

Unlike paleontologists, modern conservation biologists *are* interested in short time scales. Thus, they typically standardize extinction rates into the number of extinctions per million species per year, known as the E/MSY. There is considerable disagreement among conservationists about what constitutes a "normal" rate, but most authors use a value between 0.1 and 2 E/MSY, which seems pretty reasonable given the vagaries of the fossil record (May et al. 1995; Pimm et al. 1995; Barnosky et al. 2011; Ceballos et al. 2015). For example, standardizing our average per capita annual rate death of 2×10^{-6} above into these units yields an E/MSY of 0.5 for mammals (i.e., 1,000,000 species/2×10^6), consistent with published values.

So, how does the rate of anthropogenic extinction compare to the normal turnover in the geologic record? Are current rates so elevated that they rise to the level of a mass extinction? This is where it gets complicated because we are mixing geological and ecological records (Plotnick et al. 2016). Paleontologists generally conduct analyses at the level of genera or families because preservation becomes patchier at lower hierarchical levels, and because the necessary reliance on a morphological species concept means that we are likely underestimating species-level diversity. This can lead to problems. For example, while about 40% of modern mammal genera have a fossil record, only roughly 20% of mammal species do (Plotnick et al. 2016). Moreover, not only are we attempting to compare a baseline formulated with information collected at a geological scale with an ongoing process operating at the ecological scale, but we also have the issue of the dependence of rates on time interval (Plotnick et al. 2016). The shorter the interval, the higher the perceived rate (e.g., Gardner et al. 1987; Gingerich 1993; Foote 1994). Thus, at the very least, it is imperative to try to standardize time intervals.

With these caveats in mind, estimates in the literature suggest that the current rate of biodiversity loss far exceeds "normal" turnover. Indeed, current extinction rates are more than one hundred times, or two orders of magnitude, higher than the background level through the Phanerozoic (Myers 1990; May et al. 1995; Pimm et al. 1995; Barnosky et al. 2011). Recent work suggests a rate of 200 E/MSY, meaning it should have taken between 800 and 10,000 years for the species that have gone extinct in the last century to disappear (Ceballos et al. 2015). Moreover, the acceleration of biodiversity loss means extinction rates could exceed 1,500 E/MSY in the near future (Pimm et al. 1995). Overall, the current extinction rate *is* higher than those estimated to have led to the big five mass extinctions in the fossil record, although some caution is necessary because of differences in the temporal scale of these estimates and incompleteness of the fossil record (Barnosky et al. 2011; Plotnick et al. 2016). Better temporal calibration of previous extinctions may further refine rate estimates (Plotnick et al. 2016). Nonetheless, if sustained, it appears that

we might reach the mass extinction threshold in less than 350 years (Barnosky et al. 2011). So, while we are not *quite* at the level of a sixth mass extinction, we *are* firmly on the trajectory toward it (Jablonski 1994; Barnosky et al. 2011).

A New Driver of Extinction—Hominins

While the outcome may be the same—the catastrophic loss of life—the mechanism behind the ongoing extinction is fundamentally different than earlier events in the geological past. It is clear that the evolution of hominins was a watershed event in Earth history. With the possible exception of microbes, no other taxon has come to so profoundly dominate ecological ecosystems. How did we transform from nondescript primates into efficient, novel predators that spread rapidly around the world and reshaped virtually all terrestrial surfaces?

The Advent of Hominins

Humans belong to the family Hominidae—primates without tails—that today consist of orangutans, gorillas, chimpanzees, and humans (Wood and Richmond 2000). Hominids arose during the Miocene, and at the onset were just another animal on the landscape, probably not particularly abundant or noteworthy. Most species were probably arboreal, mostly frugivorous, and largely restricted to tropical forests; the modern strictly arboreal orangutan is probably the best extant model for what the lifestyle was like (Robson and Wood 2008). Around 12-13 Ma, the gorilla/chimp/hominin lineage spilt from the orangutans (Glazko and Nei 2003). Although still largely confined to arboreal habitat, these early primates were probably terrestrial knuckle-walkers when on the ground. They, too, were herbivores, eating mostly fruit and leaves (Wood and Richmond 2000). Somewhere around 7-8 Ma, the gorilla lineage split off from the chimpanzee/human line (Glazko and Nei 2003). By this time, there was a great diversity of primate species varying in both anatomy and ecology (Kunimatsu et al. 2007; Stringer and Andrews 2012). Shortly thereafter, about 6 Ma, hominins separated from the chimpanzees (Wood and Richmond 2000; Glazko and Nei 2003; Fig. 11.6). The principal defining characteristics of hominins include bipedality, which evolved by 4-5 Ma; the

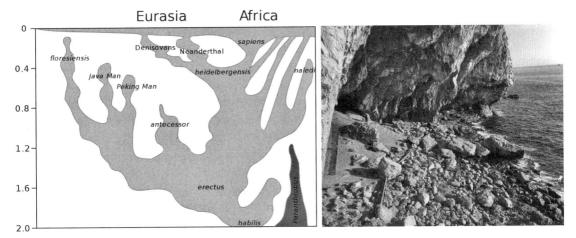

Fig. 11.6. The phylogeny of humans. *Left*: While still in debate, this figure illustrates the hypothesized relationships between various *Homo* species. Note that here *H. heidelbergensis* is shown as an ancestor of Neanderthals, Denisovans, and *H. sapiens*, which are all sister species. Recent DNA evidence suggests that after early *H. sapiens* emerged from Africa, they interbred with Neanderthals and Denisovans. The emerging consensus seems to be that the "Hobbit people," *H. floresiensis*, arose from *H. erectus*. *Right*: Gorham's Cave in the British overseas territory of Gibraltar. This UNESCO World Heritage site represents one of the last known habitations of Neanderthals in Europe. It has yielded many important fossils. "Homo lineage 2017update.svg" by Conquistador, CC BY-SA 4.0, courtesy of Wikimedia Commons; Photo by author.

progressive evolution of smaller, blunt canines; and increased brain capacity (Wood and Richmond 2000).

Bipedalism was an enormously important evolutionary innovation. By standing upright, the hands were freed for reaching and carrying food and, later, for using tools; the earliest known stone tools appeared approximately 2.6 Ma from Kada Gona at Hadar in Ethiopia (Semaw et al. 2003). Moreover, walking is energy efficient, requiring very little muscular activity (Alexander 1992). It also allows long distance running and hunting while providing an enhanced field of vision to watch for prey or predators (Kwang Hyun 2015). But by becoming good walkers, hominins probably became poorer climbers, less able to access arboreal resources or escape from terrestrial predators. What led to this critical adaption remains unclear. Was it because of the origin of monogamy, to carry or access important resources, to reduce overheating by exposing less of the body to sunlight, or in response to changing climates that necessitated the crossing of open grasslands (Rodman and McHenry 1980; Lovejoy 1988; Hunt 1994)? Nonetheless, it predated the evolution of large brains in hominins and was an important step in the transition to *Homo*.

The Genus *Homo*

Our genus arose around 2-3 Ma, likely from an australopithecine ancestor (McHenry and Coffing 2000). As yet we are not sure of our last common ancestor or closest cousins; hominin taxonomy is both complicated and controversial, with disagreement about the timing, validity, and placement of species (Wood and Richmond 2000; White et al. 2009; Stringer 2012, 2016). Our species, *Homo sapiens*, arose around 200 ka (Stringer 2016). We were one of several extant hominins on the landscape in the Pleistocene—distinct from Neanderthals and Denisovans, although not reproductively isolated. Indeed, today most humans contain genetic traces of earlier introgression from these extinct human groups (Stringer and Barnes 2015; Stringer 2012; Sankararaman et al. 2016; Vernot et al. 2016). Along with the evolution of the genus *Homo*, came changes in brain size, reductions in cheek teeth, and a change in walking and climbing behavior (Stringer 2016). Indeed, it was not until after the emergence of *H. erectus* that hominids grew tall, evolved long legs, and became completely terrestrial creatures (McHenry and Coffing 2000; Wood and Richmond 2000; Antón et al. 2014).

Somewhere around 2 Ma, an earlier member of our genus, *Homo habilis*, began using primitive stone tools to butcher animal carcasses, adding calorie-dense meat and bone marrow to their diet (Ferraro et al. 2013). Carnivory seems to have been present only in the *Homo* lineage, and especially prevalent in a later species, *Homo erectus*; earlier hominins do not display the anatomical features generally linked to a meat-based diet (McHenry 1992; Aiello and Wheeler 1995; Antón 2003, 2014; Braun et al. 2010). The transition from a plant-based diet led to further evolutionary innovations (Bunn and Ezzo 1993; Aiello and Wheeler 1995; Foley 2001; Aiello and Wells 2002; Antón 2003, Antón et al. 2014). For example, not only is meat energy-rich, but it provides many essential amino acids and micronutrients. Thus, this shift in diet may have provided a catalyst for hominin evolution, particularly the growth of the brain (Milton 1999). Certainly, the morphology of *Homo erectus* reflected the adoption of an increased consumption of animal products (Andrews and Martin 1991; Milton 1999; Antón 2003; Watts 2008; Antón et al. 2014). Brain size increased, as did body size, and both tooth and gut size decreased (McHenry 1992; Aiello and Wheeler 1995; Antón 2003; Braun et al. 2010).

There is abundant fossil evidence demonstrating the shift in hominins toward carnivory. Many *Homo habilis* remains in East Africa have been found associated with butchered animal bones and simple stone tools, and tool use became more sophisticated and regular with *Homo erectus* (Blumenschine and Pobiner 2006; Roebroeks and Villa 2011; Joordans et al. 2015; Shea 2017). Multiple sites at Olduvai Gorge, Tanzania, (~1.8 Ma) contain butchered mammal bones, ranging from hedgehogs to elephants, as well as numerous stone tools (Domínguez-Rodrigo et al. 2007; Blumenschine and Pobiner 2006).

Just how these early hominins obtained meat is a still a subject of considerable debate (Domínguez-Rodrigo 2002; Bunn et al. 2007). Did they opportunistically scavenge from other predators, or were they active hunters? Early on it was likely some combination of the two, but over time hominins probably became more efficient and effective hunters (Bunn and Ezzo 1993; Domínguez-Rodrigo et al. 2002; Luca et al. 2010). Indeed, by the mid-

Pleistocene, hominins may have been formative predators—to the point of potentially outcompeting other carnivores. A drastic decline in carnivore functional richness in African ecosystems has been reported at about this time (Werdelin and Lewis 2005). Moreover, stable isotope analysis performed on well-preserved Neanderthal fossils reveals that their bone collagen was heavily enriched in ^{15}N, consistent with a high degree of carnivory (Fizet et al. 1995; Richards et al. 2000; Bocherens et al. 2001; Sponheimer et al. 2013). And genetic analysis conducted on plaque extracted from Neanderthal teeth revealed not only that these ancient hominins had poor dental hygiene but also that they fed on woolly rhinoceros and mushrooms (Weyrich et al. 2017). Finally, the abundance of Paleolithic cave paintings depicting bison, horses, aurochs, mammoth, and reindeer—some drawn with spears or arrows through them—highlight the importance of these mammals to early hominin hunters (Whitley 2009). Found in the Dordogne region of southwestern France, cave complexes such as Chauvet, Lascaux, Les Combarelles, and Font-de-Gaume are replete with realistic depictions of megafauna once native to the region (Fig. 11.7). These are truly remarkable because of their exceptional artistry, sophistication, and antiquity (over 20,000–32,000 years old). I recently had the privilege of visiting Les Combarelles and Font-de-Gaume; the latter contains some of the last polychromatic cave paintings still open to the public. Seeing these incredible drawings of the extinct animals I study—drawn by an early hominin who had actually watched or hunted them—was exceptionally moving.

As hominins evolved, they developed increasingly successful hunting and scavenging technologies, which allowed them to more efficiently target large-bodied prey. The earliest unequivocal evidence of spear points dates from around 500 ka; more complex projectile weapons, such as the atlatl, were in use by around 71 ka (Brown et al. 2012; Wilkins et al. 2012). And, not long after adopting a meat-based diet, hominins "discovered" fire. The oldest unambiguous evidence for the controlled use of fire at hearths is from about 790 ka (Gowlett et al. 1981; Goren-Inbar et al. 2004; Wrangham 2009; Berna et al. 2012). The use of fire allowed the manufacture of more sophisticated tools; cooking also may have

Fig. 11.7. Prehistoric cave art yields details about the population ecology and dynamics of extinct species and, moreover, how early *Homo* hunted and/or viewed them. The Grotte de Font-de-Gaume, a cave near Les Eyzies-de-Tayac-Sireuil in the Dordogne department of southwest France is one of the last sites with polychromatic cave paintings that still can be viewed by the public. The beautiful and moving images include more than 240 figures of bison—the dominant motif—as well as mammoth, horses, reindeer, deer, auroch, goat, wolf, bear, and rhinoceros. There are also enigmatic hand outlines and geometric figures. Because photographs of the paintings are not permitted, sketches are posted outside the caves. **a**, Sketch of one of the most iconic paintings within Font-de-Gaume on the left wall in the Galerie des Fresques. Here, a male reindeer is leaning over, licking the forehead of a smaller, presumably female, reindeer. The tongue is engraved and only visible with oblique light. Also, note the tectiforms at the bottom left. While still debated, some believe these schematic lines and dots represent huts or larger shelters built of logs and covered with hides. **b**, Two mammoths. More than twenty images of extinct mammoth are contained within Font-de-Gaume. **c**, Horses, reindeer, ibex, and mammoth. Note the shaggy indications of fur on the mammoth. Many of the images are overlaid on older images, reflecting the long use of this cave. **d**, The entrance to the Grotte de Font-de-Gaume. Most of the paintings are situated along a 120 m gallery and were probably done around 17-25 ka. The cave mouth is concealed by rocks and trees. Once inside, you travel down a narrow twisting passage of irregular height and width, and paintings are seemingly everywhere. The number of visitors is highly regulated and dependent on environmental conditions within the cave. **e**, Part of a frieze of bison in the Galerie des Fresques; this image was not discovered until 1966. The bison are painted in reddish-brown colors. By using the natural shape of the rock to form part of the painting, the resulting image became remarkably lifelike and full of movement. All photos by the author.

allowed more energy and nutrients to be extracted from food (Wrangham 2009).

Thus, by the late Pleistocene, hominins were likely wide-ranging generalist carnivores who foraged effectively on a broad array of mammal species, including each other. In short, we were now excellent predators.

Out of Africa

During most of the early Pleistocene, hominins were restricted to Africa, where they had evolved. The wholescale migration out of the continent began about 2 Ma as *Homo erectus* moved into Eurasia (Stiner 2002; Rightmire et al. 2017). Fairly quickly thereafter, hominins had diversified and spread to many habitats within the temperate and tropical zones; indeed, the geographic distribution of *Homo* now stretched across almost forty-seven degrees of latitude (i.e., 40° N to 7° S). The exact trajectory of early hominin migration and reconstruction of the evolutionary relationships between species are still in flux. Thus, there is no scientific consensus on the routes, exact timing, or number of initial migrations as yet. However, it is likely that *Homo sapiens* left Africa in pulses somewhere around 100 ka (Walter et al. 2000; Carto et al. 2009; Stringer and Andrews 2012; Groucutt et al. 2015; Stringer 2016). What led to their emigration from Africa is unclear. Humans may have been following migrating herbivore herds into Eurasia (Carto et al. 2009; Muttoni et al. 2018). It is also possible that changing environmental conditions could have prompted or facilitated movement out of Africa (Antón 2003; Carto et al. 2009; Larrasoaña et al. 2013; Antón et al. 2014; Parton et al. 2015; Breeze et al. 2016; Timmermann and Friedrich 2016; Tierney and Zander 2017; Muttoni et al. 2018). For example, the migration from Africa occurred during a time of Heinrich events—climate episodes precipitated by the massive release of icebergs and the subsequent deposition of large volumes of freshwater in the North Atlantic (Carto et al. 2009). Along with changes in precipitation and the onset of glaciation, Heinrich events likely led to abrupt changes in climate and vegetation, which may have opened migration corridors (Carto et al., 2009; Tierney and Zander 2017). Modeling efforts support the role of climate variability and sea level changes in the

migration of humans around the globe (Timmermann and Friedrich 2016).

Regardless of the exact route(s) taken or the number of migration pulses, certainly by the late Pleistocene (~80-100 ka) several species of hominins were present in Eurasia and several more in the Philippines. Our extended family included archaic humans as well as Neanderthals, the enigmatic Denisovans, *H. floresiensis*, and *H. luzonensis* (Walter et al. 2000; Carto et al. 2009; Krause et al. 2010; Stringer 2016; Détroit et al. 2019). We know very little about some of these hominins, although DNA evidence is increasingly demonstrating that we interacted and interbred with them; most modern humans have a small amount of DNA derived from extinct Neanderthals and/or Denisovans (Stringer 2012; Stringer and Barnes 2015; Sankararaman et al. 2016; Vernot et al. 2016). The further expansion of *Homo sapiens*—first into Melanesia, and then into Australia—happened by 50-60 ka. And humans crossed the Bering Land Bridge and entered the New World around 13-15 ka (Dixon 1999; Bowler et al. 2003; O'Connell and Allen 2004; Goebel et al. 2008; Oppenheimer 2012; Timmermann and Friedrich 2016). Humans even managed to reach many of the most remote areas on the planet before the advent of modern technology; colonization of much of remote Oceania happened between 3,400 and 800 years ago (Montenegro et al. 2016). And, as humans colonized new habitats around the globe, they also rapidly increased in abundance (Carto et al. 2009; Timmermann and Friedrich 2016; Fig. 11.8). Population growth seems to have really accelerated around 50,000 years ago (Atkinson et al. 2008).

The transition of hominins to a meat-based diet, coupled with their rapid expansion into most habitats on the planet and contemporaneous precipitous increase in abundance, had detectable consequences for life on Earth. While the effects are most evident today, they actually predate modern society (Werdelin and Lewis 2005; Braje and Erlandson 2013; Lyons et al. 2016b; Malhi et al. 2016; Smith et al. 2016d, 2018, 2019). For example, *Homo erectus* was a key node in early Pleistocene food webs, significantly increasing extinction vulnerability of herbivores and competition with carnivores (Bibi et al. 2017). As the novel and efficient human

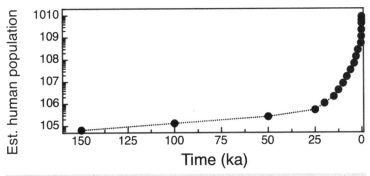

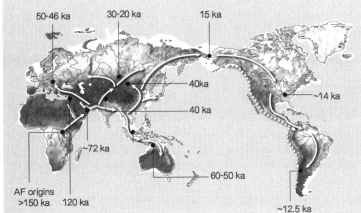

Fig. 11.8. Human population growth and migration over time. *Top*: Estimated human population over the past 150,000 years. These data include only *Homo sapiens* because information for other species of *Homo* is lacking. *Bottom*: Human migration patterns over the late Quaternary with approximate dates of arrival indicated. The routes and timing of human migration patterns are still debated. Data from Hern 1999; Map modified after Oppenheimer 2012 and Smith et al. 2019.

predators increased in abundance and dispersed across the Earth, the added mortality experienced by large-bodied mammals, in particular, led to temporally and spatially transgressive size-biased extinctions (Smith et al. 2018, 2019).

Late Quaternary Extinctions

The Earth was very different in the late Pleistocene (Fig. 11.9). The most obvious incongruity to a modern time-traveler would be the presence of numerous herds of large-bodied mammals and their associated predators. These included the familiar charismatic species, such as woolly mammoth, horses, rhinoceros, camels, llamas, saber-tooth cats, short-faced bears, and giant sloths whose fossils often grace natural history museums, as well as more obscure taxa such as *Glyptodon*—large-bodied, heavily armored mammals the size of a Volkswagen beetle (Kurtén and Anderson 1980). These extremely large-bodied megafauna were present on all the continents but Antarctica. Indeed, the Pleistocene megaherbivore community of the New World was

richer than that present in Africa today; the highest abundance occurred in South America (Lyons et al. 2004). This is the landscape and context in which hominin migrations occurred. Sometime between the late Pleistocene and today the majority of these species went extinct.

My colleagues and I have recently examined the pattern of extinction that followed the late Quaternary migration of hominins around the globe (Smith et al. 2018, 2019). This turns out to be complicated because of the nature of the fossil record. For example, ideally one would have well-dated mammal chronological collections from a number of sites that corresponded with well-dated locations of human activity—and this sort of information would be available globally. However, not only is this not available, but the trajectory and precise timing of hominin migrations are still debated. Thus, we took a large-scale macroecological approach. We compiled two spatially explicit global datasets: one for mammals across the entire Cenozoic, and a second for mammals at the late Quaternary. The first was

Fig. 11.9. A Pleistocene landscape from White Sands, New Mexico. An artist's depiction of what life in New Mexico looked like in the late Pleistocene. Shown are dire wolves, saber-tooth cats, sloth, camels, and American cave lions; a lagomorph is in the foreground. North America harbored an abundant and diverse megafauna assemblage in the Pleistocene. Painting by Karen Carr, commissioned for White Sands National Park.

constructed exclusively from fossil data and so spanned 66-1 Ma; we considered this to be largely prehominin impact. For the second late Quaternary dataset, we started with an earlier global database for modern mammals (Smith et al. 2003). To this, we added all mammals known to have gone extinct over the last 120,000 years. These efforts meant we spent a lot of time reading the literature and measuring the dimensions of old dusty fossils. Fortunately, we all share a love of fossils and tend to find these types of activities extremely rewarding.

For each mammal taxon in both datasets, we determined body mass (i.e., chapter 5), continental affiliation, extinction status (extant or extinct), and trophic group. We collapsed trophic information into four general guilds (herbivore, carnivore, insectivore, and omnivore) to reflect the major dietary modes. Here, it wasn't too important to more finely divide the type of diet, and moreover, it would have been hard to do so with the extinct species without independent isotopic or dental analyses. An important assumption we made was to ignore speciation; that is, we assumed extant mammal species were also present at the late Pleistocene. This is probably pretty reasonable given the average "life-span" for mammal species in the

fossil record is 1 million to 2 million years (Foote and Raup 1996; Alroy 2000; Vrba and DeGusta 2004).

We used these data to figure out what species were present on each continent and when. The time intervals we chose approximated the movement of hominins around the world. That is, the late Pleistocene (125-70 ka), when hominins entered Eurasia; end-Pleistocene (70-20 ka) when humans—and it was humans by this time—entered Australia; terminal Pleistocene (20-10 ka) when humans entered the Americas; and the Holocene (10-1 ka). Note that for the earlier time intervals when multiple species of hominins were present, we did not attempt to differentiate among them; we presumed that Neanderthals, Denisovans, and archaic/modern humans were all effective predators on other mammals, including each other. However, it is clear that by the terminal Pleistocene, humans were the main drivers of these extinctions. Finally, we examined the pattern of future extinctions by assuming all mammals currently listed as near-threatened, vulnerable, endangered, critically endangered, or extinct in the wild on the IUCN Red List (www.iucnredlist.org) will actually go extinct by two hundred years in the future.

We then calculated the statistical moments, magnitude of biodiversity loss, and trophic and body-size selectivity of extinction, both globally and for each continent individually, for each of our defined time intervals. Precise data for human population growth were hard to find, but we eventually settled on estimates produced by Hern (1999). While his information was somewhat imprecise, it was derived from the paleontological and archeological literature and it extended far enough into the fossil record for useful comparisons. We used the Cenozoic mammal dataset as a baseline to represent "normal" background extinctions. Data were binned into 1-million-year intervals, and we examined the pattern of extinction and survival within and between each bin.

Body Size Downgrading

Overall, we found a highly biased pattern of mammal extinction over the past 120,000 years (Smith et al. 2018, 2019). The loss of mammals was strikingly spatially and temporally transgressive, closely following the migration of humans around the globe over the late Quaternary (Fig. 11.10b). We saw a reduction in the average body mass first in Eurasia, then in Australia, and finally in the Americas. Indeed, each time hominins, and later humans, entered a new continent or landmass, a wave of mammal extinctions followed. A particularly interesting finding was that mammal mean body mass averaged 50% less on Africa than on other continents in our oldest time period. This was odd. A well-documented macroecological pattern is that both the maximum and mean body mass of mammals on a land mass is proportionate to the size of the landmass, probably reflecting the influence of resource availability (Burness et al. 2001; Smith et al. 2010a; chapter 5). Africa is the second largest continent, and thus we'd naturally expect that average mammal body mass would be higher on this continent than on North or South America. But it was not. An interesting hypothesis was that the reduced body mass reflected the long prehistory of hominins in Africa. Perhaps over the Plio-Pleistocene they had already had an impact on mammal communities before 120 ka (Smith et al. 2018)?

For the average body mass on a continent to decline means that the species extirpated at each of these intervals were disproportionately large-bodied. And they were. For example, the average victim of the late Pleistocene extinctions was more than three orders of magnitude larger in body mass than the average surviving species (Fig. 11.10f; Smith et al. 2018, 2019). They were also mostly herbivores, as would be expected if they are being exploited as a food resource (Smith et al. 2018, 2019). This was consistent throughout the late Quaternary: in each time interval, extinction probability increased significantly with larger body size, and herbivores were always particularly hard hit (Fig. 11.11). This pattern of extinction was strikingly different than the "normal" baseline of extinctions over the entire 66 million years of the Cenozoic (Fig. 11.10c; Smith et al. 2018, 2019). Being large-bodied or an herbivore did not increase extinction risk for most of mammal evolutionary history. The size and trophic bias to extinction were only present when hominins were.

The "downgrading" of mammal body mass over time was severe. For example, the average global body mass of terrestrial nonvolant mammals was around 81 kg only 120,000 years ago. Today it is almost an order of magnitude smaller, at around 16.8 kg. If extinctions continue in the future, and sadly this appears to be likely, average mass may well fall to around 6.9 kg, the lowest average body mass of mammals in the last 55 million years (Smith et al. 2010a). Moreover, when we compared this pattern of body size downgrading with the abundance of *H. sapiens* over time, we found a strong and significant correlation: the near exponential growth of human populations was associated with a near exponential decrease in mean body size over the Earth (Smith et al. 2019; Fig. 11.12). As humans became more abundant, mammals "shrunk."

These late Quaternary extinctions led to drastic changes in the *shape* of the body size distribution of mammals over time as well (Fig. 11.11). The shape of a body size distribution is important because it reflects how energy flows through the ecosystem (chapter 5). For example, a right-skewed distribution as observed for modern mammals at the regional and continental scale reflects that there are more species of small animals than large ones, suggesting that small things can or do more finely divide resources in the environment (Hutchinson and MacArthur 1959;

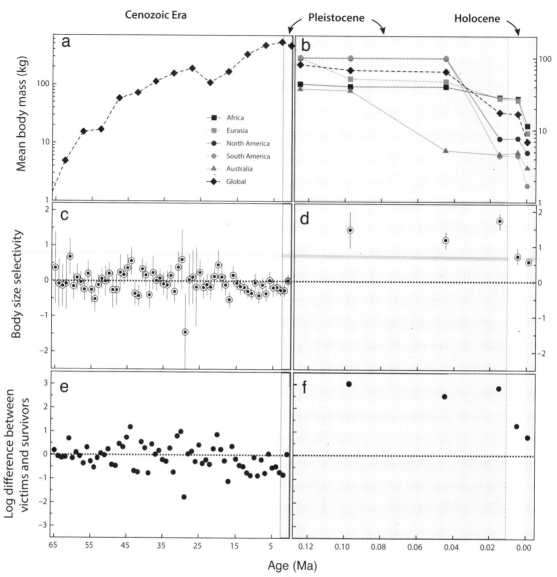

Fig. 11.10. Body size and extinction selectivity of mammals over the Cenozoic. **a,** Mean body size (kg) from 66 to 1 Ma. Notice the rapid rise after the extinction of non-avian dinosaurs at the K-Pg as mammals evolved to utilize the newly available ecological space. **b,** Mean body mass over the late Quaternary (last ~125,000 years). Here, we see that body size has declined in a spatially and temporally transgressive fashion consistent with the migration patterns of humans. **c,** Body size selectivity of mammal extinctions over the Cenozoic era. There was no bias in extinction rates over this time period; neither small- nor large-bodied animals suffered higher extinction rates. **d,** Body size selectivity over the late Quaternary. Note the highly significant bias in extinction risk for large-bodied mammals. Selectivity is in log units, so each increment represents a tenfold greater risk. **e,** Difference in body mass between victims and survivors (in log units) over the Cenozoic. Again, there is little difference between the body size of victims and survivors over most of the last 65 million years. **f,** Difference in body mass between victims and survivors (in log units) over the last 125,000 years. Over this time period, there has been a dramatic difference between the body size of survivors and victims. Redrawn from Smith et al. 2018, 2019.

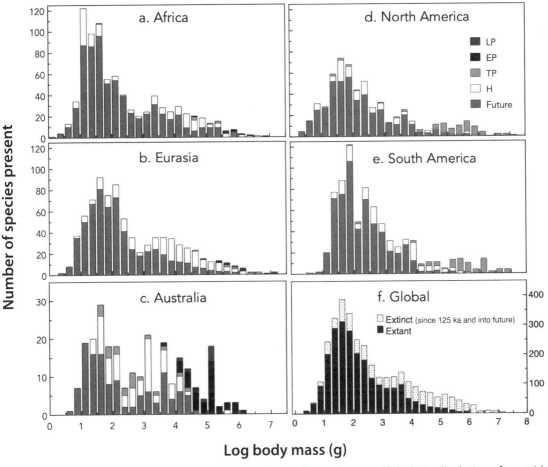

Fig. 11.11. Body size downgrading over the late Quaternary as illustrated by continent-specific body size distributions of terrestrial, nonvolant mammals (panels a–e) and the global pattern (f). Each time period is indicated by different shading. Over the past 120,000 years, the largest mammals in the world have systematically been extirpated from each continent as humans have arrived. This has led to a dramatic truncation of the right-hand portion of the distribution. Redrawn from Smith et al. 2018, 2019.

Brown and Maurer 1986; Brown and Nicoletto 1991; Fenchel 1993; Brown 1995). We found that on each continent, and globally, as hominins arrived, the right side of the body mass distribution (where all the big mammals were) was truncated (Smith et al. 2018, 2019). This change in the shape matters. It meant that energy was no longer flowing through the largest two orders of magnitude masses, reflecting lost ecological function. Was the energy that used to flow through the large-bodied mammals now being used by more smaller-bodied mammals? Was there now "excess energy" in the ecosystem that might make it easier for invasive species to establish? We tend to expect that ecosystem energy does not go unused (Ernest and Brown 2001), so such changes might be expected.

Later in this chapter, we'll come back to how the loss of these large-bodied mammals may have influenced energy flow. For now, our analyses demonstrated that starting in the middle to late Pleistocene, large-bodied mammals began to be systematically extirpated from much of the Earth's surface where they were once abundant. While all major continents once harbored giant mammals like mammoth, woolly rhinoceros, mastodon, camels, llamas, and/or horses, by the Holocene very few of these extremely large-bodied species remained, and the ones that did were mostly confined to the African continent.

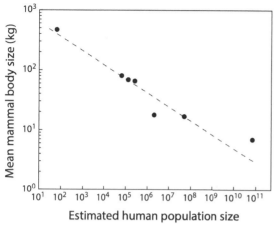

Fig. 11.12. The mean global body mass of nonvolant terrestrial mammals across the late Quaternary plotted against estimated human population size over this time period. As humans have become more populous, the mean body mass of mammals has decreased sharply. This is due to highly size-biased extinction of the largest-bodied animals. The estimate (and standard error) of the coefficients are intercept = 2.902 (0.1829), slope = −0.208 (0.0279). Multiple R^2 = 0.9171, P < 0.001, df = 5. Redrawn from Smith et al. 2019.

The End and Terminal Pleistocene Extinctions

When humans first made it to Australia around 50-60 ka, they found a continent full of unusual animals, including giant kangaroos, giant flightless birds, marsupial lions, and Tasmanian wolves. Australia's long isolation from other landmasses meant that evolution went in a novel direction: the mammals were mostly marsupials. Some of these were truly huge. Imagine if you will, a marsupial herbivore (*Diprotodon optatum*) the size of a rhinoceros (Smith et al. 2003). At around 3 tons, this oversized wombat may even have been migratory (Price et al. 2017). But, by around 45 ka, all mammals larger than 50 kg were extinct (Miller et al. 1999, 2005; Roberts et al. 2001; Lyons et al. 2004; Turney et al. 2008). The pulse of extinction dates to just after the first archeological evidence for humans in Australia (Miller et al. 2005; Turney et al. 2008). Indeed, there were negligible extinctions prior to the colonization by aboriginal humans (e.g., Figs. 11.10b, 11.11c). Megafauna persisted somewhat longer in Tasmania, reflecting the later occupation of this landmass by humans (Turney et al. 2008).

The Americas also had not experienced appreciable extinction prior to the terminal Pleistocene. The first humans entering this continent around 13-15 ka would have encountered diverse and abundant herds of naïve megafauna and many unusual predators such as saber-tooth cats and short-faced bears. This migration of humans into the New World led to the most dramatic of the late Quaternary extinction events (Figs. 11.10b, 11.11d). Not only were more species lost relative to earlier extinction events, but they also went extinct much quicker. Indeed, at least 150 mammal taxa were lost in the Americas within just a 1,000- to 1,500-year window (Faith and Surovell 2009); this may even have been as rapid as 400 years (Fiedel 2009). And it was the most size selective of the late Quaternary extinctions. Indeed, *everything* over 580 kg was lost in North America, and everything over 300 kg in South America (Lyons et al. 2004; Smith et al. 2018).

One possibility for this rapid rate of extinction was that humans had become better, and more effective, predators. By the terminal Pleistocene, humans had millennia of hunting experience. They had developed long-range weapons, which allowed targeting prey at farther and safer distances (Lyons et al. 2004). This wholescale extirpation of large-bodied mammal taxa—representing millions of large-bodied individuals—resulted in a major restructuring of the Americas. Lost were species such as the mammoth, who may have played a seminal role in establishing and maintaining grasslands; lost, too, were the dire wolf and short-faced bear, voracious predators that specialized on the herds of horses, camels, and llamas. It is sobering to consider that the imperiled apex predators we are concerned about today were only the mesocarnivores of this earlier time.

While our recent studies were the first to quantify the rate and magnitude of mammal extinctions in the late Pleistocene (Smith et al. 2018, 2019), considerable previous research has examined extinctions during the end and terminal Pleistocene in both Australia and the New World. Much of this has focused on the issue of causation—did humans, climate change, or disease cause the observed biodiversity losses (e.g., Martin 1967, 1984, 2005; Barnosky et al. 2004; Lyons et al. 2004; Miller et al. 2005; Koch and Barnosky 2006; Ripple and Van Valkenburgh 2010; Barnosky et al. 2011; Zuo et al.

2013)? This question has been debated exhaustively in the literature and, thus, is not reviewed here. However, it is fair to say that most scientists now concur that humans had a large contributing role to the end and terminal Pleistocene extinctions on all continents (Martin 1967, 1984, 2005; Martin and Steadman 1999; Roberts et al. 2001; Lyons et al. 2004; Miller et al. 2005; Koch and Barnosky 2006; Haynes 2009; Sandom et al. 2014; Smith et al. 2018, 2019).

Holocene and Historic Extinctions

By the beginning of the Holocene, humans were present on all habitable continents. They had also probably colonized much of the Australo-Oriental intercontinental mega-archipelago as they made their way into Australia. As time progressed, humans began to colonize other islands—first those close to shore, and then more and more remote islands. The most distant islands in Oceania were probably colonized in the late Holocene, somewhere around 3,500 to 500 years ago (Martin and Steadman 1999; Montenegro et al. 2016). Not surprisingly, a wave of extinctions followed (Alcover et al. 1998; Martin and Steadman 1999; Burney and Flannery 2005; Turvey 2009; Turvey and Fritz 2011; Lyons et al. 2016a). There is virtually no disagreement among scientists that humans were the cause of these more recent extinctions.

Hardest hit were flightless birds; their greater dispersal ability meant they were able to colonize a larger number of remote landmasses that were difficult for mammals to reach. Moreover, once birds made it to an island, there was strong selection to lose the ability to fly since it was energetically costly and of limited use in an insular habitat (Carlquist 1974). For example, about 35% and 25% of New Zealand and Hawaiian birds are flightless, respectively. Because remote islands largely lacked native mammals, that meant that birds could nest on the ground and not fear egg predation. Unfortunately, these characteristics made them sitting ducks (in some cases literally!) when humans, and their commensals, arrived. Hitching a ride on boats along with humans were often rats, dogs, and even cats. As many as 2,000 bird species may have gone extinct in just the Pacific islands over the Holocene, representing more than 60% of the avifauna that once existed in this region (Steadman 1995).

A number of factors influenced the likelihood that a mammal went extinct on an island. These included characteristics of the animal (i.e., Was it a large and profitable food package? Tasty?) as well as the geographic template of the island itself (i.e., How big was it? How geographically complex?) and its location and proximity to a nearby mainland. For example, 71.4% of native mammals in the West Indies went extinct after human arrival, compared to 88.9% of those found on Mediterranean islands (Alcover et al. 1998). This is despite an overall pattern of island archipelagos with a longer history of human occupation displaying lowered extinction rates. For example, only 3.8% of native mammals in Wallacea went extinct (Alcover et al. 1998). These Indonesian islands had probably been colonized in the late Pleistocene, and consequently the fauna would have coexisted with humans for a long time. But the relatively close proximity of islands in the Mediterranean to the European and African continents meant that they also supported a diverse assemblage of mammals of all sizes. Indeed, many large-bodied mammals such as elephants, deer, and hippos had successfully colonized these islands and would have been favored targets for hunting. Globally, large herbivores have been almost completely extirpated from insular habitats (Alcover et al. 1998). Other major island systems that experienced high levels of Holocene mammal extinctions were Madagascar, Cuba, and Hispaniola (Turvey and Fritz 2011). And, as island biogeography theory would predict, all else being equal, extinction rates were often higher for smaller islands than larger ones.

All in all, island ecosystems were hard hit by extinction during the Holocene. Newly arriving humans and their commensals decimated fragile and unique insular habitats. Indeed, the overall insular extinction rate for terrestrial, nonvolant mammals after the arrival of humans was around 35% (Alcover et al. 1998). Moreover, this figure is likely to be an underestimate. For example, only 38% of mammals that went extinct during the Holocene have even been formally described in the Indonesia-New Guinea region (Turvey and Fritz 2011).

While large-bodied mammals were still differentially prone to extinction (Turvey and Fritz 2011; Smith et al. 2018), the size bias decreased over the

Fig. 11.13. Examples of historic extinctions. **a,** The thylacine (*Thylacinus cynocephalus*) was one of the largest known carnivorous marsupials; the last known live animal was captured in 1933 in Tasmania. Shown here is a pair sent to the US National Zoo in the early 1900s. **b,** The quagga (*Equus quagga quagga*) is an extinct zebra-like mammal. Shown is a mare in the London Zoo in 1870. **c,** A massive pile of American bison (*Bison bison*) skulls that were ultimately ground for fertilizer, circa 1892. Bison were killed by the millions, especially after train tracks were laid across the United States. Smithsonian Institution Archives, Record Unit 95, Box 49, Folder 18; Photo of quagga by Frederick York, courtesy of Biodiversity Heritage Library and Wikimedia Commons; Photo of bison skulls modified by Chick Bowen, courtesy of Wikimedia Commons

Holocene. For example, the difference in body mass between the victims and survivors dropped to approximately one order of magnitude for extinctions during the Holocene (Fig. 11.10f). This may have been because there were fewer large-bodied mammals to go extinct, but it probably also reflected a shift in the nature of threats (Lyons et al. 2016a). While direct exploitation was likely the main driver of the late Quaternary extinctions (e.g., Alroy 2001; Zuo et al. 2013), a more complex suite of stressor—including hunting, habitat fragmentation and/or urbanization, invasive species, human conflict, and even European fashion—drove Holocene and historic extinctions (Fig. 11.13; Barnosky et al. 2004, 2011; Turvey and Fritz 2011; Dirzo et al. 2014; Lyons et al. 2016a). Future extinctions are predicted to target not only large-

bodied mammals, but also small, often highly specialized or geographically localized species (Lyons et al. 2016a; Ripple et al. 2017; Smith et al. 2018).

To date, we have lost around 8%–10% of Earth's late Quaternary mammal diversity. As mentioned earlier, without intervention the cumulative biodiversity loss may reach closer to 30% or more in a century or less, including most of the charismatic large-bodied species we all love, such as elephants, rhinos, giraffes, and lions (Cardillo et al. 2005; Schipper et al. 2008; Hoffmann et al. 2010; Barnosky et al. 2011; Dirzo et al. 2014; Ceballos et al. 2015; Ripple et al. 2015, 2016, 2017; Smith et al. 2018, 2019). This frightening estimate may even be conservative because "data deficient" mammals are generally ignored in computations—yet, they are often the species most at risk. Moreover, while the late Quaternary extinctions primarily targeted particular taxonomic groups whose biological traits made them particularly susceptible (Purvis et al. 2000; Turvey and Fritz 2011), the magnitude of future loss means extinction will necessarily become more widespread across the entire taxonomic hierarchy. This will lead to even more loss of both phylogenetic and functional diversity (Smith et al. 2019).

Does Climate Change Lead to Mammal Extinction?

Whether climate change in the late Quaternary contributed to—or even drove—extinctions in mammals has been the subject of a long and sometimes acrimonious debate in the literature. As mentioned earlier, some scholars argued that the terminal Pleistocene megafauna extinctions were the result of the end of the last ice age and the associated changes in climate and vegetation, while others argued for a seminal role for humans (e.g., Martin 1967, 1984; Graham and Lundelius 1984; Graham and Grimm 1990; Guthrie 1990; Barnosky et al. 2004; Lyons et al. 2004; Koch and Barnosky 2006; Grayson 2007; Wroe et al. 2006, 2013; Sandom et al. 2014). Today, new data and analyses have largely led paleontologists to conclude that human activities were primarily responsible for the terminal Pleistocene extinctions, although stressors induced by changing climate may have contributed (Barnosky et al. 2004; Lyons et al. 2004; Koch and Barnosky 2006; Haynes 2009; Sandom et al. 2014). However, the idea that rapid changes in climate and vegetation at the Pleistocene-Holocene transition were a primary driver of mammal extinction has not completely disappeared from the recent literature (e.g., Wroe et al. 2013).

Rather than focus solely on the terminal Pleistocene, we can ask broader questions: Over the Cenozoic has climate change *ever* led to increased rates of mammal extinction? And have shifts in climate ever led to a size-biased extinction, as might be expected on the basis of Bergmann's rule (chapter 5)? To date, few studies have directly addressed these issues. Thus, to investigate these questions—which are of obvious relevance in our modern era of climate change—my colleagues and I turned to the terrestrial mammal fossil record. We compiled a continentally explicit record of mammal presence or absence for each 1 million years over the entire 66 million years of the Cenozoic (Smith et al. 2018, 2019; Smith et al. in prep.). Mammalian turnover (i.e., extinction dynamics) over the Cenozoic was characterized using Foote's boundary crosser method (Foote 2000), which takes into account sampling intensity. These data were then directly compared with paleotemperatures over the same interval derived from the fossil foram record (chapter 9; Zachos et al. 2001). We used several metrics of temperature: (1) the global climate state; that is, the mean temperature over the interval; (2) climate variability; that is the standard deviation within each 1-million-year interval; and (3) climate variability measured as the difference in the mean temperature between 1-million-year intervals. By employing these divergent ways of looking at temperature, we were able to explore if it was the actual temperature or the fluctuations in temperature that might have influenced extinction dynamics of mammals.

While we didn't really expect to find that climate change had driven major pulses of extinction, what we did find was a bit surprising. Temperature change over the past 66 million years never led to an enhanced extinction rate (Fig. 11.14). Nor did it lead to any sort of size bias in mammal turnover (Smith et al. 2018, 2019; Smith et al. in prep). Our results were insensitive to whether we were talking about the overall temperature or fluctuations in temperature. Given the sensitivity of mammal body size to temperature over short- and long-term time scales (chapter 5), we were initially puzzled by the complete lack of selectivity in mammal extinction. Shouldn't warming climate select for smaller-bodied mammals and cooling climate larger-bodied ones? Instead, closer investigation revealed that while temperature does indeed select on mammal body size, it doesn't normally lead to clade extinction. This is because mammals experience a wide variety of temperature within their geographic range; indeed, the average temperature extent within a mammal range is about 20°C (Smith et al., in prep.). This means that selection, coupled with extirpation at the range boundaries, is how mammals coped with changing climate in the past (chapter 10). Rarely, if ever, have they experienced climatic shifts that led to extirpation of all populations, or true extinction. Thus, in

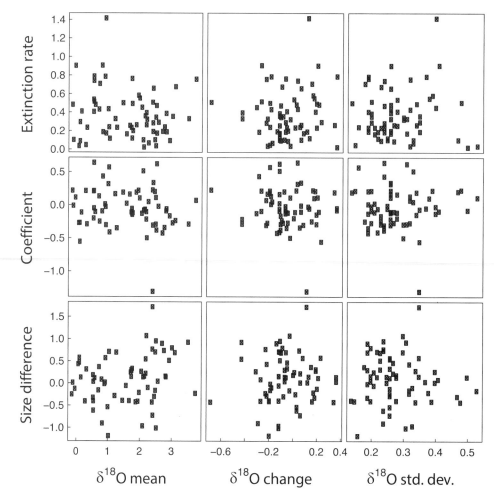

Fig. 11.14. Climate change didn't increase mammal extinction risk over the past 65 million years. Shown are scatterplots of extinction patterns versus different measures of climate across the Cenozoic at million-year intervals. Extinction is quantified as (first row) the negative of the natural log of the number of species ranging through the interval divided by the number of species entering the stage from the previous interval but not surviving into the subsequent interval (Foote, 2000); (second row) the logistic regression coefficient measuring the association between body mass and extinction; and (third row) the difference in mean size between victims and survivors. Climate is measured in terms of mean, interval-to-interval change in mean, and the standard deviation of oxygen isotope paleoclimate proxy values from planktic foraminifera. From Smith et al. 2019.

the past, climate change has not been a major factor driving biodiversity loss. That it will be in the future is likely because severe habitat fragmentation has limited the potential responses of mammals to shifts in environmental conditions (chapter 10).

What Were the Consequences of Biodiversity Loss in the Past?

Understanding the complex role of large-bodied mammals in contemporary ecosystems and the likely ecosystem consequences of their decline is essential if we are to effectively manage the remaining wild areas on Earth. Because biodiversity loss is ongoing, scientists simply do not have time to wait for the results of long-term experimental studies before proposing potential mitigation strategies. Here again is where a paleontological perspective provides unique and powerful insights for modern conservation biology. The late Quaternary extinctions are an excellent proxy for understanding how the decline and ultimate extinction of megafauna influenced ecosystem function, and for investigating how surviving mammals responded to this major loss of biodiversity. Interestingly, while considerable research has focused on the cause of mammal extinctions, particularly those of the terminal Pleistocene, we have paid much less attention to elucidating the *consequences* of biodiversity loss on the functioning of ecological communities and the environmental legacy of the loss of tens of millions of large-bodied individuals from the landscape (Smith et al. 2016d; Tomé et al. 2019). Indeed, it is only recently that we have begun appreciating just how much near-time human activities have altered fundamental macroecological patterns (Smith and Boyer 2012). Thus, we currently lack a synoptic understanding of the thresholds that may lead to the "unraveling" of ecosystems. Fortunately, this is now changing and a number of investigators are now tackling these issues, including my colleagues and I.

The Importance of Megafauna

Typically, when referring to megafauna, we mean the giant herbivores that once existed on most continents across the globe, as well as the handful that remain, and not the large-bodied predators that coexisted alongside them. The focus on megaherbivores is defensible because their impact dwarfed that of the carnivores. For example, they were more than an order of magnitude larger and more abundant than any predator within the community (Smith et al.

2010a). Megafauna were (and are) unique members of ecological communities (Fig. 11.15).

So, how and why did—or do—megafauna influence their environment? Simply put, a woolly mammoth was much more than a big elephant or a vastly supersized and fuzzy cow. Mammoths, woolly rhinos, and other megafauna were truly enormous—indeed, about two to three times as large as the average elephant today and more than an order of magnitude larger than a domestic cow (Smith et al. 2003). And while even the largest bull elephants today are smaller than their Pleistocene cousins, they are still much larger than any species of domesticated livestock (Smith et al. 2003). Sadly, given the rate of biodiversity loss today, cows are the largest-bodied mammals predicted to survive in an anthropogenic future (Smith et al. 2018). Why is this difference in size so important? It turns out that as body mass increases, there is a disproportionate, or allometric (chapter 5), increase in the influence of an animal's activities on ecosystems (Owen-Smith 1987, 1988). Thus, megafauna perform different ecological functions than do smaller herbivores, even when the biomass of each is equivalent on the landscape (Fritz 1997; Haynes 2012; Owen-Smith 1988). Indeed, virtually all aspects of ecosystems are impacted directly or indirectly by megafauna (Fig. 11.12). This cascading series of effects has led to their characterization as "ecosystem engineers" (e.g., Bakker et al.

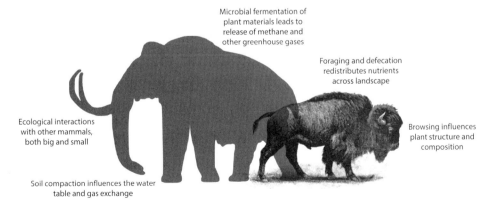

Fig. 11.15. Pivotal influences of large-bodied mammals on ecosystems. Large-bodied mammals (e.g., bison) have a disproportionate influence on their environment in almost every way, including how they eat, move, and interact with other animals (several examples are indicated here). In the past, the even-larger-bodied mammals present (shaded mammoth) likely had even larger impacts. A current focus in paleoecology is in characterizing the influence of large-bodied animals in ecosystems. Mammoth silhouette from PhyloPic (http://phylopic.org) under Creative Commons license. Bison from Beddard 1902.

2006, 2016b; Zimov et al. 1995; Johnson 2009; Haynes 2012; Asner et al. 2016). Interestingly, the insights we are arriving at today are the result of the integration of paleontological and modern perspectives on megafauna ecology; a wonderful example of how interdisciplinarity can lead to a more synoptic understanding of a major conservation issue.

The oversized influence of megafauna on ecosystem structure and function results from several salient morphological and ecological characteristics (Fig. 11.15). First, their enormous size has physical consequences. For example, the sheer weight of a 12,000 kg animal modifies soil structure, influences water tables, changes the physiographic formation of game trails, and influences what vegetation can grow. This is true of elephants today and was even truer in the past when animals were larger (Haynes 2012; Smith et al. 2016a). Second, the way megaherbivores acquire and digest food influences the structure and composition of vegetation on the landscape, as well as gas exchange with the atmosphere. Through selective foraging, herds of megafauna maintain open grasslands. Moreover, fermentation in their enormous guts releases copious amounts of methane (a potent greenhouse gas); these effects are allometrically related to their body mass (Clauss and Hummel 2005; Smith et al. 2010b, 2015). Third, large-scale movement, common to very large-bodied mammals, leads to the redistribution of nutrients across the landscape. This occurs as they walk; their dung deposits redistribute soil nutrients away from a point source (Doughty et al. 2013, 2016a, 2016c, 2016d; Doughty 2017). Multiplied by the millions of large-bodied mammals that once roamed Earth's surface, this can have sizeable consequences. For example, the reduction in large-bodied wildlife populations globally has led to a radical decrease (~90%) of the transport of nutrients across the landscape (Doughty et al. 2016a). And lastly, megafauna also have an oversized influence on ecological interactions within communities, both directly and indirectly (Smith et al. 2016a). I elaborate on each of these points in a bit more detail as follows.

Physical Influence on the Environment

Try to imagine the physical consequences on the landscape of millions of megafauna roaming freely, as they would have at the terminal Pleistocene. And indeed there were a lot of them. If we consider just the individuals over 1 ton, approximately 175 million large-bodied mammals went extinct globally in just a few thousand years (Smith et al. 2016c). While this may sound like a lot, bear in mind that today there are more than 1.5 *billion* cows on the planet. In the United States alone, we have around 100 million cows as of 2020 (Smith et al. 2016a); these figures exclude other livestock such as sheep, horses, goats, pigs, and poultry. To put this late Pleistocene scene in perspective, consider that some of these extinct megafauna were *really* large; they exceeded the mass of a small Caterpillar tractor at over 10 tons (Smith et al. 2003). If you have ever walked around a construction site, you know that the movement of even small tractors results in very hardpacked ground. In a similar way, the weight of herds of large megafauna undoubtably also led to compression of soils (Fig. 11.12). By removing air pockets, soil compaction can alter water and gas exchange with the atmosphere, which in turn can have substantial regional and even global effects on climate (Smith et al. 2016c).

But the physical impacts extend beyond soil compaction. Even the way megafauna travel through the ecosystem can alter the environment. For example, larger-bodied animals take more circuitous routes up and down mountain slopes, with the trail angle decreasing as body size increases (Reichman and Aitchison 1981). This means the game trails they establish are more shallow and curvy; such terracing of landscapes influences water movement and plant establishment. And megafauna have a unique impact on forest structure. Their massive body size means that as they move through wooded areas, they break and knock down trees (Fig. 11.15). This sort of top-down control, where megaherbivore movement opens what would otherwise be closed-canopy vegetation, is simply not replicated by smaller domestic animals (Asner and Levick 2012).

A number of recent studies have highlighted just how extensive the physical effects of large-bodied mammals can be on ecosystems. For example, a more-than-six-year survey in Kruger National Park in South Africa used advanced airborne remote-sensing techniques to examine the influence of elephants on woody plant canopies. They found extensive opening of woody vegetation by elephants; indeed, they were two times more important than fire in regulating tree fall (Asner et al. 2016). Similarly, forest structure in Gabon may be strongly shaped by elephants. There is a striking lack of small trees in Gabon forests compared to the Amazon (Terborgh et al. 2016a, 2016b). This is likely the result of elephant damage, which severely restricts tree recruitment and, thus, shapes the structure of the adult tree community (Terborgh et al. 2016b, 2015b). Because tree species diversity is concentrated in the smaller tree size classes, this means that elephant activity results in greatly reduced alpha diversity of African forests (Terborgh et al. 2016a, 2016b). Conversely, the loss or restriction of megafauna from an area can lead to an increase in woody cover (Asner et al. 2009; Gill et al. 2009; Asner and Levick 2012; Barnosky et al. 2016), although the effect is likely mediated by geography and regional climate. Such geomorphological engineering even extends to the aquatic realm, where movement by hippopota-mus leads to changes in land, vegetation composition, and hydrology at landscape scales (Bakker et al. 2016a, 2016b). These studies suggest that the wides-cale loss of large-bodied mammals from ecosystems over the Pleistocene has led to transformations that we may not have appreciated. For example, the woody savannas of South America may have been considerably more open and grassy when enormous proboscideans and ground sloths were present than are modern ones (Doughty et al. 2016a). Modeling efforts suggest some 29% of savanna woody cover in South America was lost as a result of the late Pleistocene extinction (Doughty et al. 2016a). Thus, consideration of how top-down megafaunal control influenced vegetation structure and composition in the late Pleistocene can yield new insights into the artificial, postmegafaunal nature of contemporary ecosystems.

Large-Bodied Mammals and Large-Scale Foraging

As megaherbivores move through the environment, they also forage. How they do so also has important implications in terms of ecosystem function (Fig. 11.15). For example, large-bodied browsers, such as elephants, typically do not forage at the level of an organ such as a twig or leaf, as do smaller herbivores. Instead, they feed at the level of the organism, and may consume much of a tree, including the bark. Thus, it should come as no surprise that large-bodied herbivores help maintain open grasslands and inhibit woodland regeneration (Owen-Smith 1987, 1988, 1989; Cummings et al. 1997; Whyte et al. 2003; Western and Maitumo 2004; Johnson 2009; Haynes 2012; Keesing and Young 2014; Asner et al. 2016; Bakker et al. 2016). For example, after the eradication of rinderpest in the early 1960s, native ungulates increased in abundance, leading to a significant decline in woody vegetation (Estes et al. 2011). The maintenance of open grasslands at the expense of woodlands has benefits for other herbivores in the ecosystem: by shifting carbon storage from woody tissues to the leaves and tissues of grasses, carbon becomes more freely accessible (Retallack 2001, 2013; Johnson 2009). The larger sizes of Pleistocene megafauna suggest they played an even more substantial role in regulating the structure and composition of vegetation than do modern congeners.

Of course, the type of herbivory matters as well: browsers have an inhibiting effect on woody vegetation, while the effects of grazers are mixed (Bakker et al. 2016a, 2016b). For example, grazers can have negative effects on woody cover through the trampling and destruction of seedings. But, they may also have a positive effect by reducing the plant competition between grasses and shrubs (Johnson 2009; Bakker et al. 2016a, 2016b). To complicate things, how megaherbivores foraged appears to have been variable across time and space. Stable isotopes reveal that *Palaeolama*, for example, were sometimes browsers and sometimes mixed feeders (Franca et al. 2015; chapter 7). What this means in terms of the dietary flexibility of modern mammals is unclear.

Foraging activities have other indirect effects, too, which include an influence on soil fertility and fire regimes. For example, in the absence of heavy grazing, water tables can rise. This leads to a slowdown in the rate of nutrient breakdown and recycling, which in turn increases the accumulation of organic matter and decreases soil fertility. Indeed, the large-scale transition from the vast mammoth steppe of the Pleistocene to more waterlogged habitats of the Holocene was at least partially owing to the absence of grazing and other activities by

a. Pouteria sp.

1 cm

b. Proboscidea parviflora
Devil's Claw

c. Annona holosericea

d. Maclura pomifera, commonly
known as the Osage orange

Fig. 11.16. The demise of exceptionally large-bodied mammals in the Americas led to a number of anachronisms. Shown here are four examples of fruit that were once dispersed by now-extinct megafauna. **a**, *Pouteria*, a widespread genus of flowering trees in tropical regions of the world. **b**, *Proboscidea parviflora*, or Devil's claw, is native to the desert southwest of the United States and northern Mexico, where it is found in sandy and dry habitat. It grows from a taproot and produces sprawling, spreading stems. When dry, the large fruit splits along the spine to form two large hooks (i.e., "claws"), which catch on large animals moving by that in turn disperse the fruit's seeds. **c**, *Annona holosericea*, a flowering plant in the pawpaw / sugar apple family, Annonaceae. **d**, *Maclura pomifera*, the Osage orange, which today has a limited historical range. Scientists have speculated that it was eaten by giant ground sloths, mammoths, mastodons, and/or gomphothere and that the extinction of those animals led to a decline in the fruit's geographic range (Barlow 2002).

large-bodied mammoth, camelids, and bison that went extinct (Zimov et al. 1995, 2012; Johnson 2009). The accumulation of organic matter can enhance both the frequency and intensity of fire activity; such shifts have been detected at the terminal Pleistocene in many ecosystems across the globe (Burney et al. 2003; Bond and Keeley 2005; Gill et al. 2009; Gill 2014; Bakker et al. 2016a, 2016b).

Moreover, in some ecosystems, megafauna play(ed) a significant and unique role in the dispersal of plants (Fig. 11.16). These plants are not hard to identify; they tend to produce distinct large-bodied seeds and fleshy fruit. The crucial role of megafauna in dispersal of these plants is not easily duplicated; smaller herbivores typically eat portions of the flesh and discard the seeds (Barlow 2002; Johnson 2009).

Moreover, even when smaller-bodied herbivores do ingest the seeds, they tend to discard them at much shorter distances than did larger-bodied herbivores, leading to changes in the structure of forests (Guimaraes et al. 2008; Campos-Arceiz and Blake 2011; Bueno et al. 2013). Avocados are an excellent example—a large and fatty fruit that likely evolved for dispersal through the guts of elephants and other megafauna.

The extinction of most megafauna around the globe means such plants no longer have effective dispersal agents, becoming in effect evolutionary anachronisms (Janzen and Martin 1982; Barlow 2002). Indeed, avocados are too large for dispersal by any extant herbivores in the Americas. Fortunately for avocados, humans have enthusiastically adopted them as a food source, essentially replacing the role of extinct megafauna. Other megafauna-dispersed plants have not fared as well, and many have declined in abundance and/or geographic distribution (Janzen and Martin 1982; Barlow 2002; Doughty et al. 2015c). For example, in the Congo, about 4.5% of tree species are dependent on elephants for dispersal (Beaune et al. 2013). The ongoing loss of forest elephants means that these plant taxa are not maintaining stable populations (Beaune et al. 2013). Since large-seeded fruit trees tend to have less dense wood, declines in their abundance or distribution also influences carbon flux and biomass of forests (Doughty et al. 2015c). Indeed, the total carbon content in Amazon rain forests has been reduced by about 1.5% owing to the terminal Pleistocene extinctions (Doughty et al. 2016b). The effects have been spread out over centuries or even millennia because of the slow demographics of large forest trees. Thus, in the past, many landscapes may well have been "herbivore controlled" through a combination of direct and indirect effects of megafauna (Owen-Smith 1987; Bond and Keeley 2005; Doughty 2017).

Foraging, Biogeochemical Cycles, and Earth's Climate

In addition to their crucial influence on vegetation structure and abundance, foraging by megaherbivores influences global biogeochemical cycles (Fig. 11.15). This was likely to have been more important in the past when wildlife numbers were much greater (e.g., Smith et al. 2016c). Because much animal behavior and physiology scales with body size, we can employ allometric relationships (chapter 5) to estimate the contributions of megafauna to crucial biogeochemical cycles. A good example is the greenhouse gas methane.

Megafauna are essentially giant walking fermentation vats. Because herbivores are incapable of enzymatically digesting the structural components of plants, they depend on an internal fermentation chamber stocked with symbiotic microbes to do so (Van Soest 1982). These microbes ferment the plant materials and release volatile fatty acids, which herbivores absorb in the intestines. This process also releases substantial quantities of potent greenhouse gases, particularly methane (Clauss and Hummel 2005; Smith et al. 2010b). Today, a variety of natural and anthropogenic sources produce methane, although in many countries domestic livestock are a primary source. For example, while livestock are responsible for about 20% of annual global methane emissions today (~76 to 189 million tons, Tg, per year); in pastoral New Zealand they make up around 80%–85% of the nation's contribution to the global budget (IPCC 2017; Smith et al. 2015).

How much methane herbivores produce by the anaerobic microbial digestion of plants is dependent on how big the animal is (Smith et al. 2015). Very large herbivores produce a disproportionate amount relative to smaller animals (Smith et al. 2015). Thus, it follows that the loss of millions of giant mammals in the late Quaternary impacted atmospheric composition of gases—and perhaps even altered the climate (Smith et al. 2010b, 2016c). To examine this, my colleagues and I quantified the contribution of large-bodied mammals to the global methane cycle over time in a series of papers. By constructing a series of robust allometric equations relating methane production, geographic range, and population density with body mass, we were able to compute how much methane went "missing" when hundreds of millions of megaherbivores were extirpated at the terminal Pleistocene (Smith et al. 2010b, 2016c).

We found that in the late Pleistocene, wildlife contributed about 139 Tg per year to the global methane budget (Smith et al. 2016c). Strikingly, this is approximately equivalent to contributions by livestock today (Fig. 11.17). The extinction of megafauna at the terminal Pleistocene led to the loss of around 69.6 Tg of methane annually, representing a 28%–35% reduction of the overall inputs to tropospheric input (Smith et al. 2016c). Our computations were supported by independent assessments by others. For example, using a bottom-up approach based on primary productivity, Zimov and Zimov (2014) estimated the annual contribution of wildlife to the global methane budget in the Pleistocene was 90–170 Tg; our value of 138.5 Tg is almost in the middle of this range. Furthermore, employing a top-down analysis based on the isotopic signature of methane derived from cores also yielded similar results (Zimov and Zimov 2014): during the Pleistocene, mammals were responsible for around 63%–64% of the global input; we calculated approximately 62% (Smith et al. 2016c).

An intriguing postscript to these analyses is that the terminal Pleistocene megafauna extinctions occurred simultaneously with the onset of the Younger Dryas cold interval (chapter 9). Thus, we can ask whether the extirpation of hundreds of millions of large-bodied animals contributed in any way to this somewhat enigmatic climatic event. Interestingly, the ice core record reveals a particularly precipitous drop in atmospheric methane concentration during the extinction interval and prior to the Younger Dryas; the rapidity of the methane drop is unique in the ice core record (Smith et al. 2010b, 2016c). Moreover, isotopic analysis points to a change in the sourcing of atmospheric methane. While the Pleistocene isotopic signature of methane suggests wildlife were the main contributors, during the early Holocene there is an abrupt transition to methane produced largely by boreal and tropical wetlands (Zimov and Zimov 2014). Thus, it seems likely that the drop in atmospheric methane was the result of the extinction of hundreds of millions of herbivores. How much the decrease in methane drove or contributed to the rapid drop in global temperature at the Younger Dryas remains unclear, although simplistic modeling suggests the extinction may have reduced Earth's climate by around 0.5°C

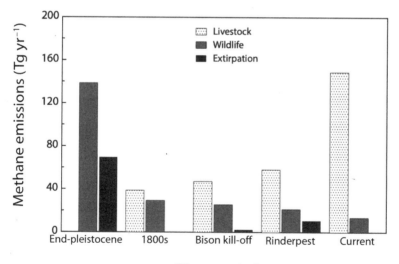

Fig. 11.17. Changes in methane emissions by wildlife over time. Methane is produced by herbivores as a by-product of the microbial fermentation of plant fiber. As wildlife (gray bars) have been extirpated across the globe and, more recently, replaced by domesticated livestock (stippled bars), there has been a corresponding shift in the sourcing of methane. Today, emissions by mammals in the United States and elsewhere are largely from domesticated livestock rather than wildlife. The black bars represent emissions by animals that were extirpated during each time period. Because methane is a greenhouse gas, the amount and sourcing of emissions is of national and global importance. Indeed, methane is often a target of mitigation efforts. The terminal Pleistocene megafauna extinction and corresponding decline in emissions was followed by an abrupt drop in global methane concentration in the atmosphere, which also closely coincided with the onset of the Younger Dryas cold episode. Thus, it is possible that megafauna declines at the terminal Pleistocene influenced global climate.

(Smith et al. 2016c). The indirect role of the extinction on methane cycles and, potentially, climate should be explored in Earth circulation models to better constrain the magnitude of the effect.

Researchers have documented other indirect effects of the megaherbivore extinction on Earth's climate. For example, the extinction of megafauna in northern high latitudes around 15,000 years ago was followed by a rapid expansion of birch (*Betula*) from a few trees to forests that covered around 25% of Siberia and Beringia (Doughty et al. 2010). This rapid shift in plant cover from grass to dwarf birch lowered the albedo, or reflectivity, of the environment (Doughty et al. 2010). Thus, more sunlight was absorbed, which may have warmed regional climate by more than 0.2°C within just a few hundred years (Doughty et al. 2010). Similarly, the type of vegetation present influences carbon flux through the environment. As we discussed earlier, megafauna often help maintain open grasslands at the expense of tree cover. But, woody biomass tends to sequester more carbon than do grasslands. Thus, in the absence of megafauna, woody encroachment can lead to a drawdown in atmospheric CO_2 concentration (Bakker et al. 2016a; Doughty et al. 2016b). However, it should be noted that carbon dioxide is much more abundant in the atmosphere than methane; indeed, it is measured in parts per million volume (ppm), rather than parts per billion (ppb) as is methane. Thus, it is unclear whether the drawdown in carbon dioxide would be of sufficient magnitude to materially influence global climate.

Large-Bodied Mammals Move Large Distances

The home range of an animal represents the environmental area where the essential activities of foraging, denning, and mating occur. As animals increase in body mass, there is a proportional increase in their use of space (Brown 1995). Thus, the very largest-bodied mammals may travel vast areas (Brown 1995), especially those taxa that undergo seasonal migrations, such as wildebeest. For example, the home range of an elephant may exceed 8,500 to 10,000 km^2 (Lindeque and Lindeque 1991; Ngene et al. 2017) with space use varying with gender, season, and geographic location (Ngene et al. 2017). Clearly elephants move around a lot. Logically,

Pleistocene megafauna also did so, and probably traveled even greater distances.

The large distances megafauna cover matter. As they traverse the landscape, megafauna engage in daily activities such as foraging, digestion, and ultimately defecation. Their extensive lateral movements coupled with long food-passage times means that megafauna often deposit nutrients or partially digested seeds a considerable distance from where they initially foraged (Doughty et al. 2016a). Thus, large-bodied mammals influence not only seed dispersal but also the biogeochemical flux of essential elements such as carbon, sodium, phosphorus, and nitrogen that are sequestered within vegetation (Schmitz et al. 2014; Doughty et al. 2013, 2016a, 2016d, 2016c). Smaller herbivores do not have the same ecological function; they eat and defecate much less and have much more restricted home ranges. Neither do domestic livestock, who while large, are typically confined to enclosed, managed pastures (Doughty et al. 2016d). This constriction of movement leads to a concentration of nutrients in one area, effectively stopping the diffusion of nutrients in the ecosystem. Moreover, in forests where large-bodied herbivores were extirpated, nutrients can be trapped within the woody tissues of trees and, hence, effectively removed from biogeochemical cycling (Owen-Smith 1988; Retallack 2001, 2013; Johnson 2009).

By reducing the spatial redistribution of nutrients on the landscape, the extinction of large-bodied herbivores led to declines in the fertility of terrestrial, aquatic, and marine environments (Doughty et al. 2013, 2016d). For example, sodium concentrations are higher on the coast today than inland; this lateral shift followed the megafauna extinction (Doughty et al. 2016c). Globally, biodiversity loss means the ability of animals to transport nutrients away from a point source has decreased to about 6% of its former capacity (Doughty et al. 2013, 2016d). The loss of nutrient transport was less in Africa, where some megafauna till remain.

The Oversized Influence of Megafauna on Community Structure

While we have thus far focused on changes in the physical and biochemical structure of the environment, megafauna also play a key role in mediating interactions among species within a community. Indeed, their loss from an ecosystem changes ecological dynamics throughout the community, so much so that the loss of top consumers is known as *trophic downgrading* (Estes et al. 2011).

There are numerous examples of the effects of trophic downgrading in both modern and paleo ecosystems. A classic example is the changes that occurred in Yellowstone National Park *after* wolves were reintroduced. Predator control policies early in the century led to the eradication of wolves in Yellowstone by the 1920s. The effects of their removal percolated throughout the ecosystem: without wolves, elk herds blossomed over the following seventy-plus years, leading to overgrazing. Elk also browsed in riparian habitat, which led to dwarfing and thinning of vegetation such as willow and aspen; this in turn led to declines in songbirds and beaver populations (Ripple and Beschta 2004). Without beaver, the morphology of streams changed; they became narrower, faster, and deeper, which was unfavorable for the establishment of riparian vegetation. The reintroduction of wolves to the park in 1995–1996 reversed some, but not all, of these effects. Elk numbers declined and they changed their foraging behavior, avoiding riparian areas where predation was higher. Without intense browsing pressure by elk, the vegetation was able to rebound in some areas of the park, leading to increases in songbirds and beavers (Ripple and Beschta 2004). However, some of the effects of seventy-plus years without wolves have not been reversed. For example, some of the changes in stream morphology were irreversible and riparian vegetation has failed to establish.

This modern example demonstrates the importance of apex carnivores. Large-bodied carnivores change prey behavior by generating "landscapes of fear," areas that prey avoid because they are particularly vulnerable to predation (Laundré et al. 2001). This can lead to reduced herbivore pressure on plants in patches of the environment. In the Yellowstone example, this was sensitive riparian areas, which are crucially important for biodiversity. Because large-bodied carnivores were particularly common in the Pleistocene, they likely exerted intense predation pressure and control on the herbivore community, leading to a landscape of fear (Van Valkenburgh et al. 2016). For example, the terminal Pleistocene extinction led to the loss of twelve of the top carnivores in North America alone, including the charismatic scimitar cat (*Homotherium serum*), American lion (*Pantherea leo atrox*), saber-tooth cat (*Smilodon fatalis*) and the short-faced bear (*Arctodus simus*). These species were considerable larger than the extant African lion, with the short-faced bear weighing about 800 kg (Smith et al. 2003). Because the Pleistocene carnivore guild was more diverse and composed of larger taxa, numerous species specialized on megafauna. Thus, even the very largest megaherbivores (>5–10 tons), often thought to be largely immune to predation, were likely influenced. Using predator/prey body mass ratios, Blaire Van Valkenburgh et al. (2016) hypothesized that the apex carnivores of the Pleistocene likely had the capacity to exert top-down control on populations by preying on juveniles. And interestingly, our stable isotope analysis supports this idea (see chapter 7, Fig. 7.7).

The extent to which the apex carnivores mediated the ecology of smaller-bodied members of their guild is unclear. For example, species distribution modeling of canids over the Pleistocene and Holocene suggests that surviving species increasingly partitioned their climatic niche; contrary to expectations, they did not expand into the niche space made newly available by the extinction of apex predators (Pardi and Smith 2016). The immigration of a new predator, humans, into North America may have outweighed the advantages of reduced intraguild competition. Thus, the negative pressure of competition with humans (and perhaps their accompanying semidomesticated dogs) outweighed the positive effects of trophic release, resulting in little range expansion in the Holocene (Pardi and Smith 2016).

While large-bodied herbivores do not create a landscape of fear, they too influence community dynamics. For example, a detailed study of a grassland community in Texas found that the terminal Pleistocene megafauna extinction was followed by significant changes in both alpha and beta diversity in that area (Smith et al. 2016d). This likely arose through encroachment by woody vegetation into what had been grassland and a subsequent increase in the number of frugivore/browser species. But the effects

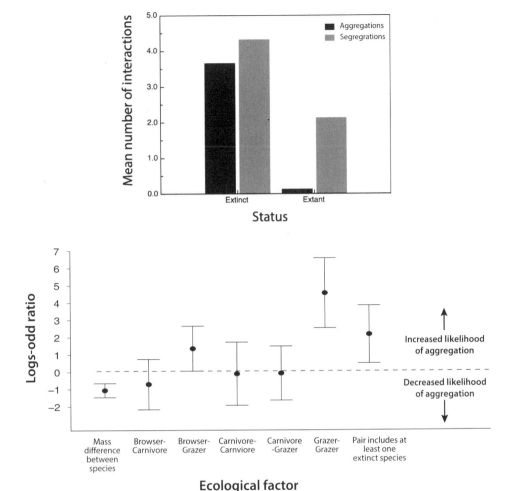

Fig. 11.18. Changes in the type and strength of species interactions over time. **a**, Mammals that went extinct (especially carnivores) formed more (and *stronger*) ecological associations with other mammals in the community during the Pleistocene. This was true of both aggregations and segregations. In contrast, extant mammals interact less strongly with potential prey or competitors. **b**, Factors promoting the formation of aggregated pairs in the late Pleistocene. Logs-odd ratios are logarithmic, thus a ratio of 2 is ten times stronger than a ratio of 1. Error bars reflect 95% confidence intervals. A dotted line reflects the null (i.e., no effect); values above the dotted line indicate factors that increase the likelihood of aggregation, while those factors below reflect a decreased likelihood of aggregation. Here, we demonstrated that species that were found together more often than by chance tended to both be grazers or grazer-browser pairs and, moreover, were pairs that included at least one species that went extinct. This, again, supports the idea that Pleistocene communities were more tightly connected than modern ones currently are. Redrawn from Smith et al. 2016d.

were more pervasive: use of PAIRS analysis, which examines species associations, revealed a decrease in the number, type, and intensity of species interactions after the extinction (Fig. 11.18; Smith et al. 2016d). Large-bodied herbivores co-occurred more often than by chance. And extinct carnivores were more likely to form positive associations with large-bodied prey than do extant carnivores, suggesting tighter predator-prey

relationships in the past. Overall, these analyses suggested that the Holocene community was simpler in ecological structure than in the Pleistocene and potentially less resilient to environmental change (Smith et al. 2016d).

Even small-bodied mammals responded to the extinction of megafauna (Tomé et al. 2019). For example, following the terminal Pleistocene extinction, cotton rats (*Sigmodon hispidus*) shifted body mass and diet—changes that were both influenced by the structure of the mammal community (Tomé et al. 2019). Such results suggest that small mammals may be sensitive to shifts in local biotic interactions within their community. Late Quaternary extinctions led to the reorganization of ecological communities and shifts in the foraging niches of surviving species, and it is likely that future extinctions will have similar effects (e.g., Gill et al. 2009; Estes et al. 2011; Gill 2014; Keesing and Young 2014; Bakker et al. 2016b; Smith et al. 2016d; Tomé et al. 2019).

What Future Do We Want?

Consequently the choice confronting humanity is not whether it affects the environment or does not. Rather the choice is about how we affect the environment, that is, in what ways and to what extent.
—Tom Lovejoy, 2000 Reith Lecture

A core insight that arises from a synthesis of ecological and paleontological perspectives is the strong top-down control that megafauna, both herbivores and carnivores, have on all aspects of ecosystem structure and function (Estes et al. 2011; Terborgh et al. 2016a, 2016b). Future work that continues to incorporate present and past extinction events can help lead to a synoptic understanding of what the decline and ultimate extinction of large-bodied animals means in terms of lost ecosystem function (e.g., Malhi et al. 2016; Smith et al. 2016a, 2016d). Clearly paleoecology is increasingly relevant to conservation biology. In particular, paleoecology helps define the range of "normal" variation of ecosystems and set baselines against which changes in species composition, abundance, and richness in modern ecosystems can be compared. While we have discussed here a number of ways that megafauna influence ecosystems, there are undoubtably other unpredictable "emergent novelties" that may arise in their absence, including changes in the transmission or frequency of disease, ecosystem homogenization, biodisparity impoverishment, and perhaps even changes in the evolutionary process itself.

Every country in the world, *except* the United States, is a signatory to the Convention on Biological Diversity (CBD). The CBD was the first binding global agreement on the conservation and sustainable use of biological diversity, and it went into effect in 1993. It was followed by the current CBD Strategic Plan for Biodiversity (2011-2020), which set out goals, polices, and obligations (WWF 2018). Unfortunately, just how the main vision of *"By 2050, biodiversity is valued, conserved, restored and wisely used, maintaining ecosystem services, sustaining a healthy planet and delivering benefits essential for all people"* is to be achieved is left to the individual countries (WWF 2018). To date, there has been only limited progress, and despite multiple international policy agreements and even extensive research, biodiversity is still in decline. This leads to the question: what future do we want on planet Earth? Without active management and conservation, the trajectory of biodiversity decline is likely to continue.

FURTHER READING

Algeo, T.J., et al. 1995. Late Devonian oceanic anoxic events and biotic crises: "rooted" in the evolution of vascular land plants? *GSA Today* 5:64-66.

Alvarez, W., et al. 1984. Impact theory of mass extinctions and the invertebrate fossil record. *Science* 223:1135-1141.

Antón, S.C. 2003. Natural history of *Homo erectus*. *American Journal of Physical Anthropology* S37:126-70.

Benton, M.J., and R.J. Twitchett. 2003. How to kill (almost) all life: the end-Permian extinction event. *Trends in Ecology and Evolution* 18:358-365.

Estes, J.A., et al. 2011. Trophic downgrading of planet Earth. *Science* 333:301-306.

Haynes, G., ed. 2009. *American Megafaunal Extinctions at the End of the Pleistocene*. Dordrecht: Springer Netherlands.

Malhi, Y., et al. 2016. Megafauna and ecosystem function from the Pleistocene to the Anthropocene. *Proceedings of the National Academy of Sciences USA* 113:838-46.

Owen-Smith, N. 1988. *Megaherbivores: The Influence of Very Large Body Size on Ecology*. Cambridge: Cambridge University Press.

Smith, F.A., et al. 2018. Body size downgrading of mammals over the late Quaternary. *Science* 360:310-313.

References

Abels, H.A., et al. 2012. Terrestrial carbon isotope excursions and biotic change during Palaeogene hyperthermals. *Nature Geoscience* 5:326-329.

Ackerly, D.D., et al. 2010. The geography of climate change: implications for conservation biogeography. *Diversity and Distributions* 16:476-487.

Agenbroad, L.D. 2005. North American proboscideans: Mammoths: the state of knowledge, 2003. *Quaternary International* 126-128:73-92.

Agricola, G. 1546. *De Natura Fossilium* [On the nature of fossils]. Translated by M.C. Bandy and J.A. Bandy. New York: Dover Publications, 2013.

Agusti, J., and M. Anton. 2002. *Mammoths, Sabretooths, and Hominids*. New York: Columbia University Press.

Aiello, L.C., and J.C. Wells. 2002. Energetics and the evolution of the genus *Homo*. *Annual Review of Anthropology* 31:323-338.

Aiello, L.C., and P. Wheeler. 1995. The expensive-tissue hypothesis: the brain and digestive system in human and primate evolution. *Current Anthropology* 36:199-221.

Albert, J.S., and D.M. Johnson. 2011. Diversity and evolution of body size in fishes. *Evolutionary Biology* 39:324-340.

Alcover, J.A., et al. 1998. The extent of extinctions of mammals on islands. *Journal of Biogeography* 25:913-918.

Ald, S.M., et al. 2007. Diversity, nomenclature, and taxonomy of protists. *Systematic Biology* 56:684-689.

Aldrich, L.T., and A.O. Nier. 1948. Argon 40 in potassium minerals. *Physical Reviews* 74:876-877.

Alexander, E.C.J., et al. 1978. MMhb-1: a new 40Ar-39Ar dating standard. *United States Geological Survey Open-File Report* 70-701:6-8.

Alexander, R.M. 1971. *Size and Shape*. London: Edward Arnold.

Alexander, R.M. 1982. *Optima for Animals*. London: Edward Arnold.

Alexander, R.M. 1989. *Dynamics of Dinosaurs and Other Extinct Giants*. New York: Columbia University Press.

Alexander, R.M. 1992. Human locomotion. In *The Cambridge Encyclopedia of Human Evolution*, edited by S. Jones, 80-85. Cambridge: Cambridge University Press.

Alexander, R.M. 2003. *Principles of Animal Locomotion*. Princeton, NJ: Princeton University Press.

Algeo, T.J., and S.E. Scheckler. 1998. Terrestrial-marine teleconnections in the Devonian: links between the evolution of land plants, weathering processes, and marine anoxic events. *Philosophical Transactions of the Royal Society of London B, Biological Sciences* 353:113-128.

Algeo, T.J., et al. 1995. Late Devonian oceanic anoxic events and biotic crises: "rooted" in the evolution of vascular land plants? *GSA Today* 5:64-66.

Allen, B.D., and R.Y. Anderson. 1993. Evidence from western North America for rapid shifts in climate during the Last Glacial Maximum. *Science* 260:1920-1923.

Allen, J. 1877. The influence of physical conditions in the genesis of species. *Radical Review* 1:108-140.

Allentoft, M.E., et al. 2012. The half-life of DNA in bone: measuring decay kinetics in 158 dated fossils. *Proceedings of the Royal Society B* 279:4724-33.

Alley, R.B. 2000. The Younger Dryas cold interval as viewed from central Greenland. *Quaternary Science Reviews* 19:213-226.

Alley, R.B. 2002. *The Two-Mile Time Machine*. Princeton, NJ: Princeton University Press.

Alley, R.B., et al. 1993. Abrupt increase in Greenland snow accumulation at the end of the Younger Dryas event. *Nature* 362:527-529.

Andrews, P. 1990. *Owls, Caves, and Fossils*. Chicago: University of Chicago Press.

Alroy, J.A. 1998. Cope's rule and the dynamics of body mass evolution in North American fossil mammals. *Science* 280:731-734.

Alroy, J.A. 1999a. The fossil record of North American mammals: evidence for a Paleocene evolutionary radiation. *Systematic Biology* 48:107-118.

Alroy, J.A. 1999b. Putting North America's end-Pleistocene megafaunal extinction in context. Large-scale analyses of spatial patterns, extinction rates, and size distributions. In *Extinctions in Near Time: Causes, Contexts, and Consequences*, edited by R.D.E. MacPhee, 105-143. New York: Kluwer.

Alroy, J.A. 2000. New methods for quantifying macroevolutionary patterns and processes. *Paleobiology* 26:707-733.

Alroy, J.A. 2001. A multispecies overkill simulation of the end-Pleistocene megafaunal mass extinction. *Science* 292:1893-1896.

Alroy, J.A., et al. 2001. Effects of sampling standardization on estimates of Phanerozoic marine diversification. *Proceedings of the National Academy of Science USA* 98:6261-6266.

Alroy, J.A., et al. 2008. Phanerozoic trends in the global diversity of marine invertebrates. *Science* 321:97-100.

Alvarez, L.W., et al. 1980. Extraterrestrial cause for the Cretaceous-Tertiary extinction. *Science* 208:1095-1108.

Alvarez, W. 1997. *T. rex and the Crater of Doom*. Princeton, NJ: Princeton University Press.

Alvarez, W., et al. 1984. Impact theory of mass extinctions and the invertebrate fossil record. *Science* 223:1135-1141.

Amelin, Y., and T.R. Ireland. 2013. Dating the oldest rocks and minerals in the solar system. *Elements* 9:39-44.

Andrewartha, H.G., and L.C. Birch 1954. *The Distribution and Abundance of Animals*. Chicago: University of Chicago Press.

Andrews, P., and L. Martin. 1991. Hominoid dietary evolution. *Philosophical Transactions of the Royal Society of London B, Biological Sciences* 334:199-209.

Angerbjorn. A. 1985. The evolution of body size in mammals on islands: some comments. *American Naturalist* 125:304-309.

Angielczch, K., et al. 2013. *Early Evolutionary History of Synapsida*. Berlin: Springer Science and Business Media.

Antón, S.C. 2003. Natural history of *Homo erectus*. *American Journal of Physical Anthropology* S37:126-70.

Antón, S.C., et al. 2014. Evolution of early *Homo*: an integrated biological perspective. *Science* 345:45. doi:1236828.

Aono, Y., and K. Kazui. 2008. Phenological data series of cherry tree flowering in Kyoto, Japan, and its application to reconstruction of springtime temperatures since the 9th century. *International Journal of Climatology* 28:905-914.

Aono, Y., and Y. Omoto. 1993. Variation in the March mean temperature deduced from cherry blossom in Kyoto since the 14th century. *Journal of Agricultural Meteorology* 48:635-638.

Appleby, V. 1979. Ladies with hammers. *New Scientist* 84:714.

Araújo, M.B., and M. Luoto. 2007. The importance of biotic interactions for modelling species distributions under climate change. *Global Ecology and Biogeography* 16:743-753.

Araújo, M.B., et al. 2006. Climate warming and the decline of amphibians and reptiles in Europe. *Journal of Biogeography* 33:1712-1728.

Arnold, J.R., and W.F. Libby. 1949. Age determinations by radiocarbon content: checks with samples of known age. *Science* 110:678-680.

Ashton, K.G., et al. 2000. Is Bergmann's rule valid for mammals? *American Naturalist* 156:390-415.

Ashworth, A.C. 1973. Fossil beetles from a fossil wood rat midden in western Texas. *Coleopterists' Bulletin* 27:139-140.

Asner, G.P., and S.R. Levick. 2012. Landscape-scale effects of herbivores on treefall in African savannas. *Ecology Letters* 15:1211-1217.

Asner, G.P., et al. 2009. A contemporary assessment of change in humid tropical forests. *Conservation Biology* 23:1386-1395.

Asner, G.P., et al. 2016. Ecosystem-scale effects of megafauna in African savannas. *Ecography* 39:240-252.

Atkinson, D., and R.M. Sibly. 1997. Why are organisms usually bigger in colder environments? Making sense of a life history puzzle. *Trends in Ecology and Evolution* 12:235-239.

Atkinson, Q.D., et al. 2008. mtDNA variation predicts population size in humans and reveals a major Southern Asian chapter in human prehistory. *Molecular Biology Evolution* 25:468-474.

Aufderheide A. 2003. *The Scientific Study of Mummies*. Cambridge: Cambridge University Press.

Avenant, N.L. 2005. Barn owl pellets: a useful tool for monitoring small mammal communities? *Belgium Journal of Zoology* 135:39-43.

Bada, J.L. 1985. Amino acid racemization dating of fossil bones. *Annual Review of Earth and Planetary Sciences* 13:241-268.

Bada, J.L., et al. 1979. Amino acid racemization dating of fossil bones, I. inter-laboratory comparison of racemization measurements. *Earth and Planetary Science Letters* 43:265-268.

Bailer-Jones, C.A.L. 2009. The evidence for and against astronomical impacts on climate change and mass extinctions: a review. *International Journal of Astrobiology* 8:213-219.

Baker, A.G., et al. 2013. Do dung fungal spores make a good proxy for past distributions of large herbivore? *Quaternary Science Reviews* 62:21-31.

Bakker, E.S., et al. 2006. Herbivore impact on grassland plant diversity depends on habitat productivity and herbivore size. *Ecology Letters* 9:780-788.

Bakker, E.S., et al. 2016a. Assessing the role of large herbivores in the structuring and functioning of freshwater and marine angiosperm ecosystems. *Ecography* 39:162-179.

Bakker, E.S., et al. 2016b. Combining paleo-data and modern exclosure experiments to assess the impact of megafauna extinctions on woody vegetation. *Proceedings of the National Academy of Science USA* 113:847-855.

Balanyá, J., et al. 2006. Global genetic change tracks global climate warming in *Drosophila subobscura*. *Science* 313:1773-1775.

Balk, M.A., et al. 2019. Investigating (a)symmetry in a small mammal's response to warming and cooling events across western North America over the late Quaternary. *Quaternary Research* 92:408-415.

Balter, V., et al. 2008. U-Pb dating of fossil enamel from the Swartkrans Pleistocene hominid site, South Africa. *Earth and Planetary Science Letters* 267:236-246.

Bambach, R.K. 1983. Ecospace utilization and guilds in marine communities through the Phanerozoic. In *Biotic Interactions in Recent and Fossil Benthic Communities,* edited by M. Tevesz and P. McCall, 719-746. New York: Plenum.

Bambach, R.K., et al. 2002. Anatomical and ecological constraints on Phanerozoic animal diversity in the marine realm. *Proceedings of the National Academy of Sciences USA* 99:6854-6859.

Bambach, R.K., et al. 2007. Autecology and the filling of ecospace: key metazoan radiations. *Palaeontology* 50:1-22.

Barbour, E.H. 1892. Notice of new gigantic fossils. *Science* 19:99-100.

Barbour, E.H. 1895. Is *Daemonelix* a burrow? A reply to Dr. Theodor Fuchs. *American Naturalist* 29:517-527.

Bargo, M.S. 2001. The ground sloth *Megatherium americanum*: skull shape, bite forces, and diet. *Acta Palaeontologica Polonica* 46:173-192.

Barlow, C. 2002. The Ghosts of Evolution: Nonsensical Fruit, Missing Partners, and Other Ecological Anachronisms. New York: Basic Books.

Barnes, I., et al. 2002. Dynamics of Pleistocene population extinctions in Beringian brown bears. *Science* 295:2267-2270.

Barnes, I., et al. 2007. Genetic structure and extinction of the woolly mammoth, *Mammuthus primigenius*. *Current Biology* 17:1072-1075.

Barnett, S.A., and R.G. Dickson 1984. Changes among wild House mice (*Mus musculus*) bred for ten generations in a cold environment, and their evolutionary implications. *Journal of Zoology* 203:163-180.

Barnosky, A.D. 2008. Megafauna biomass tradeoff as a driver of Quaternary and future extinctions. *Proceedings of the National Academy of Science USA* 105:11543-11548.

Barnosky, A.D., et al. 2003. Mammalian response to global warming on varied temporal scales. *Journal of Mammalogy* 84:354-368.

Barnosky, A.D., et al. 2004. Assessing the causes of Late Pleistocene extinctions on the continents. *Science* 306:70-75.

Barnosky, A.D., et al. 2011. Has the Earth's sixth mass extinction already arrived? *Nature* 471:51-57.

Barnosky, A.D., et al. 2016. Lasting ecological impacts of megafauna extinction. *Proceedings of the National Academy of Sciences* USA 113:856-861.

Barrell, J. 1917. Rhythms and the measurement of geologic time. *Bulletin of the Geological Society of America* 18:745-904.

Bataille, C.P., and G.J. Bowen. 2012. Mapping $^{87}Sr/^{86}Sr$ variations in bedrock and water for large scale provenance studies. *Chemical Geology* 304-305:39-52.

Baucon, A. 2010. Leonardo da Vinci, the founding father of ichnology. *Palaios* 25:361-367

Beaune, D., et al. 2013. Doom of the elephant-dependent trees in a Congo tropical forest. *Forest Ecological Management* 295:109-117.

Beddard, F. 1902. *The Cambridge Natural History. Volume X: Mammalia.* New York: MacMillan.

Beever, E., et al. 2017. Behavioral flexibility as a mechanism for coping with climate change. *Frontiers in Ecology and the Environment* 15:299-308.

Behrensmeyer, A.K. 1978. Taphonomic and ecological information from bone weathering. *Paleobiology* 4:150-162.

Behrensmeyer, A.K. 1988. Vertebrate preservation in fluvial channels. *Palaeogeography, Palaeoclimatology, Palaeoecology* 63:183-199.

Behrensmeyer, A.K. 1993. The bones of Amboseli: bone assemblages and ecological change in a modern African ecosystem. *National Geographic Research* 9(4): 402-421.

Behrensmeyer, A.K. 2007. Bonebeds through geologic time. In *Bonebeds: Genesis, Analysis, and Paleobiological Significance,* edited by R. Rogers et al., 65-102. Chicago: University of Chicago Press.

Behrensmeyer, A.K., and R.E. Chapman 1993. Models and simulations of taphonomic time-averaging in terrestrial vertebrate assemblage. *Paleontological Society Short Courses in Paleontology* 6:125-149.

Behrensmeyer, A.K., and D.E. Dechant Boaz. 1980. The recent bones of Amboseli Park, Kenya, in relation to East African paleoecology. In Behrensmeyer and Hill 1980, 72-92.

Behrensmeyer, A.K., and A.P. Hill. 1980. *Fossils in the Making: Vertebrate Taphonomy and Paleoecology,* Chicago: University of Chicago Press.

Behrensmeyer, A.K., and R.W. Hook. 1992. Paleoenvironmental contexts and taphonomic modes. In *Terrestrial Ecosystems through Time,* edited by A.K. Behrensmeyer et al., 15-136. Chicago: University of Chicago Press.

Behrensmeyer, A.K., and S.M. Kidwell. 1985. Taphonomy's contributions to paleobiology. *Paleobiology* 11:105-119.

Behrensmeyer, A.K., et al. 1997. Late Pliocene faunal turnover in the Turkana Basin, Kenya and Ethiopia. *Science* 278:1589-1594.

Behrensmeyer, A.K., et al. 2000. Taphonomy and paleobiology. *Paleobiology* 26:103-147.

Bekker, A., et al. 2004. Dating the rise of atmospheric oxygen. *Nature* 427:117-120.

Bell, A. 1983. *Dung Fungi: An Illustrated Guide to Coprophilous Fungi in New Zealand.* Wellington: Victoria University Press.

Bell, A.M. 1855. *Letters and Sounds: An Introduction to English Reading.* London: Hamilton, Adams.

Bell, P.R., et al. 2017. Tyrannosauroid integument reveals conflicting patterns of gigantism and feather evolution. *Biology Letters* 13:20170092.

Ben-David, M., and E.A. Flaherty. 2012. Stable isotopes in mammalian research: a beginner's guide. *Journal of Mammalogy* 93:312-328.

Ben-David, M., and D.M. Schell. 2001. Mixing models in analyses of diet using multiple stable isotopes: a response. *Oecologia* 127:180-184.

Bennett, K.D., et al. 1990. Fire and man in post-glacial woodlands of eastern England. *Journal of Archaeology Science* 17:635-642.

Benson, R.B.J., et al. 2014. Rates of dinosaur body mass evolution indicate 170 million years of sustained ecological innovation on the avian stem lineage. *PLoS Biology* 12 (5): e1001853. doi:10.1371/journal.pbio.1001853.

Benton, M.J. 1998. The quality of the fossil record of the vertebrates. In *The Adequacy of the Fossil Record*, edited by S.K. Donovan and C.R.C. Paul, 269-300. New York: John Wiley and Sons.

Benton, M.J. 2003. *When Life Nearly Died: The Greatest Mass Extinction of All Time.* London: Thames and Hudson.

Benton, M.J. 2005. *Vertebrate Paleontology.* Oxford: Blackwell Scientific.

Benton, M.J., and P.C.J. Donoghue. 2007. Palaeontological evidence to date the tree of life. *Molecular Biology and Evolution* 24:26-53.

Benton, M.J., and R.J. Twitchett. 2003. How to kill (almost) all life: the end-Permian extinction event. *Trends in Ecology and Evolution* 18:358-365.

Benton, M.J., et al. 2010. Dinosaurs and the island rule: The dwarfed dinosaurs from Haţeg Island. *Palaeogeography, Palaeoclimatology, Palaeoecology* 293:438-454.

Berger L.R., and R.J. Clarke. 1995. Eagle involvement in accumulation of the Taung child fauna. *Journal of Human Evolution* 29:275-299.

Berger, L.R., and W.S. McGraw. 2007. Further evidence for eagle predation of, and feeding damage on, the Taung child. *South African Journal of Science* 103:496-498.

Bergmann, C. 1847. Über die Verhältnisse der Wärmeöko-nomie der Thiere zu ihrer Grösse [About the relations of the warm economy of animals to their size]. *Göttinger Studien* 1:595-708.

Berna, F., et al. 2012. Microstratigraphic evidence of in situ fire in the Acheulean strata of Wonderwerk cave, northern Cape province, South Africa. *Proceedings of the National Academy of Sciences USA* 109:E1215-E1220.

Bernard, M., et al. 2014. The effects of mega-herbivore extinctions on seed dispersal and community structure in an East African Savanna. *Consilience* 13:312-326.

Berner, R.A. 1997. The rise of plants and their effect on weathering and atmospheric CO_2. *Science* 276:544-546.

Berry, P.M., et al. 2002. Modelling potential impacts of climate change on the bioclimatic envelope of species in Britain and Ireland. *Global Ecology and Biogeography* 11:453-462.

Best, P.B., and D.M. Schell. 1996. Stable isotopes in southern right whale (*Eubalaena australis*) baleen as indicators of seasonal movements, feeding, and growth. *Marine Biology* 124:483-494.

Betancourt, J.L., and T.R. Van Devender. 1981. Holocene vegetation in Chaco Canyon, New Mexico. *Science* 214:656-658.

Betancourt, J.L., et al. 1990. *Packrat Middens: The Last 40,000 Years of Biotic Change.* Tucson: University of Arizona Press.

Bibi, F., et al. 2017. Olduvai Gorge, Tanzania: the mammal and fish evidence. *Journal of Human Evolution* 120:48-75.

Bienvenu, T., et al. 2008. Diversity and evolution of the molar radicular complex in murine rodents (Murinae, Rodentia). *Archives of Oral Biology* 53:1030-1036.

Bininda-Emonds, O.R.P., et al. 2007. The delayed rise of present-day mammals. *Nature* 446:507-512.

Birch, L.C. 1957. The role of weather in determining the distribution and abundance of animals. Population studies: animal ecology and demography. *Cold Spring Harbor Symposia on Quantitative Biology* 22:203-218.

Birks, H.H., and B. Ammann. 2000. Two terrestrial records of rapid climatic change during the glacial-Holocene transition (14,000-9,000 calendar years B.P.) from Europe. *Proceedings of the National Academy of Science USA* 97:1390-1394.

Birks, H.J.B., and H.H. Birks. 2008. Biological responses to rapid climate change at the Younger Dryas-Holocene transition at Kråkenes, western Norway. *The Holocene* 18:19-30.

Bloch, J.I., et al. 1998. New species of *Batodonoides* (Lipotyphla, Geolabididae) from the early Eocene of Wyoming: smallest known mammal? *Journal of Mammalogy* 79:804-827.

Blois, J.L., and E.A. Hadly. 2010. Mammalian response to Cenozoic climate change. *Annual Review of Earth and Planetary Science* 37:181-208.

Blois, J.L., et al. 2008. Environmental influences on spatial and temporal patterns of body-size variation in California ground squirrels (*Spermophilus beecheyi*). *Journal of Biogeography* 35:602-613.

Blois, J.L., et al. 2010. Small mammal diversity loss in response to late-Pleistocene climatic change. *Nature* 465:771-775.

Blois J.L., et al. 2012. Modeling the climatic drivers of spatial patterns in vegetation composition since the Last Glacial Maximum. *Ecography* 36:460-473.

Blumenbach, J.F. 1799. *Handbuch der Naturgeschichte 2. durchgehends verbesserte Ausgabe* [Handbook of natural history, 2nd continuously improved edition]. Göttingen: Johann Christian Dieterich.

Blumenberg B., and A.T. Lloyd. 1983. *Australopithecus* and the origin of the genus *Homo*: aspects of biometry and systematics with accompanying catalog of tooth metric data. *Biosystems* 16:127-167.

Blumenschine, R.J., and B.L. Pobiner. 2006. Zooarchaeology and the ecology of Oldowan hominin carnivory. In Ungar 2006, 167-190. Oxford: Oxford University Press.

Bobe, R., and A.K. Behrensmeyer. 2004. The expansion of grassland ecosystems in Africa in relation to mammalian evolution and the origin of the genus *Homo*. *Palaeogeography, Palaeoclimatology, Palaeoecology* 207:399-420.

Bocherens, H., et al. 2001. New isotopic evidence for dietary habits of Neanderthals from Belgium. *Journal of Human Evolution* 40:497-505.

Boeskorov, G.G., et al. 2007. A new find of a mammoth calf. *Doklady Biological Sciences* 417:480-483.

Boeskorov, G.G., et al. 2016. The Yukagir Bison: the exterior morphology of a complete frozen mummy of the extinct steppe bison, *Bison priscus* from the early Holocene of northern Yakutia, Russia. *Quaternary International* 406:94-110.

Boeskorov, G.G., et al. 2018. A study of a frozen mummy of a wild horse from the Holocene of Yakutia, East Siberia, Russia. *Mammal Research* 63:1-8.

Boivin, N.L., et al. 2016. Ecological consequences of human niche construction: examining long-term anthropogenic shaping of global species distributions. *Proceedings of the National Academy of Sciences USA* 113:6388-6396.

Bond, G., and R. Lotti. 1995. Iceberg discharges into the North Atlantic on millennial time scales during the last glaciation. *Science* 267:1005-1010.

Bond, W.J., and J.E. Keeley. 2005. Fire as a global "herbivore": the ecology and evolution of flammable ecosystems. *Trends in Ecology and Evolution* 20:387-394.

Borths, M.R., and N.J. Stevens. 2019. *Simbakubwa kutokaafrika*, gen. et sp. nov. (Hyainailourinae, Hyaenodonta, "Creodonta," Mammalia), a gigantic carnivore from the earliest Miocene of Kenya. *Journal of Vertebrate Paleontology* 39 (1). doi:10.1080/02724634.2019.1570222.

Botkin, D. et al. 2007. Forecasting the Effects of Global Warming on Biodiversity. *Bioscience* 57:227-236.

Bourgeon, L., et al. 2017. Earliest human presence in North America dated to the Last Glacial Maximum: new radiocarbon dates from Bluefish Caves, Canada. *PLoS ONE* 12:e0169486.

Bourliere, F. 1975. Mammals, small and large: the ecological implications of size. In *Small Mammals: Their Productivity and Population Dynamics*, edited by F.B. Golley et al., 1-8. Cambridge: Cambridge University Press.

Boutin, S., and J.E. Lane. 2014. Climate change and mammals: evolutionary versus plastic responses. *Evolutionary Applications* 7:29-41.

Bowen, G.J. 2010. Isoscapes: spatial pattern in isotopic biogeochemistry. *Annual Review of Earth and Planetary Sciences* 38:161-187.

Bowen, G.J., et al. 2013. IsoMAP: Isoscapes Modeling, Analysis and Prediction (version 1.0). The IsoMAP Project. Accessed April 1, 2019. http://isomap.org.

Bowler, J.M., et al. 2003. New ages for human occupation and climatic change at Lake Mungo, Australia. *Nature* 421:837.

Bowman, S. 1995. *Radiocarbon Dating*. London: British Museum Press.

Bradley, R.S. 1999. *Paleoclimatology: Reconstructing Climates of the Quaternary*. Burlington, MA: Harcourt/Academic Press.

Bradshaw, W.E., and C.M. Holzapfel. 2001. Genetic shift in photoperiodic response correlated with global warming. *Proceedings of the National Academy of Sciences* 98:14509-14511.

Braje, T.J., and J.M. Erlandson. 2013. Human acceleration of animal and plant extinctions: a Late Pleistocene, Holocene, and Anthropocene continuum. *Anthropocene* 4:14-23.

Brand, U., et al. 2016. Methane hydrate: killer cause of Earth's greatest mass extinction. *Palaeoworld* 25:496-507.

Brasier, M.D., et al. 2002. Questioning the evidence for Earth's oldest fossils. *Nature* 416:76-81.

Brault, M.O., et al. 2013. Assessing the impact of late Pleistocene megafaunal extinctions on global vegetation and climate. *Climate of the Past* 9:1761-1771.

Braun, D.R., et al. 2010. Early hominin diet included diverse terrestrial and aquatic animals 1.95 Ma in East Turkana, Kenya. *Proceedings of the National Academy of Sciences USA* 107:10002-10007.

Brázdil, R., et al. 2010. European climate of the past 500 years: new challenges for historical climatology. *Climate Change* 101:7-40.

Breeze, P.S., et al. 2016. Palaeohydrological corridors for hominin dispersals in the Middle East similar to 250-70,000 years ago. *Quaternary Science Reviews* 144:155-185

Briggs, D.E.G. 2015. Extraordinary fossils reveal the nature of Cambrian life: a commentary on Whittington (1975) "The enigmatic animal *Opabinia regalis*, Middle Cambrian, Burgess Shale, British Columbia." *Philosophical Transactions of the Royal Society of London B, Biological Sciences* 370:20140313.

Briggs, D.E.G., et al. 1995. *Fossils of the Burgess Shale*. Washington, DC: Smithsonian Institution Press.

Brohan, P., et al. 2006. Uncertainty estimates in regional and global observed temperature changes: a new dataset from 1850. *Journal of Geophysical Research* 111:d12106.

Bromley, R.G. 1996. *Trace Fossils: Biology, Taphonomy, and Applications*. London: Chapman and Hall.

Brookfield, M.E. 2008. *Principles of Stratigraphy*. New York: John Wiley and Sons.

Brown, C.M., et al. 2017. An exceptionally preserved three-dimensional armored dinosaur reveals insights into coloration and Cretaceous predator-prey dynamics. *Current Biology* 27:P2514-2521.

Brown, J.H. 1968. Adaptation to environmental temperature in two species of woodrats, *Neotoma cinerea* and *N. albigula*. *Miscellaneous Publications of the Museum of Zoology, University of Michigan* 135:1-48.

Brown, J.H. 1995. *Macroecology*. Chicago: University of Chicago Press.

Brown, J.H., and A.K. Lee. 1969. Bergmann's rule and climatic adaptation in woodrats (*Neotoma*). *Evolution* 23:329-338.

Brown, J.H., and B.A. Maurer. 1986. Body size, ecological dominance, and Cope's rule. *Nature* 324:248-250.

Brown, J.H., and B.A. Maurer. 1989. Macroecology: the division of food and space among species on continents. *Science* 243:1145-1150.

Brown, J.H., and P.F. Nicoletto. 1991. Spatial scaling of species composition: body masses of North American land mammals. *American Naturalist* 138:1478-1512.

Brown, K.S., et al. 2012. An early and enduring advanced technology originating 71,000 years ago in South Africa. *Nature* 491:590-593.

Brown, P., et al. 2004. A new small-bodied hominin from the Late Pleistocene of Flores, Indonesia. *Nature* 431:1055-1061.

Buckland, W. 1829. On the discovery of coprolites, or fossil feces in the Lias at Lyme Regis, and in other formations. *Transactions of the Geological Society of London, Series 2* 1:223-36.

Bueno, R.S., et al. 2013. Functional redundancy and complementarities of seed dispersal by the last Neotropical megafrugivores. *PLoS ONE* 8:e56252.

Buick, R. 1990. Microfossil recognition in Archean rocks: an appraisal of spheroids and filaments from a 3500 m.y. old chert-barite unit at North Pole, Western Australia. *Palaios* 5:441-459.

Bunn, H.T. 1981. Archaeological evidence for meat-eating by Plio-Pleistocene hominids from Koobi Fora and Olduvai Gorge. *Nature* 291:547-577.

Bunn, H.T., and J.A. Ezzo. 1993. Hunting and scavenging by Plio-Pleistocene hominids: nutritional constraints, archaeological patterns, and behavioural implications. *Journal of Archaeological Science* 20:365-398.

Bunn, H.T., et al. 2007. Was FLK North levels 1-2 a classic "living floor" of Oldowan hominins or a taphonomically complex palimpsest dominated by large carnivore feeding behavior? *Quaternary Research* 74:355-362.

Bunney, K., et al. 2017. Seed dispersal kernel of the largest surviving megaherbivore—the African savanna elephant. *Biotropica* 49:395-401.

Burger, J.R., et al. 2012. The macroecology of sustainability. *PLoS Biology* 10:e1001345.

Burness, G.P., et al. 2001. Dinosaurs, dragons and dwarfs: the evolution of maximal body size. *Proceedings of the National Academy of Science USA* 98:14518-14523.

Burney, D.A., and T.F. Flannery. 2005. Fifty millennia of catastrophic extinctions after human contact. *Trends in Ecology and Evolution* 20:395-401.

Burney, D.A., et al. 2003. *Sporormiella* and the late Holocene extinctions in Madagascar. *Proceedings of the National Academy of Sciences USA* 100:10800-10805.

Burnside, W.R., et al. 2012. Human macroecology: linking pattern and process in big-picture human ecology. *Biological Reviews* 87:194-208.

Burthe, S., et al. 2011. Demographic consequences of increased winter births in a large aseasonally breeding mammal (*Bos taurus*) in response to climate change. *Journal of Animal Ecology* 80:1134-1144.

Butler, P.M. 1982. Directions of evolution in the mammalian dentition. In *Problems of Phylogenetic Reconstruction*, edited by K.A. Joysey and A.E. Friday, 235-244. London: Systematics Association.

Butler, R.F. 2004. *Paleomagnetism: Magnetic Domains to Geologic Terranes*. New York: Blackwell Science.

Butterfield, N.J. 2000. *Bangiomorpha pubescens* n. gen., n. sp.: implications for the evolution of sex, multicellularity, and the Mesoproterozoic/Neoproterozoic radiation of eukaryotes. *Paleobiology* 26:386-404.

Calder, W.A. 1984. *Size, Function, and Life History*. Cambridge, MA: Harvard University Press.

Callen, E.O., and T.W.M. Cameron. 1960. A prehistoric diet revealed in coprolites. *The New Scientist* 18:35-40.

Campos-Arceiz, A., and S. Blake. 2011. Megagardeners of the forest: the role of elephants in seed dispersal. *Acta Oecologica* 37:542-553.

Cardillo, M., et al. 2005. Multiple causes of high extinction risk in large mammal species. *Science* 309:1239-1241.

Cardillo, M., et al. 2008. The predictability of extinction: biological and external correlates of decline in mammals. *Proceedings of the Royal Society B* 275:1441-1448.

Cardonatto, M.C., and R.N. Melchor. 2018. Large mammal burrows in late Miocene calcic paleosols from central Argentina: paleoenvironment, taphonomy, and producers. *PeerJ* 6:e4787.

Carlquist, S.J. 1965. *Island Life: A Natural History of the Islands of the World*. New York: Natural History Press.

Carlquist, S.J. 1974. *Island Biology*. New York: Columbia University Press.

Carroll, R.L. 1988. *Vertebrate Paleontology and Evolution*. New York: W.H. Freeman.

Carroll, S.P. 2008. Facing change: forms and foundations of contemporary adaptation to biotic invasions. *Molecular Ecology* 17:361-372.

Carter, T.S., et al. 2016. *Priodontes maximus* (Cingulata: Chlamyphoridae). *Mammalian Species* 48:21-34.

Carto, S.L., et al. 2009. Out of Africa and into an ice age: on the role of global climate change in the late Pleistocene migration of early modern humans out of Africa. *Journal of Human Evolution* 56:139-151.

Case, T.J. 1978. A general explanation for insular body size trends in terrestrial vertebrates. *Ecology* 59:1-18.

Case, T.J., and M.L. Cody. 1983. *Island Biogeography in the Sea of Cortez*. Berkeley: University of California Press.

Caughley, G. 1966. Mortality patterns in mammals. *Ecology* 47:906-918.

Caut, S., et al. 2009. Variation in discrimination factors (δ^{15}N and δ^{13}C): the effect of diet isotopic values and applications for diet reconstruction. *Journal of Applied Ecology* 46:443-453.

Ceballos, G., and P.R. Ehrlich. 2002. Mammal population losses and the extinction crisis. *Science* 296:904-907.

Ceballos, G., et al. 2015. Accelerated modern human-induced species losses: entering the sixth mass extinction. *Science Advances* 1:e1400253.

Ceballos, G., et al. 2017. Population losses and the sixth mass extinction. *Proceedings of the National Academy of Sciences USA* 114:E6089-E6096.

Cerling, T.E., and J.M. Harris. 1999. Carbon isotope fractionation between diet and bioapatite in ungulate mammals and implications for ecological and paleoecological studies. *Oecologia* 120:347-363.

Cerling, T.E., et al. 1997. Global vegetation change through the Miocene/Pliocene boundary. *Nature* 389:153-158.

Cerling, T.E., et al. 1998. Carbon dioxide starvation, the development of C$_4$ ecosystems, and mammalian evolution. *Philosophical Transactions of the Royal Society of London B, Biological Sciences* 353:159-171.

Chamberlain, C.P., et al. 2005. Pleistocene to recent dietary shifts in California condors: implications for conservation strategies. *Proceedings of the National Academy of Sciences USA* 102:16707-16711.

Chan, Y.L., et al. 2005. Ancient DNA reveals Holocene loss of genetic diversity in a South American rodent. *Biology Letters* 1:423-426.

Chan, Y.L., et al. 2006. Bayesian estimation of the timing and severity of a population bottleneck from ancient DNA. *PLoS Genetics* 2:e59.

Chang, D., et al. 2017. The evolutionary and phylogeographic history of woolly mammoths: a comprehensive mitogenomic analysis. *Scientific Reports* 7:44585.

Chew, A.E. 2015. Mammal faunal change in the zone of the Paleogene hyperthermals ETM2 and H2. *Climate of the Past* 11:1223-1237.

Chiappe, L.M. 1995. The first 85 million years of avian evolution. *Nature* 378:349-355.

Chiyo, P.I., et al. 2015. Illegal tusk harvest and the decline of tusk size in the African elephant. *Ecology and Evolution* 5:5216-5229.

Cifelli, R.L. 2004. Marsupial mammals from the Albian-Cenomanian (Early-Late Cretaceous) boundary, Utah. *Bulletin of the American Museum of Natural History* 285:62-79.

Clark, I., and P. Fritz. 1997. *Environmental Isotopes in Hydrogeology*. New York: CRC Press.

Clark, J.S., et al. 2001. Ecological forecasts, an emerging imperative. *Science* 293:657-660.

Clary, K.H. 1984. Anasazi diet and subsistence as revealed by coprolites from Chaco Canyon. In *Recent Research on Chaco Prehistory*, Reports of the Chaco Center vol. 8, edited by W.J. Judge and J.D. Schelberg, 265-279. Albuquerque, NM: Division of Cultural Resources, National Park Service.

Clauss, M., and J. Hummel. 2005. The digestive performance of mammalian herbivores: why big may not be that much better. *Mammal Review* 35:174-187.

Clements, F.E. 1916. *Plant Succession: An Analysis of the Development of Vegetation*. Washington, DC: Carnegie Institute of Washington Publications.

Clements, F.E. 1936. Nature and structure of the climax. *Ecology* 24:252-284.

Clementz, M.T. 2012. New insight from old bones: stable isotope analysis of fossil mammals. *Journal of Mammalogy* 93:368-380.

Clementz, M.T., and P.L. Koch. 2001. Differentiating aquatic mammal habitat and foraging ecology with stable isotopes in tooth enamel. *Oecologia* 129:461-472.

Clementz, M.T., et al. 2003. A paleoecological paradox: the habitat and dietary preferences of the extinct tethythere, *Desmostylus*, inferred from stable isotope analysis. *Paleobiology* 29:506-519.

Clyde, W.C., and P.D. Gingerich. 1998. Mammalian community response to the latest Paleocene thermal maximum: an isotaphonomic study in the northern Bighorn Basin, Wyoming. *Geology* 26:1011-1014.

Colbert, E. 1984. *The Great Dinosaur Hunters and Their Discoveries*. New York: Dover.

Coltrain, J.B., et al. 2004. Rancho La Brea stable isotope biogeochemistry and its implications for the palaeoecology of late Pleistocene, coastal southern California. *Palaeogeography, Palaeoclimatology, Palaeoecology* 205:199-219.

Condon, D., et al. 2005. U-Pb ages from the Neoproterozoic Doushantuo Formation, China. *Science* 308:95-98.

Conway, M.S. 1986. The community structure of the Middle Cambrian phyllopod bed (Burgess Shale). *Palaeontology* 29:423-467.

Conybeare, W.D. 1824. On the discovery of an almost complete skeleton of the *Plesiosaurus*. *Philosophical Transactions of the Geological Society of London* 1:381-389.

Cooper, A., and H.N. Poinar. 2000. Ancient DNA: do it right or not at all. *Science* 289:1139.

Cope, E.D. 1880. On the extinct cats of North America. *American Naturalist* 14:833-858.

Cope, E.D. 1896. *The Primary Factors of Organic Evolution*. Chicago: Open Court.

Corona, A. 2019. New records and diet reconstruction using dental microwear analysis for *Neolicaphrium recens* Frenguelli, 1921 (Litopterna, Proterotheriidae). *Andean Geology* 46:153-167.

Craig, H. 1961. Isotopic variations in meteoric waters. *Science* 133:1702-1703.

Crichton, M. 1990. *Jurassic Park*. New York: Knopf.

Crompton, A.W. 1980. In *Comparative Physiology: Primitive Mammals*, edited by K. Schmidt-Nielsen et al., 1-12. Cambridge: Cambridge University Press.

Cumming, D.H., et al. 1997. Elephants, woodlands, and biodiversity in southern Africa. *South African Journal of Science* 93:231-236.

Cummins, H., et al. 1986. The size-frequency distribution in paleoecology: effects of taphonomic processes during formation of molluscan death assemblages in Texas bays. *Paleontology* 29:495-518.

Curry, D.H. 1939. Tertiary and Pleistocene mammal and bird tracks in Death Valley. *Bulletin of the Geological Society of America* 50:1971-1972.

Cuvier, G. 1796. *Memoir on the Species of Elephants, Both Living and Fossil*. Paris: Museum National d'Histoire Naturelle.

Cuvier, G. 1829. *Iconographie du Régne Animal. Tome 1: Planches des Animaux vertébrés* [Iconography of the animal kingdom. Volume 1: Plates of vertebrate animals]. Paris: Librarie de L'Académie Royale de Médecine.

Cuvier, G. 1830. Lectures on the history of the natural sciences. Lecture nine: Theophrastus. *Edinburgh New Philosophical Journal* 9:76-83.

D'Ambrosia, A.R., et al. 2017. Repetitive mammalian dwarfing during ancient greenhouse warming events. *Science Advances* 3:e1601430.

Da Vinci, L. 1817. Trattato della pittura di Lionardo da Vinci. Tratto da un codice della biblioteca Vaticana. Nella stamperia de Romanis, Roma [Treatise on painting by Lionardo da Vinci. Taken from a codex of the Vatican library. In the de Romanis printing house, Rome]. Translated and annotated by P.P. McMachon. Princeton, NJ: Princeton University Press, 1956.

Dahl-Jensen, D., et al. 1998. Past temperatures directly from the Greenland ice sheet. *Science* 282:268-271.

Dalrymple, B.G. 1991. *The Age of the Earth*. Stanford, CA: Stanford University Press.

Dalton, R. 2002. Squaring up over ancient life. *Nature* 417:782-784.

Damuth, J.D., and C.M. Janis. 2011. On the relationship between hypsodonty and feeding ecology in ungulate animals, and its utility in palaeoecology. *Biological Reviews* 86:733-758.

Damuth, J.D., and B.J. MacFadden. 1990. *Body Size in Mammalian Paleobiology: Estimation and Biological Implications*. Cambridge: Cambridge University Press.

Damuth, J.D., et al. 2002. Reconstructing mean annual precipitation, based on mammalian dental morphology and local species richness. In *EEDEN Plenary Workshop on Late Miocene to Early Pliocene Environments and Ecosystems*, edited by J. Agustí and O. Oms, 23-24. Sabadell, Spain: EEDEN Programme, European Science Foundation.

Dansgaard, W., et al. 1989. The abrupt termination of the Younger Dryas climate event. *Nature* 339:532-534.

Dansgaard, W., et al. 1993. Evidence for general instability of past climate from a 250-kyr ice-core record. *Nature* 364:218-220.

Dart, R.A. 1925. *Australopithecus africanus*: the man-ape of South Africa. *Nature* 115:195-199.

Dart, R.A., and D. Craig. 1959. *Adventures with the Missing Link*. New York: Harper Brothers.

Darwin, C. 1859. *On the origin of species by means of natural selection; or, the preservation of favoured races in the struggle for life*. London: John Murray.

Davidson, A.D., et al. 2009. Multiple ecological pathways to extinction in mammals. *Proceedings of the National Academy of Sciences USA* 106:10702-10705.

Davis, M.B. 1963. On the theory of pollen analysis. *American Journal of Science* 261:897-912.

Davis, M.B. 1969. Climatic changes in southern Connecticut recorded by pollen deposition at Rogers Lake. *Ecology* 50:409-422.

Davis, M.B. 2000. Palynology after Y2K—understanding the source area of pollen in sediments. *Annual Review of Earth and Planetary Sciences* 28:1-18.

Davis, M.B., and R.G. Shaw. 2001. Range shifts and adaptive responses to quaternary climate change. *Science* 292:673-679.

Davis, M.B., et al. 2005. Evolutionary responses to changing climate. *Ecology* 86:1704-1714.

Davis, O.K. 1987. Spores of the dung fungus *Sporormiella*: Increased abundance in historic sediments and before Pleistocene megafaunal extinction. *Quaternary Research* 28:290-294.

Davis, O.K., and D.S. Shafer. 2006. *Sporormiella* fungal spores, a palynological means of detecting herbivore density. *Palaeogeography, Palaeoclimatology, Palaeoecology* 237:40-50.

Davis, O.K., et al. 1984. The Pleistocene dung blanket of Bechan Cave, Utah. In *Contributions in Quaternary Vertebrate Paleontology: A Volume in Memorial to John E. Guilday*, edited by H.H. Genoways and M.R. Dawson, 267-282. Special Publications of the Carnegie Museum of Natural History No. 8. Pittsburgh: Carnegie Museum of Natural History.

Davis, O.K., et al. 1985. Riparian plants were a major component of the diet of mammoths of southern Utah. *Current Research in the Pleistocene* 2:81-82.

Davis, S.J. 1977. Size variation of the fox, *Vulpes vulpes*, in the Palaearctic region today, and in Israel during the late Quaternary. *Journal of the Zoological Society London* 182:343-351.

Davis, S.J. 1981. The effects of temperature change and domestication on the body size of Late Pleistocene to Holocene mammals of Israel. *Paleobiology* 7:101-114.

Davis, W.T. 1909. Owl pellets and insects. *Journal of the New York Entomological Society* 17:49-51.

Davit-Béal T., et al. 2009. Loss of teeth and enamel in tetrapods: fossil record, genetic data, and morphological adaptations. *Journal of Anatomy* 214:477-501.

Dawson, W.R. 1992. Physiological responses of animals to higher temperatures. In *Global Warming and Biological Diversity*, edited by R.L. Peters and T.E. Lovejoy, 158-170. New Haven, CT: Yale University Press.

de Amorim, M.E., et al. 2017. Lizards on newly created islands independently and rapidly adapt in morphology and diet. *Proceedings of the National Academy of Science USA* 114:8812-8816.

de Bruyn, M., et al. 2009. Rapid response of a marine mammal species to Holocene climate and habitat change. *PLoS Genetics* 5:e1000554.

Dean, G.W. 2006. The science of coprolite analysis: the view from Hinds cave. *Palaeogeography, Palaeoclimatology, Palaeoecology* 237:67-79.

Debruyne, R., et al. 2008. Out of America: ancient DNA evidence for a new world origin of late Quaternary woolly mammoths. *Current Biology* 18:1320-1326.

Decker, E.H., et al. 2000. Energy and material flow through the urban ecosystem. *Annual Review of Energy and the Environment* 25:685-740.

Deino, A.L. 1998. ^{40}Ar/^{39}Ar dating in paleoanthropology and archaeology. *Evolutionary Anthropology* 6:63-75.

Delsuc, F., et al. 2019. Ancient mitogenomes reveal the evolutionary history and biogeography of sloths. *Current Biology* 17:2031-2042.

Deméré, T.A., et al. 2008. Morphological and molecular evidence for a stepwise evolutionary transition from teeth to baleen in mysticete whales. *Systematic Biology* 57:15-37.

DeNiro, M.J., and S. Epstein. 1978. Influence of diet on the distribution of carbon isotopes in animals. *Geochimica et Cosmochimica Acta* 42:495-506.

DeNiro, M.J., and S. Epstein. 1981. Influence of diet on the distribution of nitrogen isotopes in animals. *Geochimica et Cosmochimica Acta* 45:341-351.

DeSantis, L.R.G. 2016. Dental microwear textures: reconstructing diets of fossil mammals. *Surface Topography: Metrology and Properties* 4:023002.

Détroit, F., et al. 2019. A new species of *Homo* from the Late Pleistocene of the Philippines. *Nature* 568:181-186.

Devender, T.R. 1987. Holocene vegetation and climate in the Puerto Blanco Mountains, southwestern Arizona. *Quaternary Research* 27:51-72.

Diniz-Filho, J.A.F., and P. Raia. 2017. Island Rule, quantitative genetics, and brain-body size evolution in *Homo floresiensis*. *Proceedings of the Royal Society B* 284:20171065.

Dirzo, R., et al. 2014. Defaunation in the Anthropocene. *Science* 345:401-406.

Dixon, E.J. 1999. *Bones, Boats, and Bison: Archeology and the First Colonization of Western North America*. Albuquerque: University of New Mexico Press.

Doelling, H.H., et al. 2000. Geology of Grand Staircase-Escalante National Monument, Utah. *Utah Geological Association Publication* 28:1-43.

Domínguez-Rodrigo, M. 2002. Hunting and scavenging by early humans: the state of the debate. *Journal of World Prehistory* 16:1-54.

Domínguez-Rodrigo, M., et al. 2007. *Deconstructing Olduvai: A Taphonomic Study of the Bed I Sites.* Dordrecht: Springer Netherlands.

Donlan, C.J., et al. 2005. Rewilding North America. *Nature* 436 (7053): 913-914.

dos Reis, M., et al. 2012. Phylogenomic datasets provide both precision and accuracy in estimating the timescale of placental mammal phylogeny. *Proceedings of the Royal Society B* 279:3491-3500.

Doucett, R.R., et al. 2007. Measuring terrestrial subsidies to aquatic food webs using stable isotopes of hydrogen. *Ecology* 88:1587-1592.

Doughty, C.E. 2017. Herbivores increase the global availability of nutrients over millions of years. *Nature Ecology and Evolution* 1:1820-1827.

Doughty, C.E., et al. 2010. Biophysical feedbacks between the Pleistocene megafauna extinction and climate: the first human-induced global warming? *Geophysical Research Letters* 37:1-5.

Doughty, C.E., et al. 2013. The legacy of the Pleistocene megafauna extinctions on nutrient availability in Amazonia. *Nature Geoscience* 6:761-764.

Doughty, C.E., et al. 2016a. Global nutrient transport in a world of giants. *Proceedings of the National Academy of Sciences* USA 113:868-873.

Doughty, C.E., et al. 2016b. The impact of the megafauna extinctions on savanna woody cover in South America. *Ecography* 39:213-222.

Doughty, C.E., et al. 2016c. Interdependency of plants and animals in controlling the sodium balance of ecosystems and the impacts of global defaunation. *Ecography* 39:204-212.

Doughty, C.E., et al. 2016d. Megafauna extinction, tree species range reduction, and carbon storage in Amazonian forests. *Ecography* 39:194-203.

Downhower, J.F., and L.S. Bulmer. 1988. Calculating just how small a whale can be. *Nature* 335:675.

Dubois, E. 1897. Über die Abhängigkeit des Hirngewichtes von der Körpergrösse bei den Säugetieren [About the dependence of the brain weight on the body size in mammals]. *Archives of Anthropology* 25:1-28.

Dyke, A.S., and V.K. Prest. 1987. Late Wisconsinan and Holocene History of the Laurentide Ice Sheet. *Géographie physique et Quaternaire* 41:237-263.

Eaton, J. 2003. Potent packrat leavings tempted starving 49ers. *Berkeley Daily Planet.* http://www .berkeleydailyplanet.com/article.cfm?archiveDate=12-12 -03&storyID=17931.

Economos, A.C. 1981. The largest land mammal. *Journal of Theoretical Biology* 89:211-215.

Efremov, I.A. 1940. Taphonomy: a new branch of paleontology. *Pan-American Geology* 74:81-93.

Ehleringer, J.R., and R.K. Monson. 1993. Evolutionary and ecological aspects of photosynthetic pathway variation. *Annual Review of Ecology and Systematics* 24:411-439.

Ehleringer, J.R., et al. 2008. Hydrogen and oxygen isotope ratios in human hair are related to geography. *Proceedings of the National Academy of Sciences USA* 105:2788-2793.

Elbroch, M. 2003. *Mammal Tracks and Sign: A Guide to North American Species.* Mechanicsburg, PA: Stackpole Books.

Eldredge, N. 2014. *Extinction and Evolution: What Fossils Reveal about the History of Life.* Ontario: Firefly Books.

Elias, S.A. 1990. Observations on the taphonomy of late Quaternary insect fossil remains in packrat middens of the Chihuahuan Desert. *Palaios* 5:356-363.

Emerson, S.B., and L. Radinsky. 1980. Functional analysis of sabertooth cranial morphology. *Paleobiology* 6:295-312.

Emling, S. 2009. *The Fossil Hunter: Dinosaurs, Evolution, and the Woman Whose Discoveries Changed the World.* Basingstoke, UK: Palgrave Macmillan.

Ernest, S.K.M., and J.H. Brown. 2001. Homeostasis and compensation: the role of resources in ecosystem stability. *Ecology* 82:2118-2132.

Ernest, S.K.M., et al. 2003. Thermodynamic and metabolic effects on the scaling of production and abundance. *Ecology Letters* 6:990-995.

Eronen, J.T. 2006. Eurasian Neogene large herbivorous mammals and climate. *Acta Zoologica Fennica* 216:1-72.

Eronen, J.T., et al. 2010a. Precipitation and large herbivorous mammals I: estimates from present-day communities. *Evolutionary Ecology Research* 12:217-233.

Eronen, J.T., et al. 2010b. Precipitation and large herbivorous mammals II: application to fossil data. *Evolutionary Ecology Research* 12:235-248.

Erwin, D.H. 2006. Extinction: How Life on Earth Nearly Ended 250 Million Years Ago. Princeton, NJ: Princeton University Press.

Estep, M.F., and H. Dabrowski. 1980. Tracing food webs with stable hydrogen isotopes. *Science* 209:1537-1538.

Estes, J.A., et al. 2011. Trophic downgrading of planet Earth. *Science* 333:301-306.

Evans, A.R., and G.D. Sanson. 2003. The tooth of perfection: functional and spatial constraints on mammalian tooth shape. *Biological Journal of the Linnean Society* 78:173-191.

Evans, A.R., et al. 2001. Confocal imaging, visualization, and 3-D surface measurement of small mammalian teeth. *Journal of Microscopy* 204:108-118.

Evans, A.R., et al. 2007. High-level similarity of dentitions in carnivorans and rodents. *Nature* 445:78-81.

Evans, A.R., et al. 2012. The maximum rate of mammal evolution. *Proceedings of the National Academy of Science USA* 109:4187-4190.

Ewer, R.F. 1973. *The Carnivores*. London: Weidenfeld and Nicolson.

Faegri, K., et al. 1989. *Textbook of Pollen Analysis*. Caldwell: John Wiley and Sons.

Fagan, B. 2007. *The Little Ice Age: How Climate Made History 1300-1850*. New York: Basic Books.

Faith, J.T., and A.K. Behrensmeyer. 2006. Changing patterns of carnivore modification in a landscape bone assemblage, Amboseli Park, Kenya. *Journal of Archeological Science* 33:1718-1733.

Faith, J.T., and T.A. Surovell. 2009. Synchronous extinction of North America's Pleistocene mammals. *Proceedings of the National Academy of Sciences USA* 106:20641-20645.

Falconer, D.S. 1953. Selection for large and small size in mice. *Journal of Genetics* 51:470-501.

Falconer, D.S. 1973. Replicated selection for body weight in mice. *Genetic Research* 22:291-321.

Falconer, D.S. 1989. *Introduction to Quantitative Genetics*. London: Longman.

Famoso, N.A., and E.B. Davis. 2016. On the relationship between enamel band complexity and occlusal surface area in Equids (Mammalia, Perissodactyla). *PeerJ* 4:e2181.

Famoso, N.A., et al. 2013. Occlusal enamel complexity and its implications for lophodonty, hypsodonty, body mass and diet in extinct and extant ungulates. *Palaeogeography, Palaeoclimatology, Palaeoecology* 387:211-216.

Farquhar, G.D., et al. 1989. Carbon isotope discrimination and photosynthesis. *Annual Review of Plant Physiology and Plant Molecular Biology* 40:503-537.

Faure, G., and T.M. Mensing. 2004. *Isotopes: Principles and Applications*. 3rd ed. New York: John Wiley and Sons.

Fenchel, T. 1993. There are more small than large species? *Oikos* 68:375-378.

Fennessy, J., et al. 2016. Multi-locus analyses reveal four giraffe species instead of one. *Current Biology* 26:2543-2549.

Feranec, R.S. 2003. Stable isotopes, hypsodonty, and the paleodiet of *Hemiauchenia* (Mammalia: Camelidae): a morphological specialization creating ecological generalization. *Paleobiology* 29:230-242.

Feranec, R.S., and B.J. MacFadden. 2006. Isotopic discrimination of resource partitioning among ungulates in C3-dominated communities from the Miocene of Florida and California. *Paleobiology* 32:191-205.

Feranec, R.S., et al. 2011. The *Sporormiella* proxy and end-Pleistocene megafaunal extinction: a persepective. *Quaternary International* 245:333-338.

Ferguson-Lees, J., and D.A. Christie. 2001. *Raptors of the World*. Princeton, NJ: Princeton University Press.

Ferguson, C.W., and D.A. Graybill. 1983. Dendrochronology of Bristlecone pine: a progress report. *Radiocarbon* 25:87-88.

Ferraro, J.V., et al. 2013. Earliest archaeological evidence of persistent hominin carnivory. *PLoS ONE* 8:e62174.

Fiedel, S. 2009. Sudden deaths: the chronology of terminal Pleistocene megafaunal extinction. In *American Megafaunal Extinctions at the End of the Pleistocene*, edited by G.A. Haynes, 21-37. Dordrecht: Springer Netherlands.

Figueirido, B., et al. 2012. Cenozoic climate change influences mammalian evolutionary dynamics. *Proceedings of the National Academy of Sciences* 109:722-727.

Finnegan, S., et al. 2016. Biogeographic and bathymetric determinants of brachiopod extinction and survival during the Late Ordovician mass extinction. *Proceedings of the Royal Society B* 283:20160007.

Fisher, D.C., et al. 2012. Anatomy, death, and preservation of a woolly mammoth (*Mammuthus primigenius*) calf, Yamal Peninsula, Northwest Siberia. *Quaternary International* 255:94-105.

Fizet, M., et al. 1995. Effect of diet, physiology, and climate on carbon and nitrogen isotopes of collagen in a late Pleistocene anthropic paleoecosystem. *Journal of Archaeology* 22:67-79.

Fogel, M.L. 2019. My stable isotope journey in biogeochemistry, geoecology, and astrobiology. *Geochemical Perspectives* 8:105-281.

Fogel, M.L., and N. Tuross. 2003. Extending the limits of paleodietary studies of humans with compound specific carbon isotope analysis of amino acids. *Journal of Archaeological Science* 30:535:545.

Fogel, M.L., et al. 1989. Nitrogen isotope tracers of human lactation in modern and archaeological populations. *Annual Report of the Director, Geophysical Laboratory, Carnegie Institute of Washington*, 111-117.

Foley, R.A. 2001. The evolutionary consequences of increased carnivory in hominids. In *Meat-Eating and Human Evolution*, edited by C.B. Stanford and H.T. Bunn, 305-331. Oxford: Oxford University Press.

Foote, M. 1994. Temporal variation in extinction risk and temporal scaling of extinction metrics. *Paleobiology* 20:424-444.

Foote, M. 1997. Sampling, taxonomic description, and our evolving knowledge of morphological diversity. *Paleobiology* 23:181-206.

Foote, M. 2000. Origination and extinction components of taxonomic diversity: general problems. *Paleobiology* 26:74-102.

Foote, M., and D.M. Raup. 1996. Fossil preservation and the stratigraphic ranges of taxa. *Paleobiology* 22:121-140.

Forir, M., et al. 2007. Preliminary investigation of the trackways and claw marks within the Riverbluff Cave system, Springfield, Missouri. In Lucas et al. 2007, 3-4.

Fortelius, M., and J. Kappelman. 1993. The largest land mammal ever imagined. *Zoological Journal of the Linnean Society* 107:85-101.

Fortelius, M., and N. Solounias. 2000. Functional characterization of ungulate molars using the abrasion attrition wear gradient: A new method for reconstructing paleodiets. *American Museum Novitates* 3301:1-36.

Fortelius, M., et al. 2002. Fossil mammals resolve regional patterns of Eurasian climate change during 20 million years. *Evolutionary Ecology Research* 4:1005-1016.

Fortelius, M., et al. 2006. Late Miocene and Pliocene large land mammals and climatic changes in Eurasia. *Palaeogeography, Palaeoclimatology, Palaeoecology* 238:219-227.

Fortney, R. 1982. *Fossils: The Key to the Past*. Cincinnati, OH: Van Nostrand Reinhold.

Foster, F., and M. Collard. 2013. A reassessment of Bergmann's rule in modern humans. *PLoS ONE* 8 (8): e72269.

Foster, J.B. 1964. Evolution of mammals on islands. *Nature* 202:234-235.

Fox, D.L., and P.L. Koch. 2004. Carbon and oxygen isotopic variability in Neogene paleosol carbonates: constraints on the evolution of the C4 grasslands of the Great Plains, USA. *Palaeogeography, Palaeoclimatology, Palaeoecology* 207:305-329.

Franca, L.D., et al. 2015. Review of feeding ecology data of Late Pleistocene mammalian herbivores from South America and discussions on niche differentiation. *Earth Science Reviews* 140:158-165.

Franz-Odendaal, T.A., et al. 2002. New evidence for the lack of C4 grassland expansions during the early Pliocene at Langebaanweg, South Africa. *Paleobiology* 28:378-388.

Freckleton, R.P., et al. 2003. Bergmann's rule and body size in mammals. *American Naturalist* 161:821-825.

Freeman, P.W., and C.A. Lemen. 2007. The trade-off between tooth strength and tooth penetration: predicting optimal shape of canine teeth. *Mammalogy Papers, University of Nebraska State Museum*. Paper 33.

Friedrich, M., et al. 2004. The 12,460-year Hohenheim oak and pine tree-ring chronology from central Europe—a unique annual record for radiocarbon calibration and paleoenvironment reconstructions. *Radiocarbon* 46:1111-1122.

Fritz, H. 1997. Low ungulate biomass in west African savannas: primary production or missing megaherbivores or large predator species? *Ecography* 20:417-421.

Froese, D., et al. 2017. Fossil and genomic evidence constrains the timing of bison arrival in North America. *Proceedings of the National Academy of Science USA* 114:3457-3462.

Fry, B. 1981. Natural stable isotope tag traces Texas shrimp migrations. *Fishery Bulletin* 79:337-345.

Fuchs, T. 1892. On a new order of giant fossils. *University of Nebraska University Studies* 1:301-335.

Gaines, R.R., et al. 2012. Mechanism for Burgess Shale-type preservation. *Proceedings of the National Academy of Sciences USA* 109:5180-5184.

Galetti, M., et al. 2017. Ecological and evolutionary legacy of megafauna extinctions. *Biological Reviews* 93:845-862.

Garcia-Ramos, G., and M. Kirkpatrick. 1997. Genetic models of adaptation and gene flow in peripheral populations. *Evolution* 51:21-28.

Gardner, J.L., et al. 2009. Shifting latitudinal clines in avian body size correlate with global warming in Australian passerines. *Proceedings of the Royal Society B* 276:3845-3852.

Gardner, J.L., et al. 2011. Declining body size: a third universal response to warming? *Trends in Ecology and Evolution* 26:285-291.

Gardner, T.W., et al. 1987. Geomorphic and tectonic process rates: effects of measured time interval. *Geology* 15:259-261.

Garland Jr., T. 1983. The relation between maximal running speed and body mass in terrestrial mammals. *Journal of Zoology* 199:157-170.

Garrels, R.M., and F.T. Mackenzie. 1971. *Evolution of Sedimentary Rocks*. New York: W.W. Norton.

Gavin, D.G., et al. 2007. Forest fire and climate change in western North America, insights from sediment charcoal records. *Frontiers of Ecology and the Environment* 9:499-506.

Gearty, W., et al. 2018. Energetic tradeoffs control the size distribution of aquatic mammals. *Proceedings of the National Academy of Sciences USA* 115:4194-4199.

Gee, C.T., et al. 2003. A Miocene rodent nut cache in coastal dunes of the Lower Rhine Embayment, Germany. *Palaeontology* 46:1133-1149.

Gensel, P.G. 2008. The earliest land plants. *Annual Review of Ecology, Evolution, and Systematics* 39:459-477.

Geyh, M.A. 2001. Bomb radiocarbon dating of animal tissues and hair. *Radiocarbon* 43:723-730.

Gibson, J.J., et al. 2010. Stable isotopes in large scale hydrological applications. In *Isoscapes: Understanding Movement, Pattern, and Process on Earth through Isotope Mapping*, edited by J.B. West et al., 389-406. Dordrecht: Springer Netherlands.

Gill, J.L. 2014. The ecological impacts of the late Quaternary megaherbivore extinctions. *New Phytologist* 201:1163-1169.

Gill, J.L., et al. 2009. Pleistocene megafaunal collapse, novel plant communities, and enhanced fire regimes in North America. *Science* 326:1100-1103.

Gill, J.L., et al. 2012. Climatic and megaherbivory controls on late-glacial vegetation dynamics: a new, high-resolution multi-proxy record from Silver Lake, OH. *Quaternary Science Reviews* 34:66-80.

Gill, J.L., et al. 2013. Linking abundances of the dung fungus *Sporormiella* to the density of American bison (*Bison bison*): implications for assessing grazing by megaherbivores in the paleorecord. *Journal of Ecology* 101:1125-1136.

Gill, T. 1897. Edward Drinker Cope, naturalist: a chapter in the history of science. *American Naturalist* 31:831-863.

Gingerich, P.D. 1983. Rates of evolution: effects of time and temporal scaling. *Science* 222:159-161.

Gingerich, P.D. 1989. New earliest Wasatchian mammalian fauna from the Eocene of northwestern Wyoming: composition and diversity in a rarely sampled high-floodplain assemblage. *University of Michigan Papers in Paleontology* 28:37-71.

Gingerich, P.D. 1990. Prediction of body mass in mammalian species from long bone lengths and diameters. *Contributions from the Museum of Paleontology, University of Michigan* 28:79-92.

Gingerich, P.D. 1993. Quantification and comparison of evolutionary rates. In *Functional Morphology and Evolution*, edited by P. Dodson and P. D. Gingerich, 453-478. A special volume of the *American Journal of Science*.

Gingerich, P.D. 2003. Mammalian responses to climate change at the Paleocene-Eocene boundary: Polecat Bench record in the northern Bighorn Basin, Wyoming. *Special Papers of the Geological Society of America* 369:463-478.

Gingerich P.D., et al. 1982. Allometric scaling in the dentition of primates and prediction of body weight from tooth size in fossils. *American Journal of Physical Anthropology* 58:81-100.

Glazko, G.V., and M. Nei. 2003. Estimation of divergence times for major lineages of primate species. *Molecular Biology Evolution* 20:424-434.

Gleason, H.A. 1926. The individualistic concept of plant association. *Bulletin of the Torrey Botany Club* 53:7-26.

Gobetz, K.E. 2007. New considerations for interpreting fossilized mammal burrows from observations of living species. In Lucas et al. 2007, 7-9.

Gobetz, K.E., and L.D. Martin. 2006. Burrows of a gopher-like rodent, possibly *Gregorymys* (Geomyoidea: Geomyidae: Entoptychinae), from the early Miocene Harrison Formation, Nebraska. *Palaeogeography, Palaeoclimatology, Palaeoecology* 237:305-314.

Goebel, T., et al. 2008. The late Pleistocene dispersal of modern humans. *Science* 319:497-1502.

Goldberg, M., et al. 2011. Dentin: structure, composition, and mineralization. The role of dentin ECM in dentin formation and mineralization. *Frontiers in Bioscience* 3:711-735.

Goldsmith, B. 2005. *Obsessive Genius: The Inner World of Marie Curie*. New York: W.W. Norton.

Gordon, K.D. 1988. A review of methodology and quantification in dental microwear analysis. *Scanning Microscopy* 2:1139-1147.

Goren-Inbar, N., et al. 2004. Evidence of hominin control of fire at Gesher Benot Ya'aqov, Israel. *Science* 304:725-727.

Goring, S., et al. 2015. Neotoma: A programmaic interface to the neotoma paleoecological database. *Open Quaternary* 1:1-17.

Gould, S.J. 1966. Allometry and size in ontogeny and phylogeny. *Biological Reviews of the Cambridge Philosophical Society* 41:587-640.

Gould, S.J. 1971. Geometric similarity in allometric growth: a contribution to the problem of scaling in the evolution of size. *American Naturalist* 105:113-136.

Gould, S.J. 1984. Smooth curve of evolutionary rate: a psychological and mathematical artifact. *Science* 226:994-996.

Gower, G., et al. 2019. Widespread male sex bias in mammal fossil and museum collections. *Proceedings of the National Academy of Sciences* 116:19019-19024.

Gowlett, J.A.J., et al. 1981. Early archaeological sites, hominid remains, and traces of fire from Chesowanja, Kenya. *Nature* 294:125-129.

Gradstein, F.M., et al. 2012. *The Geologic Time Scale 2012*. Waltham, MA: Elsevier.

Graham, C.T., et al. 2013. Development of non-lethal sampling of carbon and nitrogen stable isotope ratios in salmonids: effects of lipid and inorganic components of fins. *Isotopes in Environmental and Health Studies* 49:555-566.

Graham, R.W. 1986. Response of mammalian communities to environmental changes during the Late Quaternary. In *Community Ecology*, edited by J. Diamond and T.J. Case, 300-313. New York: Harper and Row.

Graham, R.W., and E.C. Grimm. 1990. Effects of global climate change on the patterns of terrestrial biological communities. *Trends in Ecology and Evolution* 5:289-292.

Graham, R.W., and E.L. Lundelius. 1984. Coevolutionary disequilibrium and Pleistocene extinctions. In Martin and Klein 1984, 223-249.

Graham, R.W., and J.I. Mead 1987. Environmental fluctuations and evolution of mammalian faunas during the last deglaciation in North America. In *North American and Adjacent Oceans during the Last Deglaciation*, edited by

W.F. Ruddiman and H.E. Wright, Jr., 372-402. Boulder, CO: Geological Society of America.

Graham, R.W., et al. 1996. Spatial response of mammals to late Quaternary environmental fluctuations. *Science* 272:1601-1606.

Grant, P.R., and B.R. Grant. 2002. Unpredictable evolution in a 30-year study of Darwin's finches. *Science* 296:707-711.

Grayson, D.K. 2007. Deciphering North American Pleistocene extinctions. *Journal of Anthropology Research* 63:185-213.

Greene, S., et al. 2018. Identifying geologically meaningful U-Pb dates in fossil teeth. *Chemical Geology* 493:1-15.

Gregory, W.K. 1920. On the structure and relations of *Notharctus*, an American Eocene primate. *Memoirs of the American Museum of Natural History* 3:49-243.

Grigoriev, S.E., et al. 2017. A woolly mammoth (*Mammuthus primigenius*) carcass from Maly Lyakhovsky Island (New Siberian Islands, Russian Federation). *Quaternary International* 445:89-103.

Grimm, E.C., et al. 2013. Pollen databases and their application. In *Encyclopaedia of Quaternary Sciences*, edited by S.A. Elias, 831-838. Amsterdam: Elsevier.

Grine, F.E. 1986. Dental evidence for dietary differences in *Australopithecus* and *Paranthropus*: a quantitative analysis of permanent molar microwear. *Journal of Human Evolution* 15:783-822.

Grine, F.E., and R.F. Kay. 1988. Early hominid diets from quantitative image analysis of dental microwear. *Nature* 333:765-768.

Groucutt, H.S., et al. 2015. Rethinking the dispersal of *Homo sapiens* out of Africa. *Evolutionary Anthropology Issues News Review* 24:149-164.

Guex, J., et al. 2012. Geochronological constraints on post-extinction recovery of the ammonoids and carbon cycle perturbations during the Early Jurassic. *Palaeogeography, Palaeoclimatology, Palaeoecology* 346-347:1-11.

Guilday, J.E. 1971. The Pleistocene history of the Appalachian mammal fauna. *Virginia Polytechnic Research Institute Monograph* 4:233-262.

Guimaraes, S., et al. 2016. A cost-effective high-throughput metabarcoding approach powerful enough to genotype ~44,000-year-old rodent remains from Northern Africa. *Molecular Ecology Resources* 17:405-417.

Guimarães, P.R., Jr., et al. 2008. Seed dispersal anachronisms: rethinking the fruits extinct megafauna ate. *PLoS ONE* 3:e1745.

Gunnell, G.F. 1998. Creodonta. In *Evolution of Tertiary Mammals of North America. Volume 1: Terrestrial Carnivores, Ungulates, and Ungulatelike Mammals*, edited by C.M. Janis et al., 91-109. Cambridge: Cambridge University Press.

Guthrie, R.D. 1990. *Frozen Fauna of the Mammoth Steppe: The Story of Blue Babe*. Chicago: University of Chicago Press.

Hadly, E.A. 1996. Influence of late Holocene climate on Northern Rocky Mountain mammals. *Quaternary Research* 46:298-310.

Hadly, E.A. 1997. Evolutionary and ecological response of pocket gophers (*Thomomys talpoides*) to late-Holocene climatic change. *Biological Journal of the Linnean Society* 60:277-296.

Hadly, E.A. 1999. Fidelity of terrestrial vertebrate fossils to a modern ecosystem. *Palaeogeography, Palaeoclimatology, Palaeoecology* 149:389-409.

Hadly, E.A., et al. 1998. A genetic record of population isolation in pocket gophers during Holocene climatic change. *Proceedings of the National Academy of Sciences* 95:6893-6896.

Hadly, E.A., et al. 2003. Ancient DNA evidence of prolonged population persistence with negligible genetic diversity in an endemic tuco-tuco (*Ctenomys sociabilis*). *Journal of Mammalogy* 84:403-417.

Hadly, E.A., et al. 2004. Genetic response to climatic change: insights from ancient DNA and phylochronology. *PLoS Biology* 2:1600-1609.

Haile, J., et al. 2009. Ancient DNA reveals late survival of mammoth and horse in interior Alaska. *Proceedings of the National Academy of Science USA* 106:22352-22357.

Haldane, J.B.S. 1949. Suggestions as to quantitative measurement of rates of evolution. *Evolution* 3:51-56.

Hall, B.L., et al. 2006. Holocene elephant seal distribution implies warmer-than-present climate in the Ross Sea. *Proceedings of the National Academy of Sciences* 103:10213-10217.

Hall, E.R. 1981. *Mammals of North America*. New York: Wiley.

Hammerslough, J. 2004. *Owl Puke*. New York: Workman.

Handley, L.L., et al. 1999. The ^{15}N natural abundance (δ^{15}N) of ecosystem samples reflects measures of water availability. *Australian Journal of Plant Physiology* 26:185-199.

Hansen, J., et al. 2013. Climate sensitivity, sea level, and atmospheric carbon dioxide. *Philosophical Transactions of the Royal Society A* 371:20120294.

Hansen, R.M. 1978. Shasta ground sloth food habits, Rampart Cave, Arizona. *Paleobiology* 4:302-319.

Harris, J.M., and T.E. Cerling. 1996. Isotopic changes in the diet of African Proboscideans. *Journal of Vertebrate Paleontology* 16:40A.

Harris, J.M., and G.T. Jefferson, eds. 1985. *Rancho La Brea: Treasures of the Tar Pits*. Los Angeles: Natural History Museum of Los Angeles County.

Hasiotis, S.T. 2003. Complex ichnofossils of solitary to social soil organisms: understanding their evolution and roles in terrestrial paleoecosystems. *Palaeogeography, Palaeoclimatology, Palaeoecology* 192:259-320.

Hasiotis, S.T., et al. 2007. The trace-fossil record of vertebrates. In *Trace Fossils: Concepts, Problems, Prospects*, edited by W. Miller III, 196-218. Amsterdam: Elsevier.

Hautier, L., et al. 2016. The hidden teeth of sloths: evolutionary vestiges and the development of a simplified dentition. *Scientific Reports* 6:27763

Haynes, C.V., Jr. 1993. Contributions of radiocarbon dating to the geochronology of the peopling of the New World. In *Radiocarbon after Four Decades: An Interdisciplinary Perspective*, edited by R.E. Taylor et al., 355-374. New York: Springer-Verlag.

Haynes, G., ed. 2009. *American Megafaunal Extinctions at the End of the Pleistocene*. Dordrecht: Springer Netherlands.

Haynes, G. 2012. Elephants (and extinct relatives) as earth-movers and ecosystem engineers. *Geomorphology* 157:99-107.

Hays, J.D., et al. 1976. Variations in the Earth's orbit: pacemaker of the ice ages. *Science* 194:1121-1132.

Heaney, L.R. 1978. Island area and body size of insular mammals: evidence from the tri-colored squirrel (*Callosciurus prevosti*) of Southeast Asia. *Evolution* 32:29-44.

Heim, N.A., et al. 2015. Cope's rule in the evolution of marine mammals. *Science* 347:867-870.

Heim, N.A., et al. 2017. Hierarchical complexity and the size limits of life. *Proceedings of the Royal Society B* 284 (1857): 20171039.

Helliker, B.R., and J.R. Ehleringer. 2000. Establishing a grassland signature in veins: ^{18}O in the leaf water of C_3 and C_4 grasses. *Proceedings of the National Academy of Sciences USA* 97:7894-7898.

Hendry, A.P., et al. 2008. Human influences on rates of phenotypic change in wild animal populations. *Molecular Ecology* 17:20-29.

Henschel, P., et al. 2014. The lion in West Africa is critically endangered. *PLoS ONE* 9:e83500.

Hern, W.H. 1999. How many times has the human population doubled? Comparisons with cancer. *Population and Environment* 21:59-80.

Hester, J.J. 1960 Late Pleistocene extinctions and radiocarbon dating. *American Antiquity* 26:58-77.

Higuchi, R., et al. 1984. DNA sequences from the quagga, an extinct member of the horse family. *Nature* 312:282-284.

Hildebrand, A.R., et al. 1991. Chicxulub crater: a possible Cretaceous-Tertiary boundary impact crater on the Yucatan Peninsula, Mexico. *Geology*:19:867-871.

Hillson, S. 2001. Recording dental caries in archaeological human remains. *International Journal of Osteoarchaeology* 11:249-289.

Hillson, S. 2002. *Dental Anthropology.* Cambridge: Cambridge University Press.

Hintze, L.F. 1988. Geologic history of Utah. *Brigham Young University Geology Studies Special Publication* 7:1-202.

Hobson, K.A. 1999. Tracing origins and migration of wildlife using stable isotopes: a review. *Oecologia* 120:314-326.

Hobson, K.A., and L.I. Wassenaar. 2008. *Tracking Animal Migration with Stable Isotopes*. London: Academic Press.

Hobson, K.A., et al. 1994. Using stable isotopes to determine seabird trophic relationships. *Journal of Ecology* 63:786-798.

Hobson, K.A., et al. 2004. Using stable hydrogen and oxygen isotope measurements of feathers to infer geographical origins of migrating European birds. *Oecologia* 141:477-488.

Hody, J.W., and R. Kays. 2018. Mapping the expansion of coyotes (*Canis latrans*) across North and Central America. *ZooKeys* 759:81-97. doi:10.3897/zookeys.759.15149.

Hoffmann, A.A., and M.W. Blows. 1994. Species borders: ecological and evolutionary perspectives. *Trends in Ecology and Evolution* 9:223-227.

Hoffmann, A.A., and C. Sgrò. 2011. Climate change and evolutionary adaptation. *Nature* 470:479-485.Hoffmann, M., et al. 2010. The impact of conservation on the status of the world's vertebrates. *Science* 330:1503-1509.

Hofreiter, M., et al. 2000. A molecular analysis of ground sloth diet through the last glaciation. *Molecular Ecology* 9:1975-1984.

Hofreiter, M., et al. 2003. Phylogeny, diet, and habitat of an extinct ground sloth from Cuchillo Curá, Neuquén Province, southwest Argentina. *Quaternary Research* 59:364-378.

Holt, K.A., and K.D. Bennett. 2014. Principles and methods for automated palynology. *New Phytologist* 203:735-742.

Hone, D.W., et al. 2005. Macroevolutionary trends in the Dinosauria: Cope's rule. *Journal of Evolutionary Biology* 18:587-595.

Hoppe, K.A., and P.L. Koch. 2007. Reconstructing the migration patterns of late Pleistocene mammals from northern Florida. *Quaternary Research* 68:347-352.

Hoppe, K.A., et al. 1999. Tracking mammoths and mastodons: Reconstruction of migratory behavior using strontium isotope ratios. *Geology* 27:439-442.

Horner-Devine, M.C., et al. 2004. An ecological perspective on bacterial biodiversity. *Proceedings of the Royal Society B* 271:113-122.

Hornsby, A. et al. In prep. Climate and competition drive abrupt species turnover in a 33,000-year record of ancient DNA. Unpublished manuscript.

Hu, Y., et al. 2005. Large Mesozoic mammals feed on young dinosaurs. *Nature* 433:149-152.

Hua, Q., and M. Barbetti. 2004. Review of tropospheric bomb ^{14}C data for carbon cycle modeling and age calibration purposes. *Radiocarbon* 46:1273-1298.

Hughes, L. 2000. Biological consequences of global warming: is the signal already apparent? *Trends in Ecology and Evolution* 15:56-61.

Hughes, M.K., et al. 2010. *Dendroclimatology: Progress and Prospects*. Berlin: Springer-Verlag.

Hunt, A.P., et al. 2005. Vertebrate trace fossils from Arizona with special reference to tracks preserved in National Park Service units and notes on the Phanerozoic distribution of fossil footprints. In *Vertebrate Paleontology in Arizona*, edited by A.B. Heckert and S.G. Lucas, 158-166. Albuquerque: New Mexico Museum of Natural History and Science.

Hunt, A.P., and S.G. Lucas 2007. Cenozoic vertebrate trace fossils of North America: ichnofaunas, ichnofacies, and biochronology. In Lucas et al. 2007, 17-41.

Hunt, G. 2006. Fitting and comparing models of phyletic evolution: random walks and beyond. *Paleobiology* 32:578-601.

Hunt, G. 2007. The relative importance of directional change, random walks, and stasis in the evolution of fossil lineages. *Proceedings of the National Academy of Science USA* 104:18404-18408.

Hunt, G. 2008. Gradual or pulsed evolution: when should punctuational explanations be preferred? *Paleobiology* 34:360-377.

Hunt, K.D. 1994. The evolution of human bipedality: ecology and functional morphology. *Journal of Human Evolution* 26:183-202.

Huntley, B. 1990. Dissimilarity mapping between fossil and contemporary pollen spectra in Europe for the past 13,000 years. *Quaternary Research* 33:360-376.

Huntley, B. 2007. Limitations on adaptation: evolutionary responses to climatic change? *Heredity* 98:247-248.

Hutchinson, G.E., and R.J. MacArthur. 1959. A theoretical ecological model of size distributions among species of animals. *American Naturalist* 93:117-125.

Huxley, J.S., and G. Teissier. 1936. Terminology of relative growth. *Nature* 137:780-781

IPBES (Intergovernmental Science-Policy Platform on Biodiversity and Ecosystem Services). 2019. *Summary for Policymakers of the Global Assessment Report on Biodiversity and Ecosystem Services of the Intergovernmental Science-Policy Platform on Biodiversity and Ecosystem Services.* Bonn, Germany: IPBES Secretariat.

IPCC (Intergovernmental Panel on Climate Change). 2007. *Climate Change 2007: The Physical Science Basis.* Contribution of Working Group I to the Fourth Assessment Report of the Intergovernmental Panel on Climate Change. Cambridge: Cambridge University Press.

IPCC (Intergovernmental Panel on Climate Change). 2014. *Climate Change 2014: Synthesis Report.* Contribution of Working Groups I, II, and III to the Fifth Assessment Report of the Intergovernmental Panel on Climate Change [Core Writing Team, R.K. Pachauri and L.A. Meyer (eds.)]. Geneva: IPCC.

Irving, E. 1988. The paleomagnetic confirmation of continental drift. *EOS: Transactions of the American Geophysical Union* 69:994-1014.

Irwin, D.M., et al. 1991. Evolution of the cytochrome *b* gene of mammals. *Journal of Molecular Evolution* 32:128-144.

IUCN (International Union for Conservation of Nature). 2020. *The IUCN Red List of Threatened Species.* Version 2020-2. www.iucnredlist.org.

Ivany, L.C., and R.J. Salawitch 1993. Carbon isotopic evidence for biomass burning at the K-T boundary. *Geology* 21:487-490.

Jablonski, D. 1994. Extinctions in the fossil record. *Philosophical Transactions of the Royal Society of London B, Biological Sciences* 344:11-17.

Jablonski, D. 1997. Body-size evolution in Cretaceous molluscs and the status of Cope's rule. *Nature* 385:250-252.

Jablonski, D. 2003. The interplay of physical and biotic factors in macroevolution. In *Evolution on Planet Earth: The Impact of the Physical Environment*, edited by A. Lister and L. Rothschild, 235-252. New York: Academic Press.

Jablonski, D., and D.M. Raup. 1995. Selectivity of end-Cretaceous marine bivalve extinctions. *Science* 268:389-391.

Jachmann H., et al. 1995. Tusklessness in African elephants: a future trend. *African Journal of Ecology* 33:230-235.

Jackson, S.T. 2007. Looking forward from the past: history, ecology and conservation. *Frontiers in Ecology and the Environment* 9:455.

Jackson, S.T., and J.T. Overpeck. 2000. Responses of plant populations and communities to environmental changes of the Late Quaternary. *Paleobiology* 26:194-220.

Jackson, S.T., et al. 2005. A 40,000-year woodrat-midden record of vegetational and biogeographic dynamics in northeastern Utah. *Journal of Biogeography* 32:1085-1106.

Jaffe, M. 2000. *The Gilded Dinosaur: The Fossil War between E. D. Cope and O. C. Marsh and the Rise of American Science.* New York: Crown.

James, F.C. 1970. Geographic size variation in birds and its relationship to climate. *Ecology* 51:385-390.

James, S.R. 1989. Hominid use of fire in the lower and middle Pleistocene: a review of the evidence. *Current Anthropology* 30:1-26.

Janczewski, D.N., et al. 1992. Molecular phylogenetic inference from saber-toothed cat fossils of Rancho La Brea. *Proceedings of the National Academy of Sciences of the USA* 89:9769-9773.

Janis, C.M. 1976. The evolutionary strategy of the Equidae and the origins of rumen and cecal digestion. *Evolution* 30:757-774.

Janis, C.M. 1988. An estimation of tooth volume and hypsodonty indices in ungulate mammals, and the correlation of these factors with dietary preferences. *Mémoirs de Musée d'Histoire Naturelle Paris* 53:367-387.

Janis, C.M., et al. 2000. Miocene ungulates and terrestrial primary productivity: where have all the browsers gone? *Proceedings of the National Academy of Sciences USA* 97:7899-7904.

Janis, C.M., et al. 2002. The origins and evolution of the North American grassland biome: the story from the hoofed mammals. *Palaeogeography, Palaeoclimatology, Palaeoecology* 117:183-198.

Janis, C.M., et al. 2004. The species richness of Miocene browsers, and implications for habitat type and primary productivity in the North American grassland biome. *Palaeogeography, Palaeoclimatology, Palaeoecology* 207:371-398.

Janzen, D.H., and P.S. Martin. 1982. Neotropical anachronisms: the fruits the gomphotheres ate. *Science* 215:19-27.

Jeanloz, R., and B. Romanowicz, 1997. Geophysical dynamics at the center of the Earth. *Physics Today* 50:22-27.

Jefferson, T. 1799. A memoir on the discovery of certain bone of a quadruped of the clawed kind in the western parts of Virginia. *Transactions of the American Philosophical Society* 4:246-260.

Jernvall, J., and M. Fortelius. 2002. Common mammals drive the evolutionary increase of hypsodonty in the Neogene. *Nature* 417:538-540.

Ji, Q., et al. 2006. A swimming mammaliaform from the Middle Jurassic and ecomorphological diversification of early mammals. *Science* 311:1123-1127.

Johns, G.C., and J.C. Avise. 1998. A comparative summary of genetic distances in the vertebrates from the mitochondrial cytochrome b gene. *Molecular Biological Evolution* 15:1481-1490.

Johnson, C.N. 2009. Ecological consequences of Late Quaternary extinctions of megafauna. *Proceedings of the Royal Society B* 276:2509-2519.

Johnston, R.F., and R.K. Selander. 1964. House sparrows: rapid evolution of races in North America. *Science* 144:548-550.

Joordans, J.C., et al. 2015. *Homo erectus* at Trinil on Java used shells for tool production and engraving. *Nature* 518:228-231.

Jouzel, J. 2013. A brief history of ice core science over the last 50 yr. *Climate of the Past* 9:2525-2547.

Jouzel, J., et al. 2007. Orbital and millennial Antarctic climate variability over the past 800,000 years. *Science* 317:793-796.

Jukar, A.M., et al. 2018. A cranial correlate of body mass in proboscideans. *Zoological Journal of the Linnean Society* 184:919-931.

Kaiser, T.M., and N. Solounias. 2003. Extending the tooth mesowear method to extinct and extant equids. *Geodiversitas* 25:321-345.

Kaufman, D.S. 2000. Amino acid racemization in ostracodes. In *Perspectives in Amino Acid and Protein Geochemistry*, edited by G. Goodfriend et al., 145-160. New York: Oxford University Press.

Kaufman, D.S., and W.F. Manley. 1998. A new procedure for determining enantiomeric (D/L) amino acid ratios in fossils using reverse phase liquid chromatography: *Quaternary Science Reviews* 17:987-1000.

Kaya, F., et al. 2018. The rise and fall of the Old World savannah fauna and the origins of the African savannah biome. *Nature Ecology and Evolution* 2:241-246.

Keeling, C.D. 1979. The Suess effect: ^{13}Carbon-^{14}Carbon interrelations. *Environment International* 2:229-300.

Keesing, F., and T.P. Young. 2014. Cascading consequences of the loss of large mammals in an African Savanna. *BioScience* 64:487-495.

Kelley, S.P. 2002. K-Ar and Ar-Ar Dating. *Reviews in Mineralogy and Geochemistry* 47:785-818.

Kemp, T.S. 2006. The origin and early radiation of the therapsid mammal-like reptiles: a palaeobiological hypothesis. *Journal of Evolutionary Biology* 19:1231-1247.

Kennedy, B.P., et al. 1997. Natural isotope marks in salmon. *Nature* 387:776-767.

Kermack K.A., and J.B.S. Haldane. 1950. Organic correlation and allometry. *Biometrika* 37:30-41.

Kermit, H. 2002. The life of Niels Stensen. In *Niccolò Stenone: anatomista, geologo, vescovo* [Niccolò Stenone: anatomist, geologist, bishop], edited by K. Ascaniv et al., 17-22. Rome: Analecta Romana Instituti Danici.

Keyes, C. 1924. Grand Staircase of Utah. *The Pan-American Geologist* 16:33-68.

Kidwell, S.M., and K.W. Flessa. 1995. The quality of the fossil record: populations, species, and communities. *Annual Review of Ecology and Systematics* 26:269-299.

Kielan-Jaworowska, Z., et al. 2004. *Mammals from the Age of Dinosaurs: Origins, Evolution, and Structure.* New York: Columbia University Press.

Kim, K.S., et al. 2019. Exquisitely preserved, high-definition skin traces in diminutive theropod tracks from the Cretaceous of Korea. *Scientific Reports* 9:2039.

King, T., et al. 1999. Effect of taphonomic processes on dental microwear. *American Journal of Physical Anthropology* 108:359-373.

Kingsolver, J.G., and R.T. Paine. 1991. Conversational biology and ecological debate. In *Foundations of Ecology*, edited by L.A. Real and J.H. Brown, 309-317. Chicago: University of Chicago Press.

Kingsolver, J.G., and D.W. Pfennig. 2004. Individual-level selection as a cause of Cope's rule of phyletic size increase. *Evolution* 58:1608-1612.

Kirk, P.M., et al. 2008. *Dictionary of the Fungi*. 10th ed. Wallingford, CT: CABI (Centre for Agriculture and Bioscience International).

Kleiber, M. 1932. Body size and metabolism. *Hilgardia* 6:315-351.

Klein, R.G., and K. Scott. 1989. Glacial/interglacial size variation in fossil spotted hyenas (*Crocuta crocuta*) from Britain. *Quaternary Research* 32:88-95.

Knapp, M., and M. Hofreiter. 2010. Next generation sequencing of ancient DNA: requirements, strategies, and perspectives. *Genes* 1:227-243.

Knoll, A.H., and M.A. Nowak. 2017. The timetable of evolution. *Science Advances* 3:e1603076.

Kobashia, T., et al. 2008. 4 ± 1.5 °C abrupt warming 11,270 years ago identified from trapped air in Greenland ice. *Earth and Planetary Science Letters* 268:397-407.

Koch, P.L. 1986. Clinal geographic variation in mammals: implications for the study of chronoclines. *Paleobiology* 12:261-181.

Koch, P.L. 2007. Isotopic study of the biology of modern and fossil vertebrates. In *Stable Isotopes in Ecology and Environmental Science*, edited by R. Michener and K. Lajtha, 99-154. Boston: Blackwell Science.

Koch, P.L., and A.D. Barnosky. 2006. Late Quaternary extinctions: state of the debate. *Annual Review of Ecology, Evolution, and Systematics* 37:215-250.

Koch, P.L., et al. 1994. Tracing the diets of fossil animals using stable isotopes. In *Stable Isotopes in Ecology and Environmental Science*, edited by R. Michener and K. Lajtha, 63-92. Boston: Blackwell Science.

Koch, P.L., et al. 1995. Isotopic tracking of change in diet and habitat use in African elephants. *Science* 267:1340-1343.

Koch, P.L., et al. 1998. The isotopic ecology of late Pleistocene mammals in North America: Part 1. Florida. *Chemical Geology* 152:119-138.

Koch, P.L., et al. 2004. The effects of late Quaternary climate and pCO_2 change on C_4 plant abundance in the south-central United States. *Palaeogeography, Palaeoclimatology, Palaeoecology* 207:331-357.

Kohn, M.J. 1996. Predicting animal δ^{18}O: accounting for diet and physiological adaptation. *Geochimica et Cosmochimica Acta* 60:4811-4829.

Kohn, M.J. 2010. Carbon isotope compositions of terrestrial C_3 plants as indicators of (paleo)ecology and (paleo) climate. *Proceedings of the National Academy of Sciences USA* 107:19691-19695.

Kojola, I., et al. 1998. Foraging conditions, tooth wear, and herbivore body reserves: a study of female reindeer. *Oecologia* 117:26-30.

Kolbert, E. 2014. The Sixth Extinction: An Unnatural History. New York: Henry Holt.

Kosintsev, P.A., et al. 2010. Intestinal contents of the baby woolly mammoth (*Mammuthus primigenius blumenbach*, 1799) from Yuribey river (Yamal Peninsula). *Doklady Akademii Nauk* 432:556-558.

Kowalewski, M., et al. 2011. The Geozoic supereon. *Palaios* 26:251-255.

Krause, J., et al. 2010. The complete mitochondrial DNA genome of an unknown hominin from southern Siberia. *Nature* 464:894-897.

Kring, D.A., and M. Boslough. 2014. Chelyabinsk: portrait of an asteroid airburst. *Physics Today* 67:32-37.

Krug, A.Z., et al. 2009. Signature of the end-Cretaceous mass extinction in the modern biota. *Science* 323:767-771.

Krug, J.C., et al. 2004. Coprophilous fungi. In *Biodiversity of Fungi: Inventory and Monitoring Methods*, edited by M. Mueller et al., 467-499. Amsterdam: Elsevier.

Kunimatsu, Y., et al. 2007. A new Late Miocene great ape from Kenya and its implications for the origins of African great apes and humans. *Proceedings of the National Academy of Sciences USA* 104:19661-19662.

Kurtén, B. 1973. Geographic variation in size in the puma (*Felis concolor*). *Commentationes Biologicae* 63:1-8.

Kurtén, B., and E. Anderson. 1980. *Pleistocene Mammals of North America*. New York: Columbia University Press.

Kurtén, B., and L. Werdelin. 1990. Relationships between North and South American *Smilodon*. *Journal of Vertebrate Paleontology* 10:158-169.

Kwang Hyun, K. 2015. Origins of bipedalism. *Brazilian Archives of Biology and Technology* 58:929-934.

Lacey, E.A., and J.R. Wieczorek. 2003. Ecology of sociality in rodents: a ctenomyid perspective. *Journal of Mammalogy* 84:1198-1211.

Lacey, E.A., and J.R. Wieczorek. 2004. Kinship in colonial tuco-tucos: evidence from group composition and population structure. *Behavioral Ecology* 15:988-996.

Lagaria, A., and D. Youlatos. 2006. Anatomical correlates to scratch digging in the forelimb of European ground squirrels (*Spermophilus citellus*). *Journal of Mammalogy* 87:563-570.

Lamb, D.M., et al. 2009. Evidence for eukaryotic diversification in the similar to ~1800 million-year-old Changzhougou Formation, North China. *Precambrian Research* 173:93-104.

Lambert, W.D. 1992. The feeding habits of the shovel-tusked gomphotheres: evidence from tusk wear patterns. *Paleobiology* 18:132-147.

Lamsdell, J.C., and S.J. Braddy 2009. Cope's rule and Romer's theory: patterns of diversity and gigantism in eurypterids and Paleozoic vertebrates. *Biology Letters* 6:265-269

Lane, N.G. 1992. *Life of the Past*. New York: Macmillan.

Larramendi, A. 2016. Shoulder height, body mass, and shape of proboscideans. *Acta Palaeontologica Polonica* 61:537-574.

Larrasoaña, J.C., et al. 2013. Dynamics of green Sahara Periods and their role in hominin evolution. *PLoS ONE* 8:e76514.

Laundré, J.W., et al. 2001. Wolves, elk, and bison: reestablishing the "landscape of fear" in Yellowstone National Park, U.S.A. *Canadian Journal of Zoology* 79:1401-1409.

Lawlor, T.E. 1982. The evolution of body size in mammals: evidence from insular populations in Mexico. *American Naturalist* 119:54-72.

Lawrence, D.R. 1968. Taphonomy and information losses in fossil communities. *Geological Society of America Bulletin* 79:1315-1330.

Leakey, L.S.B. 1971. *Olduvai Gorge, Vol. 3: Excavations in Beds I and II, 1960-1963*. London: Cambridge University Press.

Leakey, M.G., et al. 1995. New four-million-year-old hominid species from Kanapoi and Allia Bay, Kenya. *Nature* 376:565-571.

Leaky, R.E., and R. Lewin. 1992. Origins Reconsidered: In Search of What Makes Us Human. London: Abucus.

Leamy, L. 1988. Genetic and maternal influences on brain and body size in random breed house mice. *Evolution* 42:42-53.

Lee-Thorp, J., and M. Sponheimer. 2003. Three case studies used to reassess the reliability of fossil bone and enamel isotope signals for paleodietary studies. *Journal of Anthropological Archaeology* 22:208-216.

Lee-Thorp, J.A., et al. 1989. Stable carbon isotope ratio differences between bone collagen and bone apatite, and their relationship to diet. *Journal of Archaeological Science* 16:585-599.

Lee, A.K. 1963. The adaptations to arid environments in woodrats of the genus *Neotoma*. *University of California Publications in Zoology* 64:57-96.

Legendre, S. 1986. Analysis of mammalian communities from the late Eocene and Oligocene of southern France. *Palaeovertebrata* 16:191-212.

Legendre, S., and C. Roth. 1988. Correlation of carnassial tooth size and body weight in recent carnivores (Mammalia). *Historical Biology* 1:85-98.

Lejzerowicz, F., et al. 2010. Molecular evidence for widespread occurrence of Foraminifera in soils. *Environmental Microbiology* 12:2518-2526.

Leonard, J.A., et al. 2000. Population genetics of Ice Age brown bears. *Proceedings of the National Academy of Science USA* 97:1651-1654.

Leslie, P.H., et al. 1955. The longevity and fertility of the Orkney vole, *Microtus orcadensis*, as observed in the laboratory. *Proceedings of the Zoological Society London* 125:115-125.

Leunis, J., and H. Ludwig. 1891. *Schul-Naturgeschichte* [School of natural history]. Hannover: Goldgeprägte Halblederbände.

Levin, I., and B. Kromer. 2004. The tropospheric $^{14}CO_2$ level in mid-latitudes of the Northern Hemisphere (1959-2003). *Radiocarbon* 46:1261-1272.

Levin, N.E., et al. 2006. A stable isotope aridity index for terrestrial environments. *Proceedings of the National Academy of Sciences USA* 103:11201-11205.

Libby, W.F. 1946. Atmospheric helium three and radiocarbon from cosmic radiation. *Physical Review* 69:671-672.

Libby, W.F. 1965. *Radiocarbon Dating*. Chicago: Phoenix.

Liesowska, Anna. 2017. Tragic truth about two frozen 55,000-year-old cave lion cubs is revealed by scientists. *Siberian Times*. December 1, 2017. https://siberiantimes.com/science/casestudy/news/tragic-truth-about-two-frozen-55000-year-old-cave-lion-cubs-is-revealed-by-scientists.

Lillegraven, J.A., et al. 1979. *Mesozoic Mammals*: *The First Two-Thirds of Mammalian History*. Berkeley: University of California Press.

Lindahl, T. 1993. Instability and decay of the primary structure of DNA. *Nature* 362:709-715.

Lindeque, M., and P.M. Lindeque. 1991. Satellite tracking of elephants in northwestern Namibia. *African Journal of Ecology* 29:196-206.

Lindsey, E.L., and E.X. Lopez. 2015. Tanque Loma, a new late-Pleistocene megafaunal tar seep locality from southwest Ecuador. *Journal of South American Earth Sciences* 57:61-82.

Liow, L.H., et al. 2008. Higher origination and extinction rates in larger mammals. *Proceedings of the National Academy of Science USA* 105:6097-6102.

Lister, A.M. 1989. Rapid dwarfism of red deer on Jersey in the last interglacial. *Nature* 342:539-542.

Lister, A.M. 1996. Dwarfing in island elephants and deer: processes in relation to time of isolation. *Symposia of the Zoological Society of London* 69:277-292.

Lister, A.[M.], and P. Bahn. 2007. *Mammoths: Giants of the Ice Age.* Berkeley: University of California Press.

Loarie, S.R., et al. 2009. The velocity of climate change. *Nature* 462:1052-1055.

Lomolino, M.V. 1985. Body size of mammals on islands: the island rule re-examined. *American Naturalist* 125:310-316.

Lomolino, M.V. 2005. Body size evolution in insular vertebrates: generality of the island rule. *Journal of Biogeography* 32:1683-1699.

Longman, E.K., et al. 2018. Extreme homogenization: the past, present, and future of mammal assemblages on islands. *Global Ecology and Biogeography* 27:77-95.

Lourens, L.J., et al. 2005. Astronomical pacing of late Palaeocene to early Eocene global warming events. *Nature* 435:1083-1087.

Louys, J. 2012. *Palaeontology in Ecology and Conservation.* New York: Springer-Verlag.

Lovejoy, C.O. 1988. Evolution of human walking. *Scientific American* 259:82-89.

Lovejoy, T.E. 2019. Eden no more. *Science Advances* 5:eaax7492.

Luca, F., et al. 2010. Evolutionary adaptations to dietary changes. *Annual Review of Nutrition* 30:291-314.

Lucas, S.G., et al., eds. 2007. Cenozoic vertebrate tracks and traces. *Bulletin of the New Mexico Museum of Natural History and Science* 42:1-330.

Ludwig, K.R., and P.R. Renne. 2000. Geochronology on the paleoanthropological time scale. *Evolutionary Anthropology* 9:101-110.

Lundelius, E.L., et al. 1983. Terrestrial vertebrate faunas. In *Late-Quaternary Environments of the United States, Vol 1: The Late Pleistocene*, edited by S.C. Porter, 311-353. Minneapolis: University of Minnesota Press.

Lundqvist, N. 1972. Nordic Sordariaceae s. lat. Vol. 20-21 of *Acta universitatis Upsaliensis: Symbolae Botanicae Upsalienses.* Uppsala, Sweden: Almqvist & Wiksells.

Luo, Z.-X., et al. 2001. A new mammaliaform from the Early Jurassic of China and evolution of mammalian characteristics. *Science* 292:1535-1540.

Lydolph, M.C., et al. 2005. Beringian paleoecology inferred from permafrost-preserved fungal DNA. *Applied and Environmental Microbiology* 71:1012-1017.

Lyford, M.E., et al. 2003. Influence of landscape structure and climate variability in a late Holocene natural invasion. *Ecological Monographs* 73:567-583.

Lyman, R.E. 2006. Paleozoology in the service of conservation biology. *Evolutionary Anthropology* 15:11-19.

Lyons, S.K. 2003. A quantitative assessment of the range shifts of Pleistocene mammals. *Journal of Mammalogy* 84:385-402.

Lyons, S.K. 2005. A quantitative model for assessing community dynamics of Pleistocene mammals. *American Naturalist* 165:168-185.

Lyons, S.K., et al. 2004. Of mice, mastodon, and men: human-mediated extinctions on four continents. *Evolutionary Ecology Research* 6:339-358.

Lyons, S.K., et al. 2010. Ecological correlates of range shifts of Late Pleistocene mammals. *Philosophical Transactions of the Royal Society of London B, Biological Sciences* 365 3681-3693.

Lyons, S.K., et al. 2016a. The changing role of mammal life histories in late Quaternary extinction vulnerability on continents and islands. *Biology Letters* 12:20160342.

Lyons, S.K., et al. 2016b. Holocene shifts in the assembly of plant and animal communities implicate human impacts. *Nature* 529:80-U183.

Madden, R.H. 2014. Hypsodonty in mammals: evolution, geomorphology, and the role of Earth surface processes. Cambridge: Cambridge University Press.

MacArthur, R.H., and E.O. Wilson. 1967. *The Theory of Island Biogeography.* Princeton, NJ: Princeton University Press.

MacDonald, G., et al. 2008. Impacts of climate change on species, populations, and communities: palaeobiogeographical insights and frontiers. *Progress in Physical Geography* 32:139-172.

MacFadden, B.J. 1998. Tale of two rhinos: isotopic ecology, paleodiet, and niche differentiation of *Aphelops* and *Teleoceras* from the Florida Neogene. *Paleobiology* 24:274-286.

MacFadden, B.J., and P. Higgins. 2004. Ancient ecology of 15-million-year-old browsing mammals within C3 plant communities from Panama. *Oecologia* 140:169-182.

MacFadden, B.J., et al. 1994. South American fossil mammals and carbon isotopes: a 25-million-year sequence from the Bolivian Andes. *Palaeogeography, Palaeoclimatology, Palaeoecology* 107:257-268.

MacFadden, B.J., et al. 1996. Cenozoic terrestrial ecosystem evolution in Argentina: evidence from carbon isotopes of fossil mammal teeth. *Palaios* 11:319-327.

MacFadden, B.J., et al. 1999. Ancient latitudinal gradients of C3/C4 grasses interpreted from stable isotopes of New World Pleistocene horse (*Equus*) teeth. *Global Ecology and Biogeography* 8:137-149.

Maclure, W. 1809. Observations on the geology of the United States, explanatory of a geological map. *Transactions of the American Philosophical Society* 6:411-428.

Maclure, W. 1818. Observations on the geology of the United States of North America; with remarks on the probable effects that may be produced by the decomposition of the different classes of rocks on the nature and fertility of soils: applied to the different States of the Union, agreeably to the accompanying geological map. *Transactions of the American Philosophical Society* 1:1-91.

MacMillen, R.E. 1964. Population ecology, water relations, and social behavior of a southern California semi-desert rodent fauna. *University of California Publications in Zoology* 71:1-59.

Malhi, Y., et al. 2016. Megafauna and ecosystem function from the Pleistocene to the Anthropocene. *Proceedings of the National Academy of Sciences USA* 113:838-846.

Maloof, A.C., et al. 2010. Possible animal-body fossils in pre-Marinoan limestones from South Australia. *Nature Geoscience* 3:653-659.

Manley, G. 1974. Central England temperatures: monthly means 1659 to 1973. *Quarterly Journal of the Royal Meteorological Society* 100:389-405.

Manly, W.L. 1894. Death Valley in '49: important chapter of California pioneer history. San Jose: Pacific Tree and Vine.

Mann, M. 2003. Little ice age. In *Encyclopedia of Global Environmental Change, Volume 1; The Earth System: Physical and Chemical Dimensions of Global Environmental Change*, edited by M.C. Maccracken and J.S. Perry, 504-509. New York: John Wiley and Sons.

Mann, M., et al. 2008. Proxy-based reconstructions of hemispheric and global surface temperature variations over the past two millennia. *Proceedings of the National Academy of Sciences USA* 105:13252-13257.

Marcot, J.D., and D.W. McShea. 2007. Increasing hierarchical complexity throughout the history of life: phylogenetic tests of trend mechanisms. *Paleobiology* 33:182-200.

Marquet, P.A., and M.L. Taper. 1998. On size and area: patterns of mammalian body size extremes across landmasses. *Evolutionary Ecology* 12:127-139.

Marshall, C.P., et al. 2011. Haematite pseudomicrofossils present in the 3.5-billion-year-old Apex Chert. *Nature Geoscience* 4:240-243.

Martin, L.D. 1994. Devil's corkscrew. *Natural History* 4:59-60.

Martin, L.D., and D.K. Bennett. 1977. The burrows of the Miocene beaver *Palaeocaster*, western Nebraska. *Palaeogeography, Palaeoclimatology, Palaeoecology* 22:173-193.

Martin, P.S. 1967. Prehistoric overkill. In *Pleistocene Extinctions: The Search for a Cause*, edited by P.S. Martin and H.E. Wright, Jr., 75-120. New Haven, CT: Yale University Press.

Martin, P.S. 1984. Prehistoric overkill: the global model. In Martin and Klein 1984, 354-403.

Martin, P.S. 2005. *Twilight of the Mammoths: Ice Age Extinctions and Rewilding America*. Berkeley: University of California Press.

Martin, P.S., and R.G. Klein. 1984. *Quaternary Extinctions: A Prehistoric Revolution*. Tucson: University of Arizona Press.

Martin, P.S., and D.W. Steadman. 1999. Prehistoric extinctions on islands and continents. In *Extinctions in Near Time: Causes, Contexts, and Consequences*, edited by R.D.E. MacPhee, 17-53. New York: Kluwer.

Martin, P.S., and A.J. Stuart. 1995. Mammoth extinction: two continents and Wrangel Island. *Radiocarbon* 37:7-10.

Martin, P.S., et al. 1961. Rampart Cave coprolite and ecology of the Shasta ground sloth. *American Journal of Science* 259:102-127.

Martinez, L. 1996. Useful tree species for tree-ring dating. Laboratory of Tree-Ring Research. University of Arizona. Updated October 2001. https://www.ltrr.arizona.edu/lorim/good.html.

Maxwell, W.D., and M.J. Benton. 1990. Historical tests of the absolute completeness of the fossil record of tetrapods. *Paleobiology* 16:322-335.

May, R.M., et al. 1995. Assessing extinction rates. In *Extinction Rates*, edited by J.H. Lawton and R.M. May, 1-24. Oxford: Oxford University Press.

Mayor, A. 2000. *The First Fossil Hunters: Paleontology in Greek and Roman Times*. Princeton, NJ: Princeton University Press.

Mayr, E. 1956. Geographic character gradients and climatic adaptation. *Evolution* 10:105-108.

Mayr, E. 1963. *Animal Species and Evolution*. Cambridge, MA: Harvard University Press.

McCarty, J.P. 2001. Ecological consequences of recent climate change. *Conservation Biology* 15:320-331.

McDougall, I., and T.M. Harrison. 1999. *Geochronology and Thermochronology by the $^{40}Ar/^{39}Ar$ Method*. 2nd ed. Oxford: Oxford University Press.

McHenry, C.R., et al. 2007. Supermodeled sabercat, predatory behavior in *Smilodon fatalis* revealed by high-resolution 3D computer simulation. *Proceedings of the National Academy of Sciences USA* 104:16010-16015.

McHenry, H.M. 1992. Body size and proportions in early hominids. *American Journal of Physical Anthropology* 87:407-431.

McHenry, H.M., and K. Coffing. 2000. *Australopithecus* to *Homo*: transformations in body and mind. *Annual Review of Anthropology* 29:125-146.

McInerney, F.A., and S.L. Wing. 2011. The Paleocene-Eocene thermal maximum: a perturbation of carbon cycle, climate, and biosphere with implications for the future. *Annual Review of Earth and Planetary Science* 39:489-516.

McKechnie, A.E., et al. 2004. Deuterium stable isotope ratios as tracers of water resource use: an experimental test with rock doves. *Oecologia* 140:191-200.

McKirahan, R.D. 1994. *Xenophanes of Colophon: Philosophy before Socrates*. Indianapolis, IN: Hackett.

McMahon, T.A. 1973. Size and shape in biology. *Science* 179:1201-1204.

McMenamin, M.A.S., et al. 1982. Amino acid geochemistry of fossil bones from the Rancho La Brea Asphalt Deposit, California. *Quaternary Research* 18:174-183.

McNab, B.K. 1971. On the ecological significance of Bergmann's rule. *Ecology* 52:845-854.

McNeil, P.E., et al. 2005. Mammoth tracks indicate a declining late Pleistocene population in southwestern Alberta, Canada. *Quaternary Science Reviews* 24:1253-1259.

McShea, D.W. 1994. Mechanisms of large-scale evolutionary trends. *International Journal of Organic Evolution* 48:1747-1763.

McShea, D.W. 1998. Possible largest-scale trends in organismal evolution: eight "live hypotheses." *Annual Review of Ecology and Systematics* 29:293-318.

McShea, D.W. 2001. The hierarchical structure of organisms: a scale and documentation of a trend in the maximum. *Paleobiology* 27:405-423.

Meachen, J.A., et al. 2015. Carnivorian postcranial adaptations and their relationships to climate. *Ecography* 39:553-560.

Meachen-Samuels, J.A., and B. Van Valkenburgh. 2010. Radiographs reveal exceptional forelimb strength in the sabertooth cat, *Smilodon fatalis*. *PLoS ONE* 5 (7): e11412.

Mead, J.I., et al. 1986. Dung of *Mammuthus* in the arid Southwest, North America. *Quaternary Research* 25:121-127.

Meiri, S., et al. 2004. Body size of insular carnivores: little support for the island rule. *American Naturalist* 163:469-479.

Meiri, S., and T. Dayan. 2003. On the validity of Bergmann's rule. *Journal of Biogeography* 30:331-351.

Melott, A.L., and B.C. Thomas. 2009. Late Ordovician geographic patterns of extinction compared with simulations of astrophysical ionizing radiation damage. *Paleobiology* 35:311-320.

Mendoza, M., and P. Palmqvist. 2008. Hypsodonty in ungulates: an adaptation for grass consumption or for foraging in open habitat? *Journal of Zoology* 274:134-142.

Merceron, G., et al. 2016. Untangling the environmental from the dietary: dust does not matter. *Proceedings of the Royal Society B* 283:20161032.

Meredith, R.W., et al. 2009. Molecular decay of the tooth gene enamelin (ENAM) mirrors the loss of enamel in the fossil record of placental mammals. *PLoS Genetics* 5 (9): e1000634.

Merrihue, C.M., and G. Turner. 1966. Potassium-argon dating by activation with fast neutrons. *Journal of Geophysical Research* 71:2852-2857.

Meyer, H.W. 1992. Lapse rates and other variables applied to estimating paleoaltitudes from fossil floras. *Palaeogeography, Palaeoclimatology, Palaeoecology* 99:71-99.

Meyer, R.C. 1999. Helical burrows as a palaeoclimate response: *Daimonelix* by *Palaeocastor*. *Palaeogeography, Palaeoclimatology, Palaeoecology* 147:291-299.

Mikhalevich, V.I. 2013. New insight into the systematics and evolution of the foraminifera. *Micropaleontology* 59:493-527.

Miller, G.H. et al. 1999. Pleistocene extinction of *Genyornis newtoni*: human impact on Australian megafauna. *Science* 283:205-208.

Miller, G.H., et al. 2005. Ecosystem collapse in Pleistocene Australia and a human role in megafaunal extinction. *Science* 309:287-290.

Miller, J.H. 2011. Ghosts of Yellowstone: multi-decadal histories of wildlife populations captured by bones on a modern landscape. *PLoS ONE* 6:e18057.

Millien, V. 2011. Mammals evolve faster on smaller islands. *Evolution* 65:1935-1944.

Millien, V., and J. Damuth. 2004. Climate change and size evolution in an island rodent species: new perspectives on the island rule. *Evolution* 58:1353-1360.

Millien, V., et al. 2006. Ecotypic variation in the context of global climate change: revisiting the rules. *Ecology Letters* 9:853-869.

Milton, K. 1999. A hypothesis to explain the role of meat-eating in human evolution. *Evolutionary Anthropology* 8:11-21.

Mizutani, H., et al. 1990. Carbon isotope ratio of feathers reveals feeding behaviour of cormorants. *The Auk* 107:400-437.

Mol, D., et al. 2001. The Jarkov mammoth: 20,000-year-old carcass of a Siberian woolly mammoth *Mammuthus primigenius* (Blumenbach, 1799). In *The World of Elephants: Proceedings of the 1st International Congress*, edited by G. Cavarretta et al., 305-309. Rome: Consiglio Nazionale delle Ricerche.

Montenegro, A., et al. 2016. Simulations of prehistoric colonization of Oceania. *Proceedings of the National Academy of Sciences USA* 113:12685-12690.

Morales, M.F., et al. 1945. Studies on body composition. 2. Theoretical considerations regarding the major body tissue components, with suggestions for application to man. *Journal of Biological Chemistry* 158:677-684.

Moritz, C., et al. 2008. Impact of a century of climate change on small-mammal communities in Yosemite National Park, USA. *Science* 322:261-264.

Morris, S.C. 1986. The community structure of the Middle Cambrian phyllopod bed (Burgess Shale). *Paleontology* 29:423-467.

Morwood, M.J., et al. 2004. Archaeology and age of a new hominin from Flores in eastern Indonesia. *Nature* 431:1087-1091.

Mould, R.F. 1998. The discovery of radium in 1898 by Maria Sklodowska-Curie (1867-1934) and Pierre Curie (1859-1906) with commentary on their life and times. *British Journal of Radiology* 71:1229-1254.

Moyes, K., et al. 2011. Advancing breeding phenology in response to environmental change in a wild red deer population. *Global Change Biology* 17:2455-2469.

Müller, A.H. 1951. Grundlagen der Biostratonomie [Basics of biostratonomy]. *Abhandlungen der deutsche Akademische Wissenschaft Berlin* 1950:1-147.

Mummery, T.R. 1860. On the structure and adaptation of the teeth in the lower animals and their relationship to the human dentition. Presented at the Odontological Society of London meeting on May 7, 1860. *British Journal of Dental Science* 3:290-299.

Munnecke, A., et al. 2010. Ordovician and Silurian sea-water chemistry, sea level, and climate: a synopsis. *Palaeogeography, Palaeoclimatology, Palaeoecology* 296:389-413.

Murie, A. 1944. *The Wolves of Mount McKinley.* Fauna of the National Parks of the United States. Fauna Series No. 5. Washington, DC: US Government Printing Office.

Murray, I.W., and F.A. Smith. 2012. Estimating the influence of the thermal environment on activity patterns of the desert woodrat (*Neotoma lepida*) using temperature chronologies. *Canadian Journal of Zoology* 90:1171-1180.

Muttoni, G., et al. 2018. Early hominins in Europe: the Galerian migration hypothesis. *Quaternary Science Reviews* 180:1-29.

Myers, D.J., et al. 2012. Evaluation of δD and $\delta^{18}O$ as natural markers of invertebrate source environment and dispersal in the middle Mississippi River–floodplain ecosystem. *River Research Applications* 28:135-142.

Myers, N. 1990. Mass extinctions: what can the past tell us about the present and future? *Palaeogeography, Palaeoclimatology, Palaeoecology* 82:175-185.

Myers, P., et al. 2019. Animal Diversity Web (online). University of Michigan. Museum of Zoology. Accessed May 1, 2019. https://animaldiversity.org.

National Research Council. 2006. *Surface Temperature Reconstructions for the Last 2,000 Years.* Washington, DC: National Academies Press.

NCEI (National Centers for Environmental Information). 2019. World magnetic model out-of-cycle release. NCEI website. National Oceanic and Atmospheric Administration. Published February 2, 2019. https://www.ncei.noaa.gov/news/world-magnetic-model-out-cycle-release.

Nenzén, H.K., et al. 2014. The impact of 850,000 years of climate changes on the structure and dynamics of mammal food webs. *PLoS ONE* 9:e106651.

Newcombe, G., et al. 2016. Revisiting the life cycle of dung fungi, including *Sordaria fimicola. PLoS ONE* 11 (2): e0147425.

Newsome, S.D. 2007. A niche for isotopic ecology. *Frontiers in Ecology and the Environment* 5:429-436.

Newsome, S.D., et al. 2010. Using stable isotope biogeochemistry to study marine mammal ecology. *Marine Mammal Science* 26:509-572.

Newsome, S.D., et al. 2012. Tools for quantifying isotopic niche space and dietary variation at the individual and population level. *Journal of Mammalogy* 93:329-341.

Ngene, S., et al. 2017. Home range sizes and space use of African elephants (*Loxodonta africana*) in the Southern Kenya and Northern Tanzania borderland landscape. *International Journal of Biodiversity and Conservation* 9:9-26.

NGRIP (North Greenland Ice Core Project). 2004. High-resolution record of Northern Hemisphere climate extending into the last interglacial period. *Nature* 431:147-151.

Nicholls, H. 2005. Ancient DNA comes of age. *PLoS Biology* 3:e56.

Nicolo, M.J., et al. 2007. Multiple early Eocene hyperthermals: their sedimentary expression on the New Zealand continental margin and in the deep sea. *Geology* 35:699-702.

Niklas, K.J. 1984. *Plant Allometry: The Scaling of Form and Process.* Chicago: University of Chicago Press.

Nyborg, T. 2009. Copper Canyon track locality (Pliocene) conservation strategies, Death Valley National Park, USA. In *PaleoParks: The Protection and Conservation of Fossil Sites Worldwide*, edited by J.H. Lipps and B.R.C Granier, 113-119. Brest, France: Carnets de Géologie (Notebooks on Geology). http://paleopolis.rediris.es/cg/BOOKS/CG2009_B03/CG2009_B03_Chapter10.html.

Oberg, J.H. 2019. Founders of plant ecology: Frederic and Edith Clements. DigitalCommons@University of

Nebraska-Lincoln. Published October 18, 2019. https://digitalcommons.unl.edu/unsmaffil/1.

O'Connell, J., and J. Allen. 2004. Dating the colonization of Sahul (Pleistocene Australia–New Guinea): a review of recent research. *Journal of Archaeology Science* 31:835–853.

O'Leary, M.H. 1988. Carbon isotopes in photosynthesis. *Bioscience* 38:328–336.

Opdyke, N.D. 1985. Reversals of the Earth's magnetic field and the acceptance of crustal mobility in North America: a view from the trenches. *EOS: Transactions of the American Geophysical Union* 66:1177–1182.

Opdyke, N.D., and J.E.T. Channell. 1996. *Magnetic Stratigraphy.* San Diego: Academic Press.

Oppenheimer, C. 2003. Climatic, environmental, and human consequences of the largest known historic eruption: Tambora volcano (Indonesia) 1815. *Progress in Physical Geography* 27:230–259.

Oppenheimer, S. 2012. Out-of-Africa, the peopling of continents and islands: tracing uniparental gene trees across the map. *Philosophical Transactions of the Royal Society of London B, Biological Sciences* 367:770–784.

Ordonez, A., and J.W. Williams. 2013. Climatic and biotic velocities for woody taxa distributions over the last 16,000 years in eastern North America. *Ecology Letters* 16:773–781.

Orlando, L., et al. 2002. Ancient DNA and the population genetics of cave bears (*Ursus spelaeus*) through space and time. *Molecular Biology and Evolution* 19:1920–1933.

Osborn, H.F. 1907. *Evolution of Mammalian Molar Teeth.* New York: Macmillan.

Osborn, H.F. 1911. A dinosaur mummy. *American Museum Journal* 11:7–11.

Osborn, H.F. 1912. Integument of the iguanodont dinosaur *Trachodon. Memoirs of the American Museum of Natural History* 1:33–35, 46–54.

Osborn, H.F. 1930. Biographical memoir of Edward Drinker Cope 1840–1897. *Proceedings of the National Academy of Science USA* 3:127–175.

Osborn, H.F. 1936. *Proboscidea, Vol. 1: Moeritherioidea, Deinotherioidea, Mastodontoidea.* New York: American Museum of Natural History.

Ota, M.S., et al. 2009. Patterning of molar tooth roots in mammals. *Journal of Oral Biosciences* 51:193–198.

Overpeck, J., et al. 1992. Mapping eastern North American vegetation change of the past 18 ka: no-analogs and the future. *Geology* 20:1071–1074.

Owen, R. 1841. Report on British fossil reptiles. Part 2. *Report of the Eleventh Meeting of the British Association for the Advancement of Science,* 60–204.

Owen-Smith, N. 1987. Pleistocene extinctions: the pivotal role of megaherbivores. *Paleobiology* 13:351–362.

Owen-Smith, N. 1988. *Megaherbivores: The Influence of Very Large Body Size on Ecology.* Cambridge: Cambridge University Press.

Owen-Smith, N. 1989. Megafaunal extinctions: the conservation message from 11,000 years B.P. *Conservation Biology* 3:405–412.

Pääbo, S., et al. 2004. Genetic analyses from ancient DNA. *Annual Review of Genetics* 38:645–679.

Pabst, M.A., et al. 2009. The tattoos of the Tyrolean iceman: a light microscopical, ultrastructural, and element analytical study. *Journal of Archaeology Science* 36:2335–2341.

Pagani, M., et al. 2006. An ancient carbon mystery. *Science* 314:1556–1557.

Palkopoulou, E., et al. 2018. A comprehensive genomic history of extinct and living elephants. *Proceedings of the National Academy of Science USA* 115:E2566–E2574.

Palombo, M.R. 2001. Endemic elephants of the Mediterranean Islands: knowledge, problems, and perspectives. In *The World of Elephants: Proceedings of the 1st International Congress* edited by G. Cavarretta et al., 486–491. Rome: Consiglio Nazionale delle Ricerche.

Palombo, M.R. 2003. *Elephas? Mammuthus? Loxodonta?* The question of the true ancestor of the smallest dwarfed elephant of Sicily. *Deinsea* 9:273–291.

Pampush, J.D., et al. 2016. Introducing molaR : a new R package for quantitative topographic analysis of teeth (and other topographic surfaces). *Journal of Mammalian Evolution* 23:397–412.

Papageorgopoulou, C., et al. 2015. Histology of a woolly mammoth (*Mammuthus primigenius*) preserved in permafrost, Yamal Peninsula, Northwest Siberia. *Anatomical Record* 298:1059–1071.

Pardi, M.I., and F.A. Smith. 2012. Paleoecology in an era of climate change: how the past can provide insights into the future. In *Palaeontology in Ecology and Conservation,* edited by J. Louys, 93–116. New York: Springer-Verlag.

Pardi, M.I., and F.A. Smith. 2016. Biotic responses of canids to the terminal Pleistocene megafauna extinction. *Ecography* 39:141–151.

Parker, D.E., et al. 1992. A new daily central England temperature series, 1772–1991. *International Journal of Climate* 12:317–342.

Parker, S. 2015. *Evolution: The Whole Story.* Ontario: Firefly Books.

Parmesan, C. 2006. Ecological and evolutionary responses to recent climate change. *Annual Review of Ecology Evolution and Systematics* 37:637–669.

Parmesan, C., and G. Yohe. 2003. A globally coherent fingerprint of climate change impacts across natural systems. *Nature* 421:37–42.

Parton, A., et al. 2015. Orbital-scale climate variability in Arabia as a potential motor for human dispersals. *Quaternary International* 382:82-97.

Passey, B.H., et al. 2005. Carbon isotope fractionation between diet, breath CO_2, and bioapatite in different animals. *Journal of Archaeological Science* 32:1459-1470.

Patterson, B.D. 2004. *The Lions of Tsavo: Exploring the Legacy of Africa's Notorious Man-Eaters*. New York: McGraw-Hill.

Patterson, C. 1956. Age of meteorites and the Earth. *Geochimica et Cosmochimica Acta* 10:230-237.

Patterson, J.H. 1907. *The Man-Eaters of Tsavo and Other East African Adventures*. London: Macmillan.

Patton, J.L., et al. 2007. *The Evolutionary History and a Systematic Revision of Woodrats of the* Neotoma lepida *Group*. Berkeley: University of California Press.

Payne, J.L., and M.E. Clapham. 2012. End-Permian mass extinction in the oceans: an ancient analog for the twenty-first century? *Annual Review of Earth and Planetary Sciences* 40:89-111.

Payne, J.L., and S. Finnegan. 2007. The effect of geographic range on extinction risk during background and mass extinction. *Proceedings of the National Academy of Science USA* 104:10506-10511.

Payne, J.L., et al. 2009. Two-phase increase in the maximum size of life over 3.5 billion years reflects biological innovation and environmental opportunity. *Proceedings of the National Academy of Science USA* 106:24-27.

Pearson, P.O. 1948. Metabolism of small mammals, with remarks on the lower limit of mammalian size. *Science* 108:44.

Peters, R.H. 1983. *The Ecological Implications of Body Size*. Cambridge: Cambridge University Press.

Peterson, A.T., et al. 2002. Predicting distributions of Mexican birds using ecological niche modelling methods. *Ibis* 144:E27-E32.

Peterson, B.J., and B. Fry. 1987. Stable isotopes in ecosystem studies. *Annual Review of Ecology, Evolution, and Systematics* 18:293-320.

Peyer, B. 1968. *Comparative Odontology*. Chicago: University of Chicago Press.

Pfister, C., et al. 1999. Documentary evidence on climate in sixteenth-century Europe. *Climate Change* 43:55-110.

Phillips, D.L., et al. 2014. Best practices for use of stable isotope mixing models in food-web studies. *Canadian Journal of Zoology* 92:823-835.

Phillips, J. 1844. *Memoirs of William Smith*. 1st ed. London: John Murray.

Pimm, S.L., et al. 1995. The future of biodiversity. *Science* 269:347-350.

Pliny the Elder. (77) 1855. *The Natural History*. Translated by John Bostock. London: Taylor and Francis.

Plotnick, R., et al. 2016. The fossil record of the sixth extinction. *Ecology Letters* 19:546-553.

Podlesak, D.W., et al. 2008. Turnover of oxygen and hydrogen isotopes in the body water, CO_2, hair, and enamel of a small mammal. *Geochimica et Cosmochimica Acta* 72:19-35.

Poinar, H.N., et al. 1998. Molecular coproscopy: dung and diet of the extinct ground sloth *Nothrotheriops shastensis*. *Science* 281:402-406.

Poinar, H.N., et al. 2006. Metagenomics to paleogenomics: large-scale sequencing of mammoth DNA. *Science* 311:392-394.

Polly, P.D. 2010. Tiptoeing through the trophics: geographic variation in carnivoran locomotor ecomorphology in relation to environment. In *Carnivoran Evolution: New Views on Phylogeny, Form, and Function*, edited by A. Goswami and A.R. Friscia, 374-410. Cambridge: Cambridge University Press.

Popova, O.P., et al. 2013. Chelyabinsk airburst, damage assessment, meteorite recovery, and characterization. *Science* 342:1069-1073.

Post, E., and M.C. Forchhammer. 2008. Climate change reduces reproductive success of an Arctic herbivore through trophic mismatch. *Philosophical Transactions of the Royal Society of London B, Biological Sciences* 363:2369-2375.

Powell, J., et al. 2013. Results from an amino acid racemization inter-laboratory proficiency study; design and performance evaluation. *Quaternary Geochronology* 16:183-197.

Prentice, C. 1988. Records of vegetation in time and space: the principles of pollen analysis. In *Vegetation History: Handbook of Vegetation Science*, edited by B. Huntley and T. Webb, 17-42. Dordrecht: Springer Netherlands.

Presslee, S., et al. 2019. Palaeoproteomics resolve sloth relationships. *Nature Ecology and Evolution* 3:1121-1130.

Preston, F.W. 1962. The canonical distribution of commonness and rarity: part I. *Ecology* 43:185-215.

Pretorius, Y., et al. 2016. Why elephant have trunks and giraffe long tongues: how plants shape large herbivore mouth morphology. *Acta Zoologica* 97:246-254.

Price, G.J., et al. 2017. Seasonal migration of marsupial megafauna in Pleistocene Sahul (Australia-New Guinea). *Proceedings of the Royal Society B* 284:20170785.

Prost, S., et al. 2013. Effects of late quaternary climate change on Palearctic shrews. *Global Change Biology* 19:1865-1874.

Purdue, J.R. 1980. Clinal variation of some mammals during the Holocene in Missouri. *Quaternary Research* 13:242-258.

Purdue, J.R. 1989. Changes during the Holocene in the size of white-tailed deer (*Odocoileus virginianus*) from Central Illinois. *Quaternary Research* 32:307-316.

Purvis, A., et al. 2000. Nonrandom extinction and the loss of evolutionary history. *Science* 288:328-330.

Qiang, J., et al. 2006. A swimming mammaliaform from the Middle Jurassic and ecomorphological diversification of early mammals. *Science* 311:1123-1127.

Rae, A.M., and M. Ivanovich. 1986. Successful application of uranium series dating of fossil bone. *Applied Geochemistry* 1:419-426.

Raia, P., and S. Meiri. 2007. The island rule in large mammals: paleontology meets ecology. *Evolution* 60:1731-1742.

Rameaux, J.F., and F. Sarrus. 1838. Rapport sur un mémoire adressé à l'Académie royale de médecine [Report on a dissertation to the Royal Academy of Medicine]. *Bulletin de l'Académie de Médecine, Paris* 3:1094-1100.

Raper, D., and M. Bush. 2009. A test of *Sporormiella* representation as a predictor of megaherbivore presence and abundance. *Quaternary Research* 71:490-496.

Rasbury, E.T., and J.M. Cole. 2009. Directly dating geologic events: U-Pb dating of carbonates. *Reviews of Geophysics* 47:RG3001.

Raup, D.M. 1979. Biases in the fossil record of species and genera. *Bulletin of the Carnegie Museum of Natural History* 13:85-91.

Raup, D.M. 1981. A kill curve for Phanerozoic marine species. *Paleobiology* 17:37-48.

Raup, D.M. 1991. *Extinction: Bad Genes or Bad Luck?* New York: W.W. Norton.

Raup, D.M., and J.J. Sepkoski, Jr. 1982. Mass extinctions in the marine fossil record. *Science* 215:1501-1503.

Raup, D.M., and J.J. Sepkoski, Jr. 1984. Periodicity of extinctions in the geologic past. *Proceedings of the National Academy of Sciences USA* 81:801-805.

Reichman, O.J., and S. Aitchison. 1981. Mammal trails on mountain slopes: optimal paths in relation to slope angle and body weight. *American Naturalist* 117:416-420.

Reinhard, K.J., and V.M. Bryant, Jr. 1992. Coprolite analysis: a biological perspective on archaeology. *Papers in Natural Resources* 46:245-288.

Reinhard, K.J., et al. 1987. Helminth remains from prehistoric Indian coprolites on the Colorado Plateau. *Journal of Parasitology* 73:630-39.

Renne, P.R., et al. 1997. Ar-40/Ar-39 dating into the historical realm: calibration against Pliny the Younger. *Science* 277:1279-1280.

Rensch, B. 1938. Some problems of geographical variation and species-formation. *Proceedings of the Linnean Society of London* 150:275-285.

Retallack, G.J. 2001. Cenozoic expansion of grasslands and climatic cooling. *Journal of Geology* 109:407-426.

Retallack, G.J. 2013. Global cooling by grassland soils of the geological past and near future. *Annual Review of Earth and Planetary Science* 41:69-86.

Richards, M.R., et al. 2000. Neanderthal diet at Vindija and Neatherthal predation: the evidence from stable isotopes. *Proceedings of the National Academy of Sciences USA* 97:7663-7666.

Richardson, M.J., 2001. Diversity and occurrence of coprophilous fungi. *Mycological Research* 105:387-402.

Rightmire, G.P., et al. 2017. Skull 5 from Dmanisi: descriptive anatomy, comparative studies, and evolutionary significance. *Journal of Human Evolution* 104:50-79.

Ripple, W.J., and R.L. Beschta. 2004. Wolves and the ecology of fear: can predation risk structure ecosystems? *Bioscience* 54:755-766.

Ripple, W.J., and B. Van Valkenburgh. 2010. Linking top-down forces to the Pleistocene megafaunal extinctions. *Bioscience* 60:516-526.

Ripple, W.J., et al. 2015. Collapse of the world's largest herbivores. *Science Advances* 1:31400103.

Ripple, W.J., et al. 2016. Saving the world's terrestrial megafauna. *Bioscience* 66:807-812.

Ripple, W.J., et al. 2017. Extinction risk is most acute for the world's largest and smallest vertebrates. *Proceedings of the National Academy of Sciences USA* 114:10678-10683.

Rivals, F., and G.M. Semprebon. 2011. Dietary plasticity in ungulates: insight from tooth microwear analysis. *Quaternary International* 245:279-284.

Roberts, D.F. 1953. Body weight, race, and climate. *American Journal of Physical Anthropology* 11:533-558.

Roberts, J.W., ed. 2007. Pliny the Younger. In *Oxford Dictionary of the Classical World.* Oxford: Oxford University Press.

Roberts, R.G., et al. 2001. New ages for the last Australian megafauna: continent-wide extinction about 46,000 years ago. *Science* 292:1888-1892.

Robertson, D.S., et al. 2013. K-Pg extinction patterns in marine and freshwater environments: the impact winter model. *Journal of Geophysical Research: Biogeoscience* 118:1006-1014.

Robson, S.L., and B. Wood. 2008. Hominin life history: reconstruction and evolution. *Journal of Anatomy* 212:394-425.

Rodman, P.S., and H.M. McHenry. 1980. Bioenergetics and the origin of hominid bipedalism. *American Journal of Physical Anthropology* 52:103-106.

Rodríguez, M.Á., et al. 2008. Bergmann's rule and the geography of mammal body size in the Western Hemisphere. *Global Ecology and Biogeography* 17:274-283.

Roebroeks, W., and P. Villa 2011. On the earliest evidence for habitual use of fire in Europe. *Proceedings of the National Academy of Sciences USA* 108:5209-5214.

Rogers, S.O., and A.J. Bendich. 1985. Extraction of DNA from milligram amounts of fresh herbarium and mummified plant tissues. *Plant Molecular Biology* 5:69-76.

Rohde, R., and R. Muller. 2005. Cycles in fossil diversity. *Nature* 434:208-210.

Rohland, N., and M. Hofreiter. 2007. Ancient DNA extraction from bones and teeth. *Nature Protocols* 2:1756-1762.

Rohland, N., et al. 2018. Extraction of highly degraded DNA from ancient bones, teeth, and sediments for high-throughput sequencing. *Nature Protocols* 13:2447-2461.

Rohling, E.J., and H. Pälike. 2005. Centennial-scale climate cooling with a sudden cold event around 8,200 years ago. *Nature* 434:975-979.

Romer, A.S. 1964. Cope versus Marsh. *Systematic Zoology* 13:201-207.

Roopnarine, P.D. 2006. Extinction cascades and catastrophe in ancient food webs. *Paleobiology* 32:1-19.

Root, T.L., et al. 2003. Fingerprints of global warming on wild animals and plants. *Nature* 421:57-60.

Rosenberger, F.C. 1953. *Jefferson Reader*. New York: E.P. Dutton.

Rosenzweig, M.L. 1968. Net primary productivity of terrestrial environments: predictions from climatological data. *American Naturalist* 102:67-74.

Rosenzweig, M.L. 1995. *Species Diversity in Space and Time*. Cambridge: Cambridge University Press.

Roth, V.L. 1990. Insular dwarf elephants: a case study in body mass estimation and ecological inference. In Damuth and MacFadden 1990, 151-180.

Roth, V.L. 1992. Inference from allometry and fossils: dwarfing of elephants on islands. *Oxford Surveys in Evolutionary Biology* 8:259-288.

Rudwick, M.J.S. 1997. *Georges Cuvier, Fossil Bones, and Geological Catastrophes*. Chicago: University of Chicago Press.

Rudwick, M.J.S. 2008. *Worlds before Adam: The Reconstruction of Geohistory in the Age of Reform*. Chicago: University of Chicago Press.

Ruff, C.B. 1990. Body mass and hindlimb bone cross-sectional and articular dimensions in anthropoid primates. In Damuth and MacFadden 1990, 119-150.

Ruff, C.B. 1994. Morphological adaptation to climate in modern and fossil hominids. *American Journal of Physical Anthropology* 37:65-107.

Rule, S., et al. 2012. The aftermath of megafaunal extinction: ecosystem transformation in Pleistocene Australia. *Science* 335:1483-1486.

Rull, V. 2014. Time continuum and true long-term ecology: from theory to practice. *Frontiers in Ecology and Evolution* 2:75.

Rutledge, J.J., et al. 1973. An experimental evaluation of genetic correlation. *Genetics* 75:709-726.

Ryder, M.L. 1974. Hair of the mammoth. *Nature* 249:190-192.

Saarinen, J.J., et al. 2014. Patterns of maximum body size evolution in Cenozoic land mammals: eco-evolutionary processes and abiotic forcing. *Proceedings of the Royal Society B* 281 (1784): 20132049

Salazar-Ciudad, I., and J. Jernvall. 2002. A gene network model accounting for development and evolution of mammalian teeth. *Proceedings of the National Academy of Sciences USA* 99:8116-8120.

Salzer, M.W., et al. 2009. Recent unprecedented tree-ring growth in bristlecone pine at the highest elevations and possible causes. *Proceedings of the National Academy of Sciences USA* 106:20348-20353.

Salzer, M.W., et al. 2019. Dating the Methuselah walk bristlecone pine floating chronologies. *Tree-Ring Research* 75:61-66.

Samuels, J.X., and B. Van Valkenburgh. 2008. Skeletal indicators of locomotor adaptations in living and extinct rodents. *Journal of Morphology* 269:1387-1411.

Sandom, C., et al. 2014. Global late Quaternary megafauna extinctions linked to humans, not climate change. *Proceedings of the Royal Society B* 281:20133254.

Sankararaman, S., et al. 2016. The combined landscape of Denisovan and Neanderthal ancestry in present-day humans. *Current Biology* 26:1241-1247.

Sanson, G.D., et al. 2007. Do silica phytoliths really wear mammalian teeth? *Journal of Archaeological Science* 34:526-531.

Savage, D.E. 1951. Late Cenozoic vertebrates of the San Francisco Bay region. *University of California Publications in Geological Science* 28:215-314.

Savage, R.J.G. 1977. Evolution in carnivorous mammals. *Palaeontology* 20:237-271.

Scerri, E. 2007. *The Periodic Table: Its Story and Its Significance*. Oxford: Oxford University Press.

Schaller, G.B. 1972. *The Serengeti Lion: A Study of Predator-Prey Relations*. Chicago: University of Chicago Press.

Schell, D.M., et al. 1989. Bowhead whale (*Balaena mysticetus*) growth and feeding as estimated by C techniques. *Marine Biology* 103:433-443.

Schipper, J., et al. 2008. The status of the world's land and marine mammals: diversity, threat, and knowledge. *Science* 322:225-230.

Schmidt-Nielsen, K. 1984. *Scaling: Why Is Animal Size So Important?* Cambridge: Cambridge University Press.

Schmitz, O.J., et al. 2014. Animating the carbon cycle. *Ecosystems* 17:344–359.

Schoene, B. 2014. U-Th-Pb geochronology. *Treatise on Geochemistry* 4:341–378.

Scholander, P.F. 1955. Evolution of climatic adaptation in homeotherms. *Evolution* 9:15–26.

Schopf, J.W. 1993. Microfossils of the early Archean Apex Chert: new evidence for the antiquity of life. *Science* 260:640–646.

Schopf, J.W. 2001. *Cradle of Life: The Discovery of Earth's Earliest Fossils*. Princeton, NJ: Princeton University Press.

Schopf, J.W. 2006. Fossil evidence of Archaean life. *Philosophical Transactions of the Royal Society of London B, Biological Sciences* 361:869–885.

Schulte, P., et al. 2010. The Chicxulub asteroid impact and mass extinction at the Cretaceous- Paleogene boundary. *Science* 327:1214–1218.

Schwarcz, H.P. 1982. Applications of U-series dating to archaeometry. In *Uranium Series Disequilibrium: Applications to Environmental Problems*, edited by M. Ivanovich and R.S. Harmon, 302–325. Oxford: Clarendon Press.

Scott, R.S., et al. 2005. Dental microwear texture analysis reflects diets of living primates and fossil hominins. *Nature* 436:693–695.

Scott, R.S., et al. 2006. Dental microwear texture analysis: technical considerations. *Journal of Human Evolution* 51:339–349.

Scrivner, P.J., and D.J. Bottjer. 1986. Neogene avian and mammalian tracks from Death Valley National Monument, California: their context, classification, and preservation. *Palaeogeography, Palaeoclimatology, Palaeoecology* 57:285–331.

Secord, R., et al. 2012. Evolution of the earliest horses driven by climate change in the Paleocene-Eocene Thermal Maximum. *Science* 335:959–962.

Seilacher, A. 1967. Bathymetry of trace fossils. *Marine Geology* 5:413–428.

Seilacher, A., et al. 1985. Sedimentological, ecological, and temporal patterns of fossil Lagerstätten. *Philosophical Transactions of the Royal Society of London B, Biological Sciences* 311:5–23.

Selden, P., and J. Nudds. 2012. *Evolution of Fossil Ecosystems*. London: Manson.

Semaw, S., et al. 2003. 2.6-million-year-old stone tools and associated bones from OGS-6 and OGS-7, Gona, Afar, Ethiopia. *Journal of Human Evolution* 45:169–177.

Semprebon, G.M., et al. 2011. Potential bark and fruit browsing as revealed by stereomicrowear analysis of the peculiar clawed herbivores known as chalicotheres (Perissodactyla, Chalicotherioidea). *Journal of Mammalian Evolution* 18:33–55.

Semprebon, G.M., et al. 2016. An examination of the dietary habits of *Platybelodon grangeri* from the Linxia Basin of China: evidence from dental microwear of molar teeth and tusks. *Palaeogeography, Palaeoclimatology, Palaeoecology* 457:109–116.

Seo, H., et al. 2017. Regulation of root patterns in mammalian teeth. *Scientific Reports* 7:12714.

Sepkoski, J.J., Jr. 1978. A kinetic model of Phanerozoic taxonomic diversity I: analysis of marine orders. *Paleobiology* 4:223–251.

Sepkoski, J.J., Jr. 1979. A kinetic model of Phanerozoic taxonomic diversity II: early Phanerozoic families and multiple equilibria. *Paleobiology* 5:222–251.

Sepkoski, J.J., Jr. 1981. A factor analytic description of the Phanerozoic marine fossil record. *Paleobiology* 7:36–53.

Sepkoski, J.J., Jr. 1984. A kinetic model of Phanerozoic taxonomic diversity III: post-Paleozoic families and mass extinctions. *Paleobiology* 10:246–267.

Sepkoski, J.J., Jr. 1993. Ten years in the library: new data confirm paleontological patterns. *Paleobiology* 19:43–51.

Sepkoski, J.J., Jr., et al. 1981. Phanerozoic marine diversity and the fossil record. *Nature* 293:435–437.

Serduk, N., et al. 2014. The morphology and internal anatomy of the frozen mummy of the extinct steppe bison, *Bison priscus*, from Yakutia, Russia. In *Meeting Program and Abstracts of the 74th Annual SVP Meeting*, 228. Berlin: Society of Vertebrate Paleontology (SVP).

Shapiro, B. 2015. *How to Clone a Mammoth*. Princeton, NJ: Princeton University Press.

Shapiro, B., et al. 2004. Rise and fall of the Beringian steppe bison. *Science* 306:1561–1565.

Sharp, Z. 2017. *Principles of Stable Isotope Geochemistry*. 2nd ed. University of New Mexico Digital Repository. Open Textbooks. Accessed April 1, 2019. doi:10.5072/FK2GB24S9F.

Sharpe, T. 2016. William Smith's 1815 map, a delineation of the strata of England and Wales: its production, distribution, variants, and survival. *Earth Sciences History* 35:47–61.

Shea, J.J. 2017. Occasional, obligatory, and habitual stone tool use in hominin evolution. *Evolutionary Anthropology* 26:200–217.

Shillito, L., et al. 2020. The what, how, and why of archaeological coprolite analysis. *Earth-Science Reviews* 207:103196.

Shipman, P. 1981. *Life History of a Fossil: An Introduction to Taphonomy and Paleoecology*. Cambridge, MA: Harvard University Press.

Shoshani, J. 1998. Understanding proboscidean evolution: a formidable task. *Trends in Ecology and Evolution* 13:480–487.

Shoshani, J., ed. 2000. *Elephants: Majestic Creatures of the Wild*. New York: Facts On File.

Shoshani, J., and P. Tassy. 1996. *The Proboscidea: Evolution and Palaeoecology of Elephants and Their Relatives.* Oxford: Oxford University Press.

Simpson, G.G. 1941. The function of saber-like canines in carnivorous mammals. *American Museum Novitates* 1130:1-12.

Slater, G.J., and B. Van Valkenburgh. 2008. Long in the tooth: evolution of sabertooth cat cranial shape. *Paleobiology* 34:403-419.

Slobodkin, L.B. 1962. *Growth and Regulation of Animal Populations.* New York: Holt, Rinehart, and Winston.

Smith, A.T. 1974. The distribution and dispersal of pikas: influences of behavior and climate. *Ecology* 55:1368-1376.

Smith, F.A. 1992. Evolution of body size among woodrats from Baja California, Mexico. *Functional Ecology* 6:265-273.

Smith, F.A. 1995. Scaling of digestive efficiency with body size in *Neotoma*. *Functional Ecology* 9:299-305.

Smith, F.A. 1997. *Neotoma cinerea. Mammalian Species* 564:1-8.

Smith, F.A., and J.L. Betancourt. 1998. Response of bushy-tailed woodrats (*Neotoma cinerea*) to late Quaternary climatic change in the Colorado Plateau. *Quaternary Research* 50:1-11.

Smith, F.A., and J.L. Betancourt. 2003. The effect of Holocene temperature fluctuations on the evolution and ecology of *Neotoma* (woodrats) in Idaho and northwestern Utah. *Quaternary Research* 59:160-171.

Smith, F.A., and J.L. Betancourt. 2006. Predicting woodrat (*Neotoma*) responses to anthropogenic warming from studies of the palaeomidden record. *Journal of Biogeography* 33:2061-2076.

Smith, F.A., and A.G. Boyer. 2012. Losing time? Incorporating a deeper temporal perspective into modern ecology. *Frontiers of Biogeography* 4:26-39.

Smith, F.A., and E.L. Charnov. 2001. Fitness trade-offs select for semelparous reproduction in an extreme environment. *Evolutionary Ecology Research* 3:595-602.

Smith, F.A., and S.K. Lyons. 2011. How big should a mammal be? A macroecological look at mammalian body size over space and time. *Proceedings of the Royal Society B* 366:2364-2378.

Smith, F.A., and S.K. Lyons. 2013. *Animal Body Size: Linking across Space, Time, and Taxonomy.* Chicago: University of Chicago Press.

Smith, F.A., et al. 1995. Evolution of body size in the woodrat over the past 25,000 years of climate change. *Science* 270:2012-2014.

Smith, F.A., et al. 1998. The influence of climate change on the body mass of woodrats (*Neotoma*) in an arid region of New Mexico, USA. *Ecography* 21:140-148.

Smith, F.A., et al. 2003. Body mass of late Quaternary mammals. *Ecology* 84:3402-3403.

Smith, F.A., et al. 2004. Similarity of mammalian body size across the taxonomic hierarchy and across space and time. *American Naturalist* 163:672-691.

Smith, F.A., et al. 2009. A tale of two species: extirpation and range expansion during the late Quaternary in an extreme environment. *Global and Planetary Change* 65:122-133.

Smith, F.A., et al. 2010a. The evolution of maximum body size of terrestrial mammals. *Science* 330:1216-1219.

Smith, F.A., et al. 2010b. Methane emissions from extinct megafauna. *Nature Geoscience* 3:374-375.

Smith, F.A., et al. 2014. Life in an extreme environment: a historical perspective on the influence of temperature on the ecology and evolution of woodrats. *Journal of Mammalogy* 95:1128-1143.

Smith, F.A., et al. 2015. The importance of considering animal body mass in IPCC greenhouse inventories and the underappreciated role of wild herbivores. *Global Change Biology* 21:3880-3888.

Smith, F.A., et al. 2016a. Body size evolution across the Geozoic. *Annual Review of Earth and Planetary Sciences* 44:523-553.

Smith, F.A., et al. 2016b. Exploring the influence of ancient and historic megaherbivore extirpations on the global methane budget. *Proceedings of the National Academy of Sciences USA* 113:874-879.

Smith, F.A., et al. 2016c. Megafauna in the Earth system. *Ecography* 39:99-108.

Smith, F.A., et al. 2016d. Unraveling the consequences of the terminal Pleistocene megafauna extinction on mammal community assembly. *Ecography* 39:223-239.

Smith, F.A., et al. 2018. Body size downgrading over the late Quaternary. *Science* 360:310-313.

Smith, F.A., et al. 2019. The accelerating influence of hominins on mammalian macroecological patterns over the late Quaternary. *Quaternary Science Reviews* 211:1-16.

Smith, F.A. et al. In prep. Climate change did not drive extinction in mammals. Unpublished manuscript.

Smith, F.A. et al. In prep. The missing piece: changes in the ecological niche of a mammal community in North America over the late Quaternary. Unpublished manuscript.

Smith, R.J. 1984. Allometric scaling in comparative biology: problems of concept and method. *American Journal of Physiology-Regulatory, Integrative and Comparative Physiology* 246:152-160.

Smith, R.J., et al. 1996. Distinguishing between populations of fresh- and salt-water harbour seals (*Phoca vitulina*) using stable-isotope ratios and fatty acids. *Canadian Journal of Fisheries and Aquatic Sciences* 53:272-279.

Smith, T., et al. 2006. Rapid Asia-Europe-North America geographic dispersal of earliest Eocene primate *Teilhardina* during the Paleocene-Eocene Thermal Maximum. *Proceedings of the National Academy of Science* 103:11223-11227.

Smith, W. 1816-19. *Strata Identified by Organized Fossils, Containing Prints on Coloured Paper of the Most Characteristic Specimens in Each Stratum.* London: W. Arding.

Solounias, N., and G. Semprebon. 2002. Advances in the reconstruction of ungulate ecomorphology with application to early fossil equids. *American Museum Novitates* 3366:1-49.

Sondaar, P.Y. 1977. Insularity and its effect on mammal evolution. In *Patterns in Vertebrate Evolution*, edited by M.K. Hecht and P.C. Goody, 671-705. New York: Plenum.

Soto, D.X., et al. 2011. Effects of size and diet on stable hydrogen isotope values (δD) in fish: implications for tracing origins of individuals and their food sources. *Canadian Journal of Fishery and Aquatic Science* 68:2011-2019.

Soto, D.X., et al. 2013. The influence of metabolic effects on stable hydrogen isotopes in tissues of aquatic organisms. *Isotopes Environmental Health Studies* 49:305-311.

Spalding, K.L., et al. 2005. Forensics: age written in teeth by nuclear tests. *Nature* 437:333-334.

Spaulding, W.G., et al. 1990. Packrat middens: their composition and methods of analysis. In Betancourt et al. 1991, 59-84.

Spiker, C.J. 1933. Analysis of two hundred long-eared owl pellets. *Wilson Bulletin* 45:198.

Sponheimer, M., et al. 2013. Isotopic evidence of early hominin diets. *Proceedings of the National Academy of Sciences USA* 110:10513-10518.

Stankiewicz, B.A., et al. 1998. Chemical preservation of plants and insects in natural resins. *Proceedings of the Royal Society B* 265:641-647.

Stanley, S.M. 1973. An explanation for Cope's rule. *Evolution* 27:1-26.

Stap, L.L., et al. 2010. High-resolution deep-sea carbon and oxygen isotope records of Eocene Thermal Maximum 2 and H2. *Geology* 38:607-610.

Steadman, D.W. 1995. Prehistoric extinctions of Pacific island birds: biodiversity meets zooarchaeology. *Science* 267:1123-1131.

Stearn, W.C., et al. 1989. *Paleontology: The Record of Life.* New York: John Wiley and Sons.

Stiner, M.C. 2002. Carnivory, coevolution, and the geographic spread of the genus *Homo. Journal of Archaeology Research* 10:1-63.

Stock, C., and J.M. Harris. 1992. *Rancho La Brea: A Record of Pleistocene Life in California.* Science Series No. 37. Los Angeles: Natural History Museum of Los Angeles County.

Stothers, R.B. 1984. The great Tambora eruption in 1815 and its aftermath. *Science* 224:1191-1198.

Strani, F., et al. 2018. MicroWeaR: a new R package for dental microwear analysis. *Ecology and Evolution* 8:7022-7030. doi:10.1002/ece3.4222.

Stringer, C. 2012. Evolution: what makes a modern human. *Nature* 485:33-35.

Stringer, C. 2016. The origin and evolution of Homo sapiens. *Philosophical Transactions of the Royal Society of London B, Biological Sciences* 371:20150237.

Stringer, C., and P. Andrews. 2012. *The Complete World of Human Evolution.* London: Thames and Hudson.

Stringer, C., and I. Barnes. 2015. Deciphering the Denisovans. *Proceedings of the National Academy of Sciences USA* 112:15542-15543.

Strömberg, C.A.E., et al. 2013. Decoupling the spread of grasslands from the evolution of grazer-type herbivores in South America. *Nature Communications* 4:1478. doi:10.1038/ncomms2508.

Stuart, A.J. 1991. Mammalian extinctions in the late Pleistocene of northern Eurasia and North America. *Biological Reviews of the Cambridge Philosophical Society* 66:453-562.

Suess, H.E. 1955. Radiocarbon concentration in modern wood. *Science* 122:415-417.

Surovell, T.A., et al. 2016. Test of Martin's overkill hypothesis using radiocarbon dates on extinct megafauna. *Proceedings of the National Academy of Sciences USA* 113:886-891.

Sutikna, T., et al. 2016. Revised stratigraphy and chronology for *Homo floresiensis* at Liang Bua in Indonesia. *Nature* 532:366-369.

Svensen, H., et al. 2009. Siberian gas venting and the end-Permian environmental crisis. *Earth and Planetary Science Letters* 277:490-500.

Swift, J.A., et al. 2019. Micro methods for megafauna: novel approaches to late Quaternary extinctions and their contributions to faunal conservation in the Anthropocene. *BioScience* 69:877-887.

Talalay, P.G. 2016. *Mechanical Ice Drilling Technology.* Beijing: Springer.

Tammone, M.N., et al. 2012. Habitat use by colonial tuco-tucos (*Ctenomys sociabilis*): specialization, variation, and sociality. *Journal of Mammalogy* 93:1409-1419.

Tans, P.P., et al. 1979. Natural atmospheric 14C variation and the Suess effect. *Nature* 280:826-828.

Tansley, A.G. 1935. The use and abuse of vegetational concepts and terms. *Ecology* 16:284-307.

Tapaltsyan, V., et al. 2015. Continuously growing rodent molars result from a predictable quantitative evolutionary change over 50 million years. *Cell Reports* 11:673-680.

Tappen, M.J. 1994. Savanna ecology and natural bone deposition: implications for early hominid site formation, hunting, and scavenging. *Current Anthropology* 36:223-260.

Tauxe, L. 2010. *Essentials of Paleomagnetism*. Berkeley: University of California Press.

Teplitsky, C., and V. Millien. 2014. Climate warming and Bergmann's rule through time: is there any evidence? *Ecological Applications* 7:156-168.

Terborgh, J., et al. 2016a. Foraging impacts of Asian megafauna on tropical rain forest structure and biodiversity. *Biotropica* 50:84-89.

Terborgh, J., et al. 2016b. Megafaunal influences on tree recruitment in African equatorial forests. *Ecography* 39:180-186.

Terry, R.C. 2004. Owl pellet taphonomy: a preliminary study of the post regurgitation taphonomic history of pellets in a temperate forest. *Palaios* 19:497-506.

Terry. R.C. 2010. On raptors and rodents: testing the ecological fidelity and spatiotemporal resolution of cave death assemblages. *Paleobiology* 36:137-160.

Terry, R.C., and R. Rowe. 2015. Energy flow and functional compensation in Great Basin small mammals under natural and anthropogenic environmental change. *Proceedings of the National Academy of Science* 112:9656-9661.

Terwilliger, V.J., et al. 2008. Reconstructing palaeoenvironment from δ13C and δ15N values of soil organic matter: a calibration from arid and wetter elevation transects in Ethiopia. *Geoderma* 147:197-210.

Thali, M.J., et al. 2011. Brienzi—the blue vivianite man of Switzerland: time since death estimation of an adipocere body. *Forensic Science International* 211:34-40.

Thomas, B.A., and R.A. Spicer. 1987. *The Evolution and Palaebiology of Land Plants*. London: Croom Helm.

Thomas, C.D., et al. 2004. Extinction risk from climate change. *Nature* 427:145-148.

Thompson, D.W., trans. 1910. *The Works of Aristotle, Vol. 4: Historia Animalium*. Edited by J.A. Smith and W.D. Ross. Oxford: Clarendon Press.

Thompson, D.W. 1942. *On Growth and Form*. New York: Dover.

Thomsen, P.F., and E. Willerslev. 2015. Environmental DNA: an emerging tool in conservation for monitoring past and present biodiversity. *Biological Conservation* 183:4-18.

Thuiller, W., et al. 2004. Patterns and uncertainties of species' range shifts under climate change. *Global Change Biology* 10:2020-2027.

Tierney, J.E., and P.D. Zander. 2017. A climatic context for the out-of-Africa migration. *Geology* 45:1023-1026.

Tieszen, L.L., et al. 1983. Fractionation and turnover of stable carbon isotopes in animal tissues: implications for δ13C analysis of diet. *Oecologia* 57:32-37.

Timmermann, A., and T. Friedrich. 2016. Late Pleistocene climate drivers of early human migration. *Nature* 538:92-95.

Tomé, C.P., et al. 2020. Changes in the diet and body size of a small herbivorous mammal (hispid cotton rat, *Sigmodon hispidus*) following the late Pleistocene megafauna extinction. *Ecography* 43:604-619.

Toon, O.B., et al. 1997. Environmental perturbations caused by the impacts of asteroids and comets. *Reviews of Geophysics* 35:41-78.

Torrens, H. 1995. Mary Anning (1799-1847) of Lyme; "the greatest fossilist the world ever knew." *British Journal for the History of Science* 28:257-284.

Torrens, H. 2016. William Smith (1769-1839): his struggles as a consultant, in both geology and engineering, to simultaneously earn a living and finance his scientific projects, to 1820. *Earth Sciences History* 35:1-46.

Tracy, R.C. 1992. Ecological responses of animals to climate. In *Global Warming and Biological Diversity*, edited by R.L. Peters and T.E. Lovejoy, 171-179. New Haven, CT: Yale University Press.

Tracy, R.L., and G.E. Walsberg. 2002. Kangaroo rats revisited: re-evaluating a classic case of desert survival. *Oecologia* 133:449-457.

Traverse, A. 1988. *Paleopalynology*. Winchester, MA: Allen & Unwin.

Turner, G. 1970. Argon-40/argon-39 dating of lunar rock samples. *Science* 167:466-468.

Turney, C.S., et al. 2008. Late-surviving megafauna in Tasmania, Australia, implicate human involvement in their extinction. *Proceedings of the National Academy of Sciences USA* 105:12150-12153.

Turvey, S.T. 2009. *Holocene Extinctions*. Oxford: Oxford University Press.

Turvey, S.T., and S.A. Fritz. 2011. The ghosts of mammals past: biological and geographical patterns of global mammalian extinction across the Holocene. *Philosophical Transactions of the Royal Society of London B, Biological Sciences* 366:2564-2576.

Tykot, R.H. 2004. Stable isotopes and diet: you are what you eat. In *Proceedings of the International School of Physics "Enrico Fermi." Course CLIV: Physics Methods in Archaeometry*, edited by M. Martini et al., 433-444. Amsterdam: IOS Press.

Tyler J.M., et al. 2019. PalaeoChip Arctic1.0: an optimised eDNA targeted enrichment approach to reconstructing past environments. *bioRxiv*. August 26, 2019. doi:10.1101/730440.

Uhen, M.D., et al. In prep. The trajectory of marine mammal body mass over the Cenozoic. Unpublished manuscript.

Umina, P.A., et al. 2005. A rapid shift in a classic clinal pattern in *Drosophila* reflecting climate change. *Science* 308:691-693.

Ungar, P.S., ed. 2006. Dental functional morphology: the known, the unknown, and the unknowable. In *Evolution of the Human Diet: The Known, the Unknown, and the Unknowable*, 39-55. Oxford: Oxford University Press.

Ungar, P.S. 2010. *Mammal Teeth: Origin, Evolution, and Diversity*. Baltimore: Johns Hopkins University Press.

Ungar, P.S., and M. Williamson. 2000. Exploring the effects of tooth wear on functional morphology: a preliminary study using dental topographical analysis. *Palaeontologia Electronica* 3 (1): 1-18. https://palaeo-electronica.org/2000 _1/gorilla/issue1_00.htm.

Ungar, P.S., et al. 1991. A semiautomated image analysis procedure for the quantification of dental microwear. *Scanning* 13:31-36.

Uno, K.T., et al. 2013. Bomb-curve radiocarbon measurement of recent biologic tissues and applications to wildlife forensics and stable isotope (paleo)ecology. *Proceedings of the National Academy of Sciences USA* 110:11736-11741.

USDA-NRCS PLANTS Database/Hitchcock, A.S. (rev. A. Chase). 1950. *Manual of the Grasses of the United States.* USDA Miscellaneous Publication No. 200. Washington, DC: US Department of Agriculture, Natural Resources Conservation Service. https://plants.usda.gov/java /usageGuidelines?imageID=phcol5_001_avd.tif.

Usoskin, I.G. 2017. A history of solar activity over millennia. *Living Reviews in Solar Physics* 14:3.

Valentine, J.W. 1989. How good was the fossil record? Clues from the Californian Pleistocene. *Paleobiology* 15:83-94.

Vallano, D.M., and J.P. Sparks. 2013. Foliar $\delta^{15}N$ is affected by foliar nitrogen uptake, soil nitrogen, and mycorrhizae along a nitrogen deposition gradient. *Oecologia* 172:47-58.

Van de Water, P.K., et al. 1994. Trends in stomatal density and $^{13}C/^{12}C$ ratios of *Pinus flexilis* needles during Last Glacial-Interglacial cycle. *Science* 264:239-243.

et al..Van der Merwe, N.J., and J.C. Vogel. 1978. ^{13}C content of human collagen as a measure of preshistoric diet in woodland North America. *Nature* 276:815-816.

Van der Plicht, J., et al. 1989. Uranium and thorium in fossil bones: activity ratios and dating. *Applied Geochemistry* 4:339-342.

Van Devender, T.R. 1977. Holocene woodlands in southwestern deserts. *Science* 198:189-192.

Van Devender, T.R. 1983. Our first curators. *Sonorensis* 5:4-10.

Van Devender, T.R. 1986. Climatic cadences and the composition of Chihuahuan Desert communities: the late Pleistocene packrat midden record. In *Community Ecology*, edited by J. Diamond and T.J. Case, 285-299. New York: Harper and Row.

Van Devender, T.R. 1988. Pollen in packrat (*Neotoma*) middens: pollen transport and the relationship of pollen to vegetation. *Palynology* 12:221-229.

Van Devender, T.R., et al. 1985. Fossil packrat middens and the tandem accelerator mass spectrometer. *Nature* 317:610-613.

Van Soest, P.J. 1982. *Nutritional Ecology of the Ruminant.* Corvallis, OR: O & B Books.

Van Soest, P.J. 1994. *The Nutritional Ecology of the Ruminant.* 2nd ed. Ithaca, NY: Cornell University Press.

Van Valen, L. 1973. Pattern and the balance of nature. *Evolutionary Theory* 1:31-49.

Van Valkenburgh, B. 1987. Skeletal indicators of locomotor behavior in living and extinct carnivores. *Journal of Vertebrate Paleontology* 7:162-182.

Van Valkenburgh, B. 1989. Carnivore dental adaptations and diet: a study of trophic diversity within guilds. In *Carnivore Behavior, Ecology, and Evolution*, vol. 1, edited by J.L. Gittleman, 410-436. Ithaca, NY: Cornell University Press.

Van Valkenburgh, B. 1995. Tracking ecology over geological time: evolution within guilds of vertebrates. *Trends in Ecology and Evolution* 10:71-76.

Van Valkenburgh, B. 1999. Major patterns in the history of carnivorous mammals. *Annual Review of Earth and Planetary Sciences* 27:463-493.

Van Valkenburgh, B. 2001. Predation in sabre-tooth cats. In *Palaeobiology II*, edited by D.E.G. Briggs and P.R. Crowther, 420-423. Oxford: Blackwell Science.

Van Valkenburgh, B., and F. Hertel. 1993. Tough times at La Brea: tooth breakage in large carnivores of the late Pleistocene. *Science* 261:456-459.

Van Valkenburgh, B., and K.-P. Koepfli. 1993. Cranial and dental adaptations to predation in canids. *Symposium of the Zoological Society of London* 65:15-37.

Van Valkenburgh, B., and C.B. Ruff. 1987. Canine tooth strength and killing behaviour in large carnivores. *Journal of Zoology* 212:379-397.

Van Valkenburgh, B., and T. Sacco. 2002. Sexual dimorphism and intra-sexual competition in large Pleistocene carnivores. *Journal of Vertebrate Paleontology* 22:164-169.

Van Valkenburgh, B., et al. 2004. Cope's rule, hypercarnivory, and extinction in North American canids. *Science* 306:101-104.

Van Valkenburgh, B., et al. 2016. The impact of large terrestrial carnivores on Pleistocene ecosystems.

Proceedings of the National Academy of Sciences USA 113:862–867.

Vander Zanden, H.B., et al. 2016. Expanding the isotopic toolbox: applications of hydrogen and oxygen stable isotope ratios to food web studies. *Frontiers in Ecology and Evolution* 4:20. doi:10.3389/fevo.2016.00020.

Vera Torres, J.A. 1994. *Estratigrafía: principios y métodos* [Stratigraphy: principles and methods]. Madrid: Editorial Rueda.

Vernot, B., et al. 2016. Excavating Neanderthal and Denisovan DNA from the genomes of Melanesian individuals. *Science* 352:235–239.

Villier, B., and G. Carnevale. 2013. A new skeleton of the giant hedgehog *Deinogalerix* from the Miocene of Gargano, southern Italy. *Journal of Vertebrate Paleontology* 33:902–923.

Visser, M.E., et al. 1998. Warmer springs lead to mistimed reproduction in great tits (*Parus major*). *Proceedings of the Royal Society B* 265:1867–1870.

Vitousek, P.M., et al. 1997. Human domination of Earth's ecosystems. *Science* 277:494–499.

Vogel, J.C., and N.J. van der Merwe. 1977. Isotopic evidence for early maize cultivation in New York State. *American Antiquity* 42:238–242.

Vogel, J.C., et al. 2002. Accurate dating with radiocarbon from the atom bomb tests. *South African Journal of Science* 98:437–438.

Voigt, C.C., et al. 2015. Stable isotope ratios of hydrogen separate mammals of aquatic and terrestrial food webs. *Methods Ecology and Evolution* 6:1332–1340.

Vorhies, C.T. 1945. Water requirements of desert animals in the Southwest. *University of Arizona Technical Bulletin* 107:487–525.

Vrba, E.S., and D. DeGusta. 2004. Do species populations really start small? New perspectives from the Late Neogene fossil record of African mammals. *Philosophical Transactions of the Royal Society of London B, Biological Sciences* 359:285–292.

Wacey, D., et al. 2011. Microfossils of sulphur-metabolizing cells in 3.4-billion-year-old rocks of Western Australia. *Nature Geoscience* 4:698–702.

Walker, A.C., et al. 1978. Microwear of mammalian teeth as an indicator of diet. *Science* 201:808–810.

Walker, K.R., and R.K. Bambach. 1971. The significance of fossil assemblages from fine-grained sediments: time-averaged communities. *Geological Society of America Abstracts with Program* 3:783–784.

Walker, M. 2005. *Quaternary Dating Methods*. New York: John Wiley and Sons.

Wallace, D.R. 2000. *The Bonehunters' Revenge: Dinosaurs and Fate in the Gilded Age*. New York: Houghton Mifflin.

Walter, R.C., et al. 2000. Early human occupation of the Red Sea coast of Eritrea during the last interglacial. *Nature* 405:65–69.

Walters, J.R., et al. 2010. Status of the California condor (*Gymnogyps californianus*) and efforts to achieve its recovery. *The Auk* 127:969–1001.

Wang, S.C. 2001. Quantifying passive and driven evolutionary trends. *Evolution* 55:849–858.

Ward, P.D. 2007. *Under a Green Sky: Global Warming, the Mass Extinctions of the Past, and What They Can Tell Us about Our Future*. New York: HarperCollins.

Ward, P.[D.], et al. 2006. Confirmation of Romer's Gap as a low oxygen interval constraining the timing of initial arthropod and vertebrate terrestrialization. *Proceedings of the National Academy of Science USA* 103:16818–16822.

Wasilewski, P., and G. Kletetschka. 1999. Lodestone: nature's only permanent magnet. What it is and how it gets charged. *Geophysical Research Letters* 26:2275–2278.

Watson, P. 2006. *Ideas: A History of Thought and Invention, from Fire to Freud*. New York: Harper Perennial.

Watts, D.P. 2008. Scavenging by chimpanzees at Ngogo and the relevance of chimpanzee scavenging to early hominin behavioral ecology. *Journal of Human Evolution* 54:125–133.

Webb, S.D., and N.D. Opdyke. 1995. Global climate influence on Cenozoic land mammal faunas. In *Effects of Past Global Change on Life*, edited by the National Research Council, 184–208. New York: National Academies of Science Press.

Webb, T., III. 1974a. Corresponding distributions of modern pollen and vegetation in lower Michigan. *Ecology* 55:17–28.

Webb, T., III. 1974b. A vegetational history from northern Wisconsin: evidence from modern and fossil pollen. *American Midland Naturalist* 92:12–34.

Weigelt, J. 1927. Über Biostratonomie [About biostratonomy]. *Der Geologe* 42:1069–1076.

Wells, P.V. 1966. Late Pleistocene vegetation and degree of pluvial climatic change in the Chihuahuan Desert. *Science* 153:970–975.

Wells, P.V. 1976. Macrofossil analysis of woodrat (*Neotoma*) middens as a key to the Quaternary vegetational history of arid America. *Quaternary Research* 6:223–248.

Wells, P.V. 1977. Reply to comments by Van Devender. *Quaternary Research* 8:238–239.

Wells, P.V. 1979. An equable glaciopluvial in the west: periglacial evidence of increased precipitation on a gradient from the Great Basin to the Sonoran and Chihuahuan deserts. *Quaternary Research* 12:311–325.

Wells, P.V., and R. Berger. 1967. Late Pleistocene history of coniferous woodland in the Mohave Desert. *Science* 155:1640–1647.

Wells, P.V., and C.D. Jorgensen 1964. Pleistocene woodrat middens and climatic change in the Mohave Desert: a record of juniper woodlands. *Science* 143:1171-1174.

Wells, P.V., and D. Woodcock. 1985. Full-glacial vegetation of Death Valley, California: juniper woodland opening to Yucca semidesert. *Madrono* 32:11-23.

Werdelin, L. 1983. Morphological patterns in the skulls of cats. *Biological Journal of the Linnean Society* 19:375-391.

Werdelin, L., and M. Lewis. 2005. Plio-Pleistocene Carnivora of eastern Africa: species richness and turnover patterns. *Zoological Journal of the Linnean Society* 144:121-144.

West, J.B., et al. 2006. Stable isotopes as one of nature's ecological recorders. *Trends in Ecology and Evolution* 21:408-414.

Western, D., and D. Maitumo. 2004. Woodland loss and restoration in a savanna park: a 20-year experiment. *African Journal of Ecology* 42:111-121.

Wetherill, G.W. 1956. Discordant uranium-lead ages. *Transactions of the American Geophysical Union* 37:320-326.

Weyrich, L., et al. 2017. Neanderthal behaviour, diet, and disease inferred from ancient DNA in dental calculus. *Nature* 544:357-361.

White, R. 1993. The dawn of adornment. *Natural History* 102:60-68.

White, T.D., et al. 1994. *Australopithecus ramidus*, a new species of early hominid from Aramis, Ethiopia. *Nature* 371:306-312.

White, T.D., et al. 2009. *Ardipithecus ramidus* and the paleobiology of early hominids. *Science* 326:75-86.

Whitehouse A.M. 2002. Tusklessness in the elephant population of the Addo Elephant National Park, South Africa. *Journal of Zoology* 257:249-254.

Whitley, D.S. 2009. Cave Paintings and the Human Spirit: The Origin of Creativity and Belief. Amherst, MA: Prometheus.

Whittaker, R.H. 1953. A consideration of climax theory: the climax as a population and pattern. *Ecological Monographs* 23:41-78.

Whittaker, R.J. 1998. *Island Biogeography: Ecology, Evolution, and Conservation*. London: Oxford University Press.

Whittington, H.B. 1975. The enigmatic animal *Opabinia regalis*, Middle Cambrian Burgess Shale, British Columbia. *Philosophical Transactions of the Royal Society of London B, Biological Sciences* 271:5-43.

Whyte, I.J., et al. 2003. Kruger's elephant population: its size and consequences for ecosystem heterogeneity. In *The Kruger Experience: Ecology and Management of Savanna Heterogeneity*, edited by J.T. Du Toit et al., 332-348. Washington, DC: Island Press.

Wible, J.R., et al. 2007. Cretaceous Eutherians and Laurasian origin for placental mammals near the K/T boundary. *Nature* 447:1003-1006.

Wilde, S.A., et al. 2001. Evidence from detrital zircons for the existence of continental crust and oceans on the Earth 4.4 Gyr ago. *Nature* 409:175-178.

Wilkins, J., et al. 2012. Evidence for early hafted hunting technology. *Science* 338:942-946.

Willerslev, E., and A. Cooper 2005. Ancient DNA. *Proceeding of the Royal Society B* 272:3-16.

Willerslev, E., et al. 2003. Diverse plant and animal genetic records from Holocene and Pleistocene sediments. *Science* 300:791-795.

Williams, J.E., and J.L. Blois. 2018. Range shifts in response to past and future climate change: can climate velocities and species' dispersal capabilities explain variation in mammalian range shifts? *Journal of Biogeography* 45:2175-2189.

Williams, J.W., and S.T. Jackson. 2007. Novel climates, no-analog communities, and ecological surprises. *Frontiers of Ecology and the Environment* 9:475-482.

Williams, J.W., et al. 2001. Dissimilarity analyses of late Quaternary vegetation and climate in eastern North America. *Ecology* 82:3346-3362.

Williams, J.W., et al. 2002. Rapid and widespread vegetation responses to past climate change in the North Atlantic region. *Geology* 30:971-974.

Williams, J.W., et al. 2009. Rapid responses of the midwestern prairie-forest ecotone to early Holocene aridity. *Global and Planetary Change* 66:195-207.

Williams, J.W., et al. 2010. Rapid, time-transgressive, and variable responses to early-Holocene midcontinental drying in North America. *Geology* 38:135-138.

Williams, S.H., and R.F. Kay. 2001. A comparative test of adaptive explanations for hypsodonty in ungulates and rodents. *Journal of Mammalian Evolution* 8:207-229.

Williamson, T.E., et al. 2014. The origin and early evolution of metatherian mammals: the Cretaceous record. *ZooKeys* 465:1-76.

Winchester, S. 2001. *The Map That Changed the World: William Smith and the Birth of Modern Geology*. New York: HarperCollins.

Winick, S. 2017. She sells seashells and Mary Anning: metafolklore with a twist. *Folklife Today*. July 26, 2017. https://blogs.loc.gov/folklife/2017/07/she-sells-seashells -and-mary-anning-metafolklore-with-a-twist.

Wistar, C. 1799. A description of the bones deposited, by the president, in the museum of the society, and represented in the annexed plates. *Transactions of the American Philosophical Society* 4:526-531.

Wolf, N., et al. 2012. An experimental exploration of the incorporation of hydrogen isotopes from dietary sources into avian tissues. *Journal of Experimental Biology* 215:1915-1922.

Wood, B., and B.G. Richmond. 2000. Human evolution: taxonomy and paleobiology. *Journal of Anatomy* 1966:19-60.

Wood, J., et al. 2011. *Sporormiella* as a proxy for non-mammalian herbivores in island ecosystems. *Quaternary Science Reviews* 30:915-920.

Woodard, G.D., and L.F. Marcus. 1973. Rancho La Brea fossil deposits: A re-evaluation from stratigraphic and geological evidence. *Journal of Paleontology* 47:54-69.

Woodburne, M.O. 2004. *Late Cretaceous and Cenozoic Mammals of North America: Biostratigraphy and Geochronology*. New York: Columbia University Press.

Woodburne, M.O., et al. 2009. Climate directly influences Eocene mammal faunal dynamics in North America. *Proceedings of the National Academy of Science* 106:13399-13403.

Woods, R., et al. 2017. The small and the dead: a review of ancient DNA studies analyzing micromammal species. *Genes* 8:312-326.

Woodward, S.R., et al. 1994. DNA sequence from Cretaceous period bone fragments. *Science* 266:1229-1232.

Wrangham, R. 2009. *Catching Fire: How Cooking Made Us Human*. New York: Basic Books.

Wroe, S., et al. 2005. Bite club: comparative bite force in big biting mammals and the prediction of predatory behaviour in fossil taxa. *Proceedings of the Royal Society B* 272:619-625.

Wroe, S., et al. 2006. Megafaunal extinction: climate, humans, and assumptions. *Trends in Ecology and Evolution* 21:61-62.

Wroe, S., et al. 2013. Climate change frames debate over the extinction of megafauna in Sahul (Pleistocene Australia-New Guinea). *Proceedings of the National Academy of Sciences USA* 110:8777-8781.

WWF (World Wildlife Fund). 2018. *Living Planet Report 2018: Aiming Higher*. Edited by M. Grooten and R.E.A. Almond. Gland, Switzerland: WWF.

Wyatt, K.B., et al. 2008. Historical mammal extinction on Christmas Island (Indian Ocean) correlates with introduced infectious disease. *PLoS ONE* 3 (11): e3602.

Yalden, D.W. 2009. The analysis of owl pellets. *Mammal Society Occasional Publications* 13:1-28.

Yan, Y., et al. 2019. Two-million-year-old snapshots of atmospheric gases from Antarctic ice. *Nature* 574:663-666.

Yeakel, J.D., et al. 2009. Cooperation and individuality among man-eating lions. *Proceedings of the National Academy of Sciences USA* 106:19040-19043.

Yom-Tov, Y., and E. Geffen. 2006. Geographic variation in body size: the effects of ambient temperature and precipitation. *Oecologia* 148:213-218.

Yom-Tov, Y., and H. Nix. 1986. Climatological correlate for body size of five species of Australian mammals. *Biological Journal of the Linnean Society* 29:245-262.

Yom-Tov, Y., and D. Wool. 1997. Do the contents of barn owl pellets accurately represent the proportion of prey species in the field? *The Condor* 99:972-976.

Yom-Tov, Y., et al. 2006. Recent changes in body weight and wing length among some British passerine birds. *Oikos* 112:91-101.

Yom-Tov, Y., et al. 2008. Recent increase in body size of the American marten (*Martes Americana*) in Alaska. *Biological Journal of the Linnean Society* 93:701-707.

Zachos, J.C., et al. 2001. Trends, rhythms, and aberrations in global climate 65 Ma to present. *Science* 292:686-693.

Zachos, J.C., et al. 2008. An early Cenozoic perspective on greenhouse warming and carbon-cycle dynamics. *Nature* 451:279-283.

Zhao, L.X., and L.Z. Zhang. 2013. New fossil evidence and diet analysis of *Gigantopithecus blacki* and its distribution and extinction in South China. *Quaternary International* 286:69-74.

Zhonghe, Z., et al. 2003. An exceptionally preserved Lower Cretaceous ecosystem. *Nature* 421:807-811.

Zhou, C.-F., et al. 2013. A Jurassic mammaliaform and the earliest mammalian evolutionary adaptations. *Nature* 500:163-168.

Zimov, S., and N. Zimov. 2014. Role of megafauna and frozen soil in the atmospheric CH_4 dynamics. *PLoS ONE* 9:e9333.

Zimov, S.A. 2005. Pleistocene park: return of the mammoth's ecosystem. *Science* 308:796-798.

Zimov, S.A., et al. 1995. Effects of mammals on ecosystem change at the Pleistocene-Holocene boundary. In *Arctic and Alpine Biodiversity: Patterns, Causes, and Ecosystem Consequences*, edited by F.S. Chapin III and C. Korner, 127-135. Berlin: Springer-Verlag.

Zimov, S.A., et al. 2012. Mammoth steppe: a high-productivity phenomenon. *Quaternary Science Reviews* 57:26-45.

Zischler, H., et al. 1995. Detecting dinosaur DNA. *Science* 268:1192-1193.

Zuo, W., et al. 2013. A life-history approach to the Late Pleistocene megafaunal extinction. *American Naturalist* 182:524-531.

index

Boxes, figures, and tables are indicated by b, f, and t following page numbers respectively.

More Books in Mammalian Paleontology from Hopkins Press

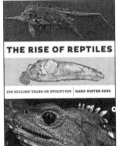

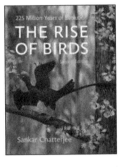
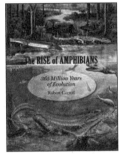
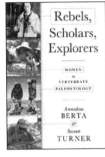